Enantioselective Homogeneous Supported Catalysis

RSC Green Chemistry

Series Editors:
James H. Clark, *Department of Chemistry, University of York, UK*
George A. Kraus, *Department of Chemistry, Iowa State University, Ames, Iowa, USA*
Andrzej Stankiewicz, *Delft University of Technology, The Netherlands*
Peter Siedl, *Federal University of Rio de Janeiro, Brazil*
Yuan Kou, *Peking University, People's Republic of China*

Titles in the Series:
1: The Future of Glycerol: New Uses of a Versatile Raw Material
2: Alternative Solvents for Green Chemistry
3: Eco-Friendly Synthesis of Fine Chemicals
4: Sustainable Solutions for Modern Economies
5: Chemical Reactions and Processes under Flow Conditions
6: Radical Reactions in Aqueous Media
7: Aqueous Microwave Chemistry
8: The Future of Glycerol: 2nd Edition
9: Transportation Biofuels: Novel Pathways for the Production of Ethanol, Biogas and Biodiesel
10: Alternatives to Conventional Food Processing
11: Green Trends in Insect Control
12: A Handbook of Applied Biopolymer Technology: Synthesis, Degradation and Applications
13: Challenges in Green Analytical Chemistry
14: Advanced Oil Crop Biorefineries
15: Enantioselective Homogeneous Supported Catalysis

How to obtain future titles on publication:
A standing order plan is available for this series. A standing order will bring delivery of each new volume immediately on publication.

For further information please contact:
Book Sales Department, Royal Society of Chemistry, Thomas Graham House, Science Park, Milton Road, Cambridge, CB4 0WF, UK
Telephone: +44 (0)1223 420066, Fax: +44 (0)1223 420247,
Email: books@rsc.org
Visit our website at http://www.rsc.org/Shop/Books/

Enantioselective Homogeneous Supported Catalysis

Edited by

Radovan Šebesta
Department of Organic Chemistry, Comenius University, Bratislava, Slovakia

RSC Publishing

RSC Green Chemistry No. 15

ISBN: 978-1-84973-176-8
ISSN: 1757-7039

A catalogue record for this book is available from the British Library

Published by The Royal Society of Chemistry,
Thomas Graham House, Science Park, Milton Road,
Cambridge CB4 0WF, UK

Registered Charity Number 207890

For further information see our web site at www.rsc.org

Printed in Great Britain by CPI Group (UK) Ltd, Croydon, CR0 4YY

Preface

Asymmetric catalysis is arguably the most promising way to synthesize chiral compounds. The notion to use the stereoinducing component of a reaction, *i.e.* a chiral catalyst, in sub-stochiometric amounts is important, because it is the catalyst which is usually the most expensive, in terms of time, labour or money. The research community therefore devotes considerable effort towards catalyst immobilization and recycling. Immobilization of a chiral catalyst is typically realized by means of grafting the catalyst onto a support. Catalysts can be supported either on solids or in liquid phase. The term 'heterogeneous catalysis', thus quite clearly describes all catalysts supported on solids. A broad range of solid supports have been described, especially inorganic materials, such as SiO_2, Al_2O_3, ZrO_2, zeolites, clays or charcoal. Another important group of solid supports are organic polymers. Catalysts have been immobilized on these supporting materials through a number of chemical or physical interactions. Out of this flourishing field of research, a relatively new concept of liquid support for catalysts came out. Catalytic systems which act in liquid phase but can be easily recovered because of specifically tailored structure form the main part of this book. The title of this book: '**Enantioselective homogeneous supported catalysis**' thus represents recoverable catalysts operating in liquid phase. This book aims at reviewing the main concepts of this area. The emphasis is on catalysts modified with ionic, polymer and fluorous tags. New ideas in catalysis in ionic liquids and water are also discussed. The distinction between heterogeneous and supported homogeneous catalysis is not always clear, so the book also deals with dendrimeric catalysts and catalysts immobilized through weak forces on solid supports. They were included because their properties and activities often resemble catalysts in homogeneous phase.

The book is composed of eight chapters, each devoted to particular concept of enantioselective homogeneous supported catalysis. The first chapter deals with catalysts with ionic tags, especially in connection with original idea of this

RSC Green Chemistry No. 15
Enantioselective Homogeneous Supported Catalysis
Edited by Radovan Šebesta
© Royal Society of Chemistry 2012
Published by the Royal Society of Chemistry, www.rsc.org

concept, ionic liquids. The original idea of ionic tagging was connected with transition metal catalysts, but rapidly gained popularity also in metal-free catalysts. Therefore Chapter 2 is devoted to organocatalysts immobilized in homogeneous phases. Štefan Toma (Comenius University in Bratislava) describes not only ionically-tagged organocatalysts, but presents also some newer ideas in supported organocatalysts. Within last the few years a great number of supported organocatalysts have been developed, therefore some other chapter of the book also present additional metal-free catalytic systems.

Chapter 3, by P. Goodrich, C. Paun and C. Hardacre at Queen's University in Belfast deals with catalysis by unmodified catalyst in ionic liquids. The focus of the chapter, however, is on newer and more elaborate concepts of ionic liquid supported phases, as simple catalysis in ionic liquids have been amply described by several reviews and book chapters.

Chapter 4, written by Marco Bandini at the University of Bologna focuses on metal-based catalysts immobilized on soluble polymers.

In Chapter 5, Robertus J. M. Klein Gebbink and Morgane A. N. Virboul from Utrecht University write about catalytic dendrimers.

Chapter 6, written by Gianluca Pozzi (The National Research Council in Milan), deals with fluorous catalyst.

In Chapter 7, Szymon Buda, Monika Pasternak and Jacek Mlynarski (Jagiellonian University in Krakow) give an overview of asymmetric catalysis in aqueous media.

Chapter 8, written by J. M. Fraile, J. I. García, C. I. Herrerias, J. A. Mayoral, and E. Pires (University of Zaragoza), has as its focus non-covalent immobilization.

Not all aspects of asymmetric catalysis with homogeneously supported catalysts could have been covered in this book. Because of space constraints, some new and interesting ideas have not been included, such as self-supported catalysts. An interested reader can find more details in a review by Ding and co-workers.[1] Topics presented in this book have been discussed in other more general volumes before.[2,3] However, the field of asymmetric catalysis is developing fast with a number of new discoveries and ideas appearing rapidly. Therefore I believe that this focused collection of comprehensive reviews will be of interest to many researchers and advanced students in the field of asymmetric catalysis.

Radovan Šebesta
Bratislava

References

1. Z. Wang, G. Chen and K. Ding, *Chem. Rev.*, 2008, **109**, 322.
2. M. Benaglia, *Recoverable and Recyclable Catalysts*, Wiley, 2009.
3. K. Ding and Y. Uozumi, *Handbook of Asymmetric Heterogeneous Catalysis*, Wiley-VCH, 2008.

Contents

RSC Green Chemistry No. 15
Enantioselective Homogeneous Supported Catalysis
Edited by Radovan Šebesta
© Royal Society of Chemistry 2012
Published by the Royal Society of Chemistry, www.rsc.org

CHAPTER 1

Ionically-tagged Transition Metal Catalysts

RADOVAN ŠEBESTA

Department of Organic Chemistry, Faculty of Natural Sciences, Comenius University in Bratislava, Mlynska dolina CH-2, SK-84215 Bratislava, Slovakia

1.1 Introduction

Ionic liquids are liquids consisting solely of ions. Those with melting points below 100 °C are often called room temperature ionic liquids but within this chapter simply the term 'ionic liquids' will be used. Their interesting physical/chemical properties stimulated the investigation of a great number of research and even technical applications. Their use in synthesis has been summarized in an excellent two-volume book edited by Wasserscheid and Welton.[1]

The concept of ionically-tagged transition metal catalysts develops on the interface of homogeneous and heterogeneous catalysis. It aims to combine positive features of both worlds. The primary function of ionic tags is to enable catalyst recycling. On the other hand, appropriate reaction media offer the possibility of maintaining catalysis in the liquid phase and so benefit from high catalytic activities of homogeneous catalysts. Furthermore, the effect of ionic tags often goes beyond a simple anchor for immobilization and recycling of catalyst. Many reactions are enhanced as a result of an ionic tag installed into the catalyst structure compared to the unmodified catalyst. This effect led Lombardo and Trombini to postulate a concept of electrosteric activation.[2]

RSC Green Chemistry No. 15
Enantioselective Homogeneous Supported Catalysis
Edited by Radovan Šebesta
© Royal Society of Chemistry 2012
Published by the Royal Society of Chemistry, www.rsc.org

This means the stabilization of a transition state of a catalytic reaction by electrostatic and steric interaction with ionic tags.

Suitable reaction media for ionically-tagged catalysts can be ionic liquids, supercritical liquids, water or combinations of these. Use of biphasic set-ups is also attractive. This chapter gives an overview of enantioselective catalysis with transition metal based catalysts having an ionic moiety. Ionic catalysts used in water are, however, excluded as they are extensively covered in Chapter 7.

In this chapter, 1-butyl-3-methylimidazolium and 1-ethyl-3-methylimidazolium cations are denoted as [bmim] and [emim]; 1-butyl-2,3-dimethylimidazolium and 1-hexyl-2,3-dimethylimidazolium as [bdmim] and [hdmim]; 1-butylpyridinium cation is denoted as [bpyr]; *N*-ethyl-3-methylpiccolinium cation is denoted as [epic]. The bis(trifluoromethyl sulfonyl)imide anion is denoted as NTf_2, and triflate and tosylate as OTf and OTs.

1.2   Achiral Ionically-tagged Transition Metal Complexes

Highly polar ligands are well known in the coordination chemistry of transition metals. Such complexes were, at first, only used in water. Several reviews and books give a good overview of synthesis and applications of these hydrophilic ligands.[3–6] However, it was only in 1996 when Chauvin and co-workers suggested that ligands bearing ionic moieties can be used for immobilization of metal complexes in ionic liquids.[7] They noticed that rhodium-catalyzed hydrogenation of pent-1-ene was five times faster in [bmim]PF_6 than in acetone. Cationic rhodium complexes were well retained in ionic liquids, which could be reused. However, in rhodium-catalyzed hydroformylation, catalyst leaching was observed. After extraction of organic products, the ionic liquid was reused but part of the rhodium complex was extracted to the organic phase too. Use of mono- and trisulfonated triphenylphosphines as ligands to rhodium completely prevented catalyst leaching to the organic phase.

The concept was then further developed in a number of ways. Various cationic phosphine ligands were tested in rhodium-catalyzed hydrogenations and hydroformylations in ionic liquids or biphasic conditions. Several, structurally different, cations were successfully applied (Figure 1.1). Wasserscheid and Olivier-Bourbigou showed that guanidinium ion is particularly suitable for immobilization of rhodium complexes.[8,9] van Leeuwen and co-workers attached two imidazolium moieties to Xantphos skeleton.[10,11] In this way they created ligand **3**, which showed excellent recycling results and very little leaching.

Ruthenium-arene catalysts were modified by appending imidazolium moiety as well (Scheme 1.1).[12] Imidazolium ionic tags are designed by their similarity with the most common type of ionic liquids. Therefore, it is also the most frequent type of ionic tag. However, it has also a disadvantage, because the proton in position 2 is acidic; therefore, an imidazolium ion with an alkyl group substituted position 2 is often used.

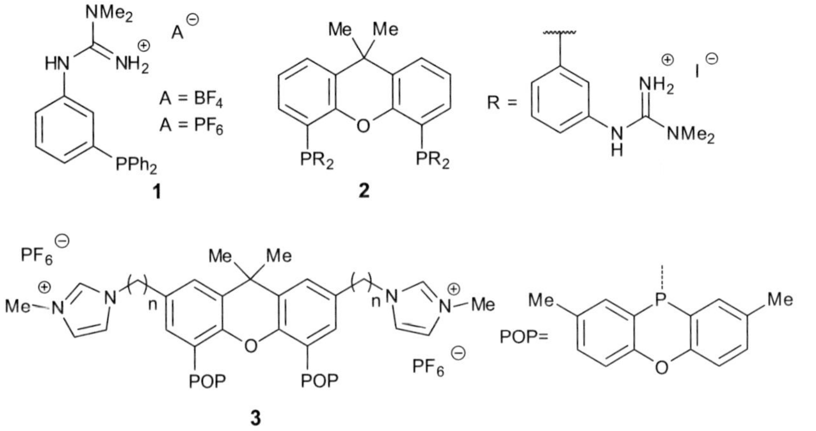

Figure 1.1 Achiral phosphine ligands with ionic tags for hydrogenations and hydroformylations.

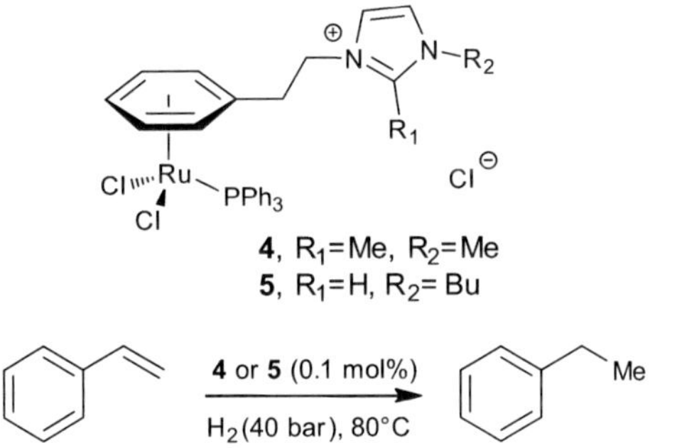

Scheme 1.1

Palladium-catalyzed Suzuki and Stille couplings[13] as well as Heck reaction[14] were successfully performed with various ionically-tagged catalysts (Figure 1.2). These catalysts have good catalytic activities and at the same time significantly reduced metal leaching from ionic liquid.

The usefulness of the ionic-tagging strategy for catalyst immobilization was also amply demonstrated in ruthenium-catalyzed olefin metathesis.[15–20] Both Grubbs catalyst of the first and second generation (**8–13**) have been modified with ionic tags. Usually imidazolium ions have been selected. Figure 1.3 shows various ways in which ionic tags can be introduced into metathesis catalysts.

Achiral Mn-Schiff base **14** was used as an epoxidation catalyst,[21] or porphyrin **15** derivatized with four pyridinium moieties was used for the oxidation of styrene (Figure 1.4).[22]

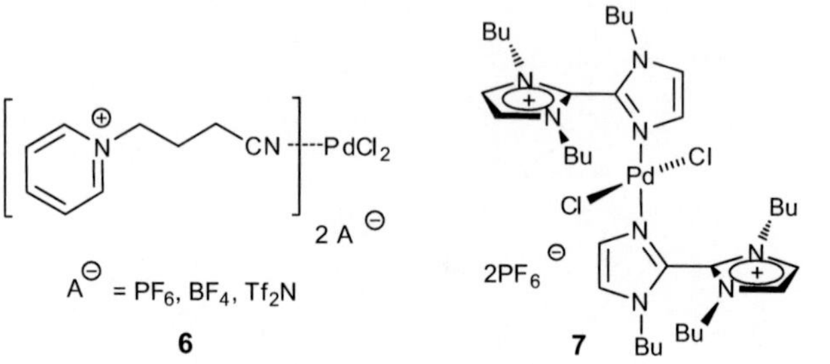

Figure 1.2 Ionically modified ligands for palladium-catalyzed cross-couplings.

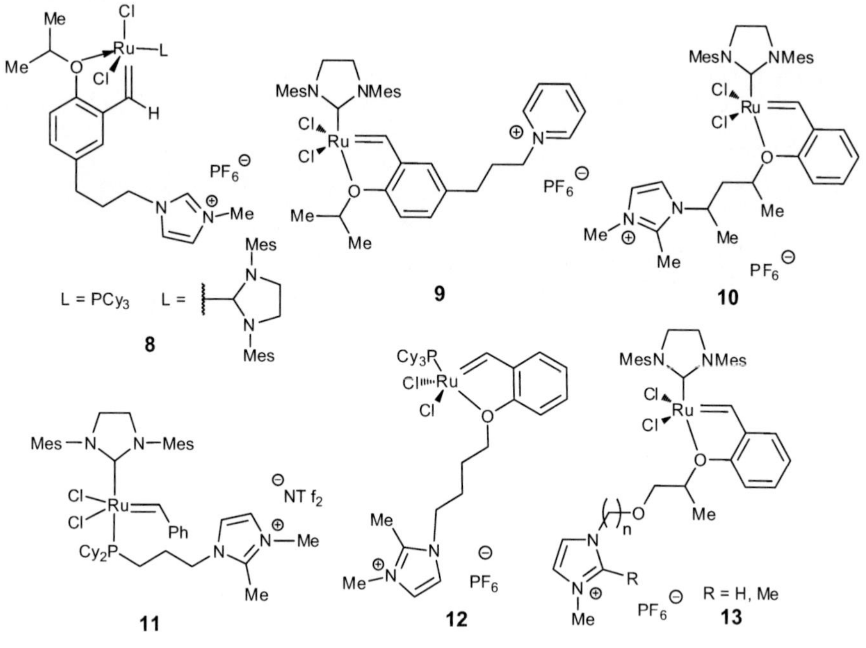

Figure 1.3 Ru complexes with ionic tags for olefin metathesis.

1.3 Chiral Transition Metal Catalysts with Ionic Tag

Chiral transition metal catalysts can be immobilized through ionic forces in ionic liquids. This is enabled by charge carried by a metal ion. Numerous examples of this approach have been described.[23–27] Such systems, however, often suffer from considerable metal leaching during work-up operations. The introduction of an auxiliary ionic moiety into the ligand has been suggested as a possible solution to this problem. The installation of an ionic tag into the ligand structure, indeed, helps prevent metal leaching. Catalytic activities are usually comparable or better than that of unmodified catalysts.

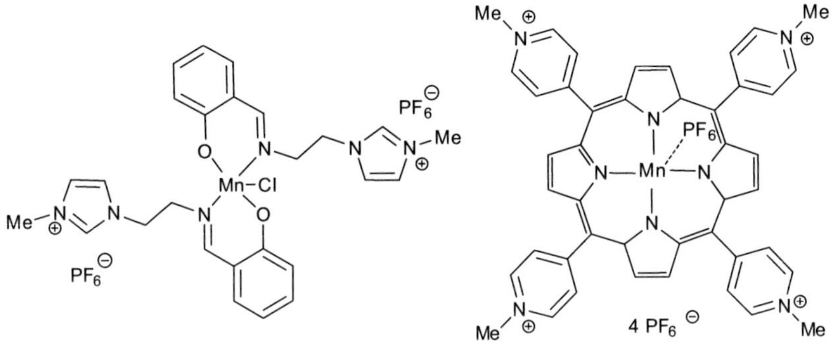

Figure 1.4 Achiral Mn-based oxidation catalysts.

1.3.1 Catalysts for Enantioselective Reduction

Enantioselective hydrogenations are one of the most widely employed asymmetric transformations. Expensive rhodium, ruthenium and iridium complexes are the most active hydrogenation catalysts. This motivated great effort for the synthesis of immobilized catalysts. Also, the concept of ionically-tagged catalysts has been developed and tested on hydrogenation reactions. Several researchers therefore tried to immobilize rhodium and ruthenium complexes with the help of ionic tags. An important factor is also good solubility of hydrogen ionic liquids. However, prototypical hydrogenation catalysts, Rh-BINAP complexes, are too air-sensitive and therefore have not been used in ionic liquids. Rhodium complexes with DIOP[7] or DUPHOS[28] have been successfully used in ionic liquids for hydrogenation of functionalized olefins.

The first rhodium catalyst with an ionic tag was prepared by Lee and co-workers. They attached two imidazolium moieties onto a rhodium-disphosphine complex.[29] The resulting catalyst **16** showed excellent catalytic activities in asymmetric hydrogenation of enamides. In the two-phase system [bmim]SbF$_6$/i-PrOH, enamide was hydrogenated with 97% ee with full conversion within 1 hour. The catalyst **16** was recycled four times with very little loss of catalytic activity (95% ee, 85% conversion in 1 h). In comparison, hydrogenation with unmodified complex **17** resulted in a similarly enantioselective reaction in the first run, but in subsequent runs enantioselectivity and conversion decreased considerably (4th run, 88% ee, 85% conversion in 12 h). After the first run the i-PrOH layer was investigated by inductively coupled plasma atomic emission spectroscopy (ICP-AES). In experiment with complex **16** no rhodium (<1 ppm) or phosphorus (<3 ppm) leaching from the ionic liquid was detected, whereas with complex **17** it was found that 2% rhodium and 6% phosphorus was leached from the ionic liquid. A control experiment with low catalyst loading (0.5 mol% of complex **17**) showed that rhodium leaching is not the sole reason for decreased catalytic performance of the catalyst. Improved stability of the catalyst **16** in ionic liquid was suggested as a possible explanation of its superior catalytic activity (Scheme 1.2).

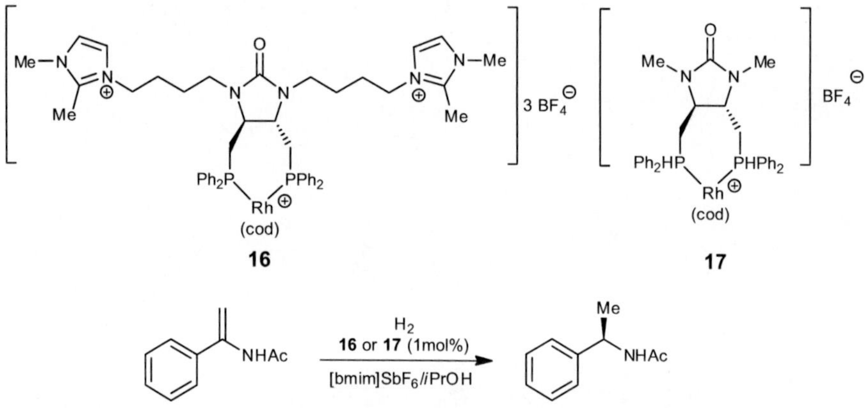

Scheme 1.2

Ferrocenyl diphosphine ligands serve in a number of excellent hydrogenation systems, including industrial applications.[30,31] Solvias and Novartis researchers tested ferrocenyl diphosphines, such as Taniaphos, Josiphos, Walphos or Mandyphos, in rhodium-catalyzed hydrogenation in ionic liquids as well as mixtures of ionic liquid with organic solvents or water.[32] The most promising results were obtained in mixtures of ionic liquids with water – wet ionic liquid. For instance, in a mixture of [omim]BF_4/H_2O ligands Taniaphos and Josiphos afforded enantioselectivities of 99% ee (Scheme 1.3). Josiphos was reused six times with constant ee and only in the seventh run did the reaction time have to

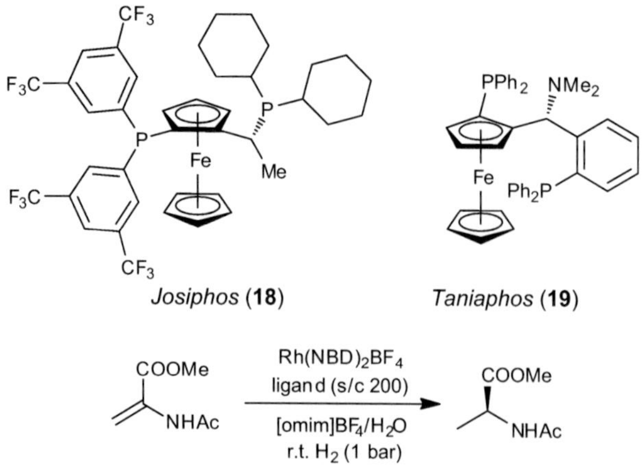

Josiphos (**18**)   Taniaphos (**19**)

Scheme 1.3

be prolonged from 20 to 40 min to reach full conversion. Enantioselectivity, however, remained unchanged (99% ee). Inductively coupled plasma mass spectrometry (ICP-MS) found only 0.9 ppm Rh in the water phase, so the decrease in activity was attributed to catalyst decomposition, rather than to leaching of the catalyst.

Inspired by these promising results, they also synthesized modified Josiphos ligands with imidazolium tag **20** and **21**.[33] These diphosphanes were successfully employed in rhodium-catalyzed asymmetric hydrogenation of methyl acetamidoacrylate (MAA) and dimethyl itaconate (DMI) (Scheme 1.4).

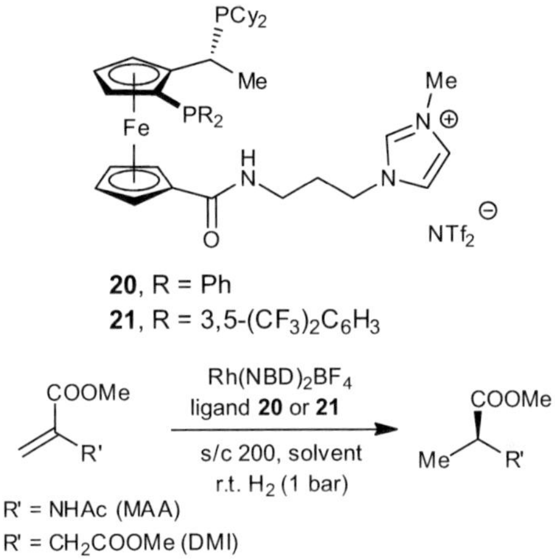
PCy2
Me
Fe
PR2
H
N
Me
N
N
NTf2
O
20, R = Ph
21, R = 3,5-(CF3)2C6H3
COOMe
R'
Rh(NBD)2BF4
ligand 20 or 21
s/c 200, solvent
r.t. H2 (1 bar)
COOMe
Me
R'
R' = NHAc (MAA)
R' = CH2COOMe (DMI)

Scheme 1.4

Modified ligands **20** and **21** paralleled catalytic behaviour of unmodified ligands in the asymmetric hydrogenation in classical organic solvents as well as under biphasic conditions. Hydrogenation of MAA using ligand **21** proceeded with 99% ee, and for reduction of DMI ligand **20** was superior (99% ee). Imidazolium tagged ligands showed much better reusability in *t*-BuOMe/[bmim]BF_4 biphasic system compared with unmodified Josiphos ligands (approx. 15% decrease of TOF for the 8th cycle, with constant enantiomeric purity of the product).

Ruthenium-catalyzed transfer hydrogenation of ketones is an important alternative to classical hydrogenation methods.[34] Several attempts were made to prepare reusable ruthenium catalysts with ionic tags. Geldbach and Dyson succeeded in the introduction of the imidazolium moiety onto η^6-arene of the Ru complex.[35] Catalyst **22** was slightly less active than unmodified catalyst **23** in the absence of ionic liquid. More importantly, enantioselectivity was significantly lower (58% ee) compared to *p*-cymene ligand **23** (98% ee). On the other hand, with catalyst **22**, dissolved in 1-butyl-2,3-dimethylimidazolium hexafluorophosphate [bdmim]PF_6, hydrogenation of acetophenone in a biphasic system (ionic liquid and propan-2-ol/KOH) proceeded well (Scheme 1.5).

Using ionic catalyst **22**, enantioselectivity was as high as with traditional *p*-cymene ligand **23** (98% ee). Recycling of the catalytic system led to decreased conversion (52% in the 4th run compared with 99% in the 1st run) of the reaction, but enantioselectivity remained the same. Decrease in catalytic activity was explained by partial leaching of the catalyst from ionic liquid due to decomposition of the ruthenium complex. As an alternative hydride donor to propan-2-ol/KOH, formic acid/triethylamine azeotrop was also used. Enantioselectivities were even higher (99% ee) but again a decrease in conversion between recycling experiments was observed.

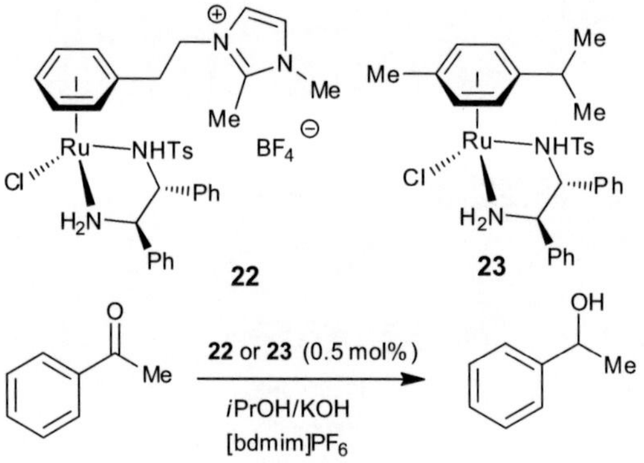

Scheme 1.5

A different approach to functionalization of Ru-TsDPEN catalyst **25** was chosen by Ohta and co-workers.[36] They introduced the imidazolium moiety into the diamine part of the ruthenium complex. Using the catalyst **24**, transfer hydrogenation of acetophenone with formic acid/triethylamine in [bmim]PF₆ proceeded with high enantioselectivity (92% ee) and high conversion (98%). Unmodified Ru-TsDPEN catalyst **25** gave similar results (93% ee, 96% conversion) but after the 4th cycle conversion decreased gradually (4th cycle, 88%; 5th cycle, 63%). In comparison, catalyst **24** showed slightly better results (4th cycle, 92%; 5th cycle, 75%) (Figure 1.5).

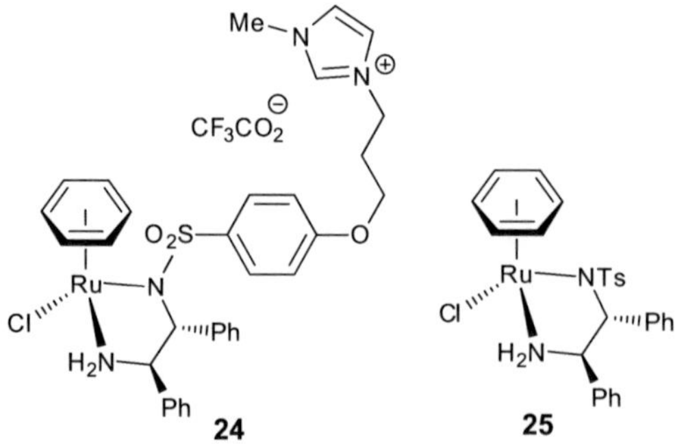

Figure 1.5   Ru-TsDPEN catalysts for transfer hydrogenation.

Ru-BINAP complexes are important alternatives to their rhodium counterparts. They can be efficiently used for hydrogenation of β-keto esters in ionic liquids.[37] Recycling of Ru-BINAP catalyst in ionic liquid was possible but a significant decrease in the conversion after the fourth cycle was observed. This decrease was attributed to leaching of the Ru complex, which was also confirmed by atomic absorption analysis. To prevent leaching and thus improve recyclability, Lin and co-workers prepared two polar phosphonic acid-derived Ru-BINAP systems 26 and 27 for asymmetric hydrogenation of β-keto esters in ionic liquids (Scheme 1.6).[38]

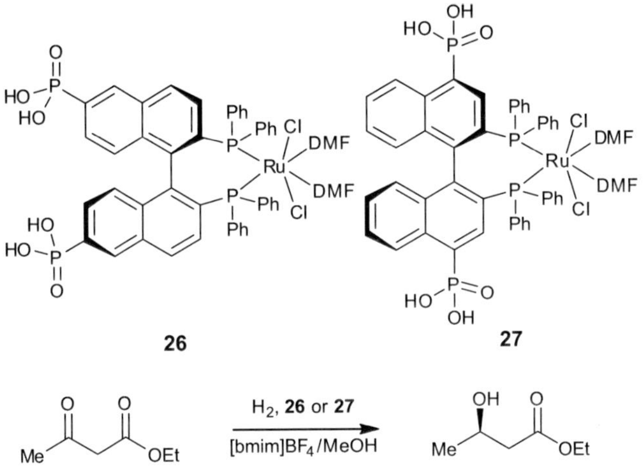

Scheme 1.6

Several β-keto esters were hydrogenated with high enantioselectivities (up to 99.3% ee) using both catalysts 26 and 27 in [bmim]BF$_4$ and [bmim]PF$_6$ as well as in [bdmim]NTf$_2$. The catalysts performed best in a mixture of ionic liquid and methanol. Products were simply extracted with degassed hexane and the ionic liquid was reused in the next experiment, after drying. Direct current plasma spectroscopy showed that only very little ruthenium leeched into the organic phase (between 0.01 and 0.04%, depending on ionic liquid used). It was possible to recycle the catalyst in ionic liquid four times with only small or no decrease in enantioselectivity and conversion. Enantioselectivities with polar catalyst were comparable with unmodified Ru-BINAP system in ionic liquids.

Lin and co-workers later studied the effect of substitution of Ru-BINAP catalytic system in more detail.[39] They found that polar substituents led to a decrease in enantioselectivity in methanol, but the P(O)(OH)$_2$ group was more enantioselective in a mixture of [bmim]BF$_4$/MeOH.

Modified Ru-BINAP catalyst 28 and 29 for hydrogenation of β-keto esters was also developed by Lemaire and co-workers (Figure 1.6).[40] These catalysts, however, were slightly less active. Hydrogenation of ethyl acetoacetate proceeded with enantioselectivity (90% ee) and only one recycling was performed.

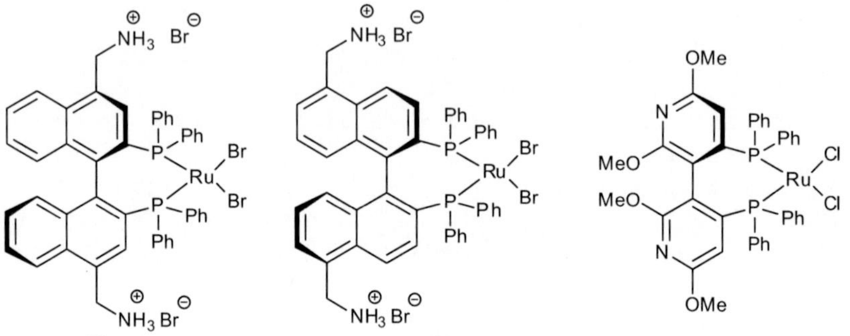

Figure 1.6   Ru-biaryl hydrogenation catalysts.

Chan and co-workers showed that ionic tagging is not absolutely necessary for highly enantioselective and recyclable ruthenium catalysts.[41] They showed that even the polar bipyridine derivative **30** can be efficiently recycled four times without loss of activity (Figure 1.6). In the mixture [bmim]PF$_6$/MeOH, methyl acetoacetonate was reduced with 99% conversion and 98% ee and in the fifth cycle conversion was still 96% and enantioselectivity 93%.

1.3.2   Catalysts for Enantioselective C–C and C–Heteroatom Bond Formation

A chiral vanadyl salen complex was tagged with the imidazolium ion. Complex **31** catalyzed the asymmetric cyanosilylation of benzaldehyde in [bmim]PF$_6$. The enantioselectivity of the reaction (57% ee) was significantly lower than that with unmodified vanadyl ligand (89% ee). However, the catalytic activity of complex **31** was considerably higher than that of the unmodified catalyst as well as the catalyst anchored on silica, activated carbon and single wall carbon nanotube (Scheme 1.7).[42]

Moreau and co-workers prepared imidazolium-tagged camphorsulfonamide ligands for titanium-promoted asymmetric Et$_2$Zn addition to benzaldehyde (Scheme 1.8).[43] Alkylation of benzaldehyde in the presence of the titanium complex of **32** was performed in CH$_2$Cl$_2$. Conversion were over 99% in all cases, but enantioselectivities were only moderate (65% ee). Separation of the product from the catalyst was achieved by evaporation and hydrolysis followed by extraction with diethyl ether, in which the catalyst was not soluble. The catalytic system was used four times without any change in catalytic activity and enantioselectivity.

Moreau and co-workers continued the development of titanium-mediated addition of diethylzinc to aldehydes by preparation of ion-tagged BINOL ligand **33** (Scheme 1.8).[44] The ligand **33** afforded Et$_2$Zn addition with 82% ee. The same result was obtained also with unmodified BINOL. The reaction was

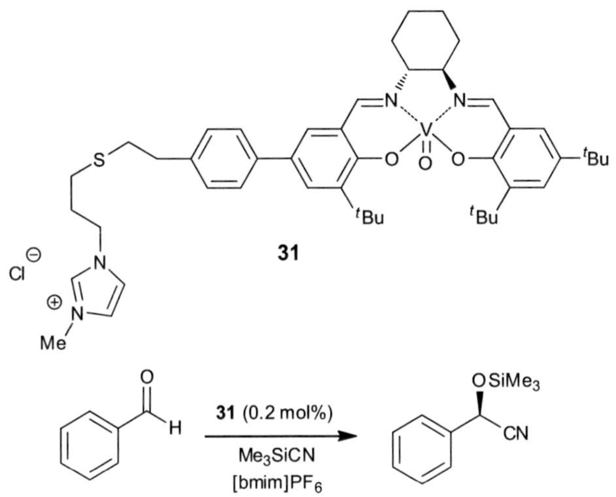

31

Scheme 1.7

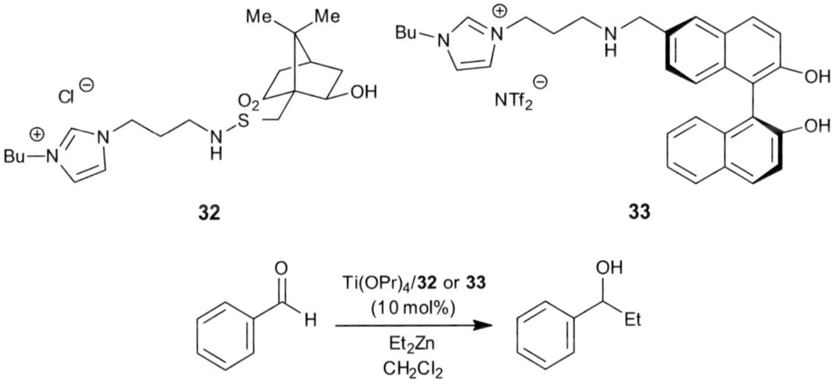

32   **33**

Scheme 1.8

performed in CH_2Cl_2 but the catalyst was recyclable three times with same activity and enantioselectivity.

The fact that dialkylzinc additions to aldehydes can also be catalyzed by simple amino alcohols motivated Lombardo and co-workers to design an amino alcohol catalyst with an ionic tag.[45] They selected diphenylprolinol for its availability and ease of modification. An ionic tag was introduced by alkylation of pyrrolidine nitrogen. Envisaged problems with carbene formation in imidazolium cations prompted them to use the simple trialkylammonium cation instead. These considerations led to three amino alcohols **34** with different spacer length. Choice of ionic liquid was also motivated by higher stability towards basic diethylzinc. Thus addition of diethylzinc to benzaldehyde proceeded best in [bmpy]NTf$_2$ giving product in 80% yield and with 89% ee.

The different length of alkyl spacer did not affect chemical yields but it had a profound effect on enantioselectivity. Catalysts **34a** ($n = 3$) and **34c** ($n = 6$) were less selective (72 and 80% ee, respectively). The striking effect of the ionic tag was demonstrated by comparison with *N*-methyldiphenylprolinol, with which the reaction proceeded with good conversion but only with 38% ee. Ethylation of benzaldehyde was run 10 times without any decrease in chemical yield or enantioselectivity (Scheme 1.9). The authors also developed an efficient recycling procedure. Zinc salts were removed from the ionic liquid by aqueous EDTA solution. The product was extracted with diethyl ether and the ionic liquid with catalyst was reused after drying. A range of aldehydes were ethylated by this method, giving products typically in over 90% yield and 90% ee.

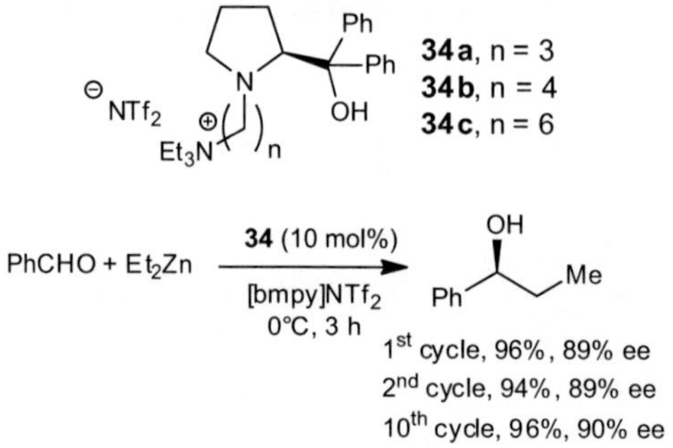

Scheme 1.9

Palladium-catalyzed allylic alkylation can be performed with ferrocenyl ligands in ionic liquids.[46] Enantioselectivities are higher than in organic solvents, but reduced yields and enantioselectivities were observed upon recycling. This motivated us to prepare imidazolium-tagged ferrocene ligands.[47] Ligands **35** and **36** were used in Pd-catalyzed allylic alkylation of 1,3-diphenylprop-2-enyl acetate with dimethyl malonate (Scheme 1.10).

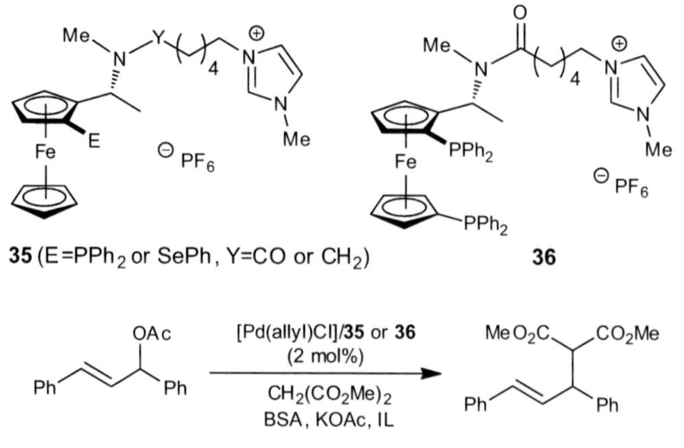

Scheme 1.10

Highest enantioselectivity (92% ee) was achieved with diphosphine **36** in [bmim]PF_6. Interestingly, this enantioselectivity is higher than that of the corresponding ligand without the imidazolium moiety (**37**, BPPFA) when used in THF as well as in ionic liquid [bmim]PF_6. This is another example of electrosteric activation mediated by an ionic tag. Other screened ligands were inferior. Recycling experiments led to a slight decrease of enantioselectivity and considerable decrease of chemical yield (1st run, 94%, 89% ee; 2nd run, 57%, 77% ee; 3rd run 16%, 69% ee in [emim]$EtOSO_3$).

Similar behaviour of the ion-tagged catalyst was also observed in allylic substitution with heteroatom nucleophiles.[48] With phthalimide, high enantioselectivities (up to 92% ee) were observed but the catalyst was again poorly recyclable. Attempts on catalyst improvement by the introduction of a methyl group in position 2 on the imidazolium moiety (ligand **39**) did not bring the desired positive effect on recycling (Scheme 1.11).

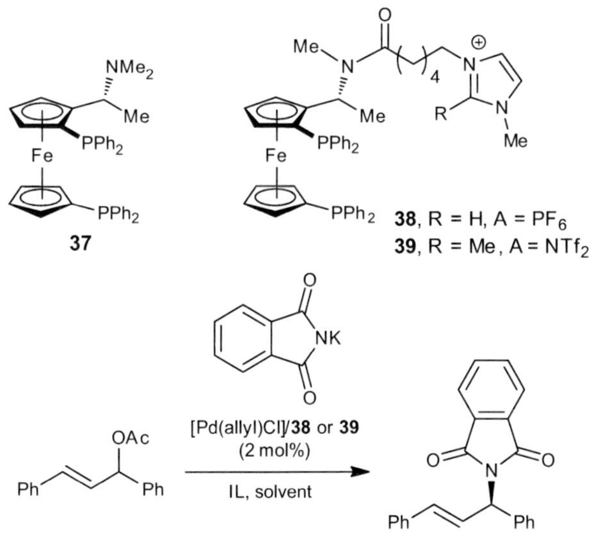

38, R = H, A = PF_6
39, R = Me, A = NTf_2

[Pd(allyl)Cl]/**38** or **39**
(2 mol%)

IL, solvent

Scheme 1.11

We investigated the reaction with respect to catalyst behaviour and its stability by NMR. ^{31}P-NMR of the Pd complex of ligand **39** in ionic liquid clearly showed complexation of Pd and ligand. However, NMR of the recovered ionic liquid after reaction suggested catalyst decomposition. The catalyst is probably unstable in the inherently basic conditions of allylic substitution reactions.

Imidazolium-tagged bis(oxazoline) **41** has been prepared and used as a chiral ligand in copper-catalyzed Diels–Alter reactions of *N*-acryloyl and *N*-crotonoyloxazolidinones with cyclopentadiene.[49] Using the ligand **41**, significant enhancement in rate and enantioselectivity of the reaction in ionic liquid was achieved compared with dichloromethane. Compared with unfunctionalized bis(oxazoline) ligand **40**, similar enantioselectivities were obtained, but with marked rate enhancement in ionic liquid in the case of the imidazolium-tagged ligand. Catalyst **41** could also be recycled 10 times without loss of its activity and enantioselectivity (Scheme 1.12). ICP-MS analysis of extracts after reaction showed that only up to 0.03% of the copper was leached to the organic phase. Ligand leaching is an even more important parameter for Lewis acid catalyzed reactions, because of the usually rapid background reaction of uncomplexed metal. HPLC analysis could not detect any leaching of immidazolium-tagged ligand. In contrast, 6–8% of uncharged oxazoline was extracted after each reaction cycle.

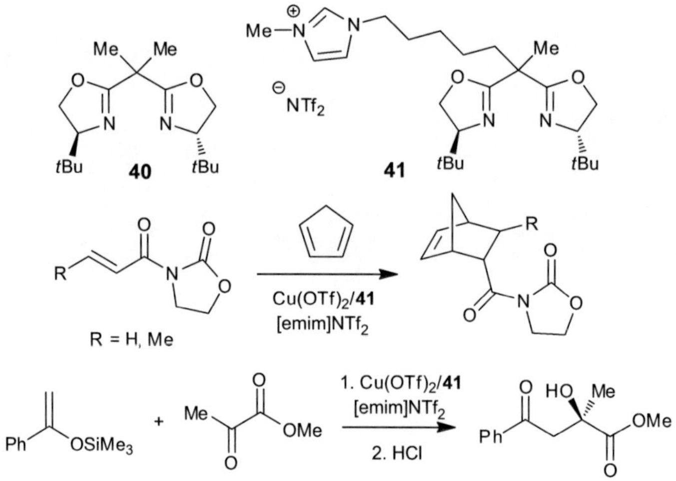

Scheme 1.12

The same catalyst was used also in a copper-catalyzed Mukaiyama aldol reaction.[50] The addition of 1-phenyl-1-trimethylsiloxyethene to methyl pyruvate was highly enantioselective (91% ee) and conversion and enantiomeric purity of the product remained the same during three reaction cycles (Scheme 1.12).

Santini and Galland designed and synthesized ionic phosphite ligands for a hydrocyanation reaction.[51] Later, Gavrilov and Reetz prepared chiral imidazolium-tagged phosphite **42** and diamidophosphite ligand **43**.[52] Their rhodium and palladium complexes were efficient catalysts for asymmetric hydrogenation and allylic substitution in organic solvents (Figure 1.7). Interestingly, hydrogenation in [bdmim]BF$_4$ was less selective than in dichloromethane (79% ee versus 94% ee). These ligands were also immobilized onto a solid support through ionic interaction. Their direct recycling, however, was not studied.

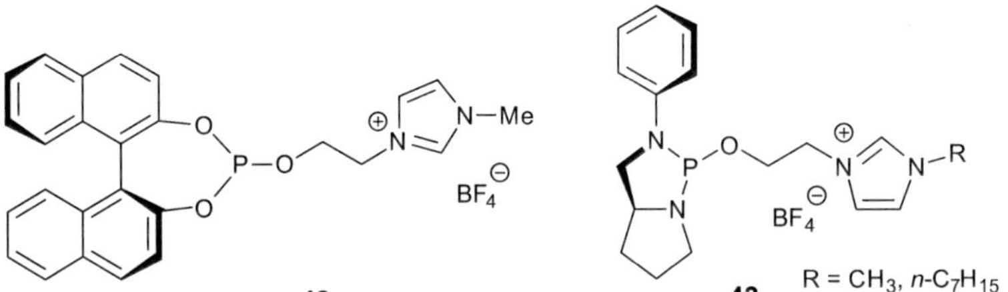

Figure 1.7 Ionically tagged phosphite and diamidophosphite ligands.

1.4 Conclusions

Catalytic systems with ionic tags proved to be useful alternatives to unmodified catalysts. The catalytic activities of these systems often parallel or even surpass those of classical catalysts with the added benefit of easy product isolation and possible catalyst reuse. Ionically tagged catalysts, however, sometimes suffer from other problems. Modified catalysts can be chemically less stable than their unmodified counterparts. Published data suggest that careful design and optimization of catalyst structure in connection with reaction parameters are crucial for success. There is a great potential for improvements of catalyst physical and chemical properties by altering both the cation and the anion part of such molecules. Although great progress has already been achieved in this field, there is still much to be discovered. One of the most important issues seems to be chemical stability of the ionically-tagged catalytic system in the desired reaction. Another issue is a limited number of enantioselective transformations studied so far that use transition metal catalysts with ionic tags. The notion of electrosteric activation, introduced on the basis of empirical observations by Lombardo and Trombini, needs more thorough theoretical investigation. This would enable further improvements and hopefully rational design of ionically-tagged catalysts in the future.

References

1. P. Wasserscheid and T. Welton, *Ionic Liquids in Synthesis*, 2nd Edition, Wiley-VCH, Weinheim, 2008.
2. M. Lombardo and C. Trombini, *ChemCatChem*, 2010, **2**, 135–145.
3. W. A. Herrmann and C. W. Kohlpaintner, *Angew. Chem., Int. Ed. Engl.*, 1993, **32**, 1524–1544.
4. K. H. Shaughnessy, *Chem. Rev.*, 2009, **109**, 643.
5. B. Cornils and W. A. Herrmann (eds.), *Aqueous Phase Organometallic Catalysis*, 2nd edn, Wiley-VCH, Weinheim, 2004.
6. D. Sinou, *Adv. Synth. Catal.*, 2002, **344**, 221–237.
7. Y. Chauvin, L. Mussmann and H. Olivier, *Angew. Chem. Int. Ed. Engl.*, 1996, **34**, 2698–2700.

8. P. Wasserscheid, H. Waffenschmidt, P. Machnitzki, K. W. Kottsieper and O. Stelzer, *Chem. Commun.*, 2001, 451–452.

9. F. Favre, H. Olivier-Bourbigou, D. Commereuc and L. Saussine, *Chem. Commun.*, 2001, 1360–1361.

10. R. P. J. Bronger, S. M. Silva, P. C. J. Kamer and P. W. N. M. van Leeuwen, *Chem. Commun.*, 2002, 3044–3045.

11. R. P. J. Bronger, S. M. Silva, P. C. J. Kamer and P. W. N. M. v. Leeuwen, *Dalton Trans.*, 2004, 1590–1596.

12. T. J. Geldbach, G. Laurenczy, R. Scopelliti and P. J. Dyson, *Organometallics*, 2006, **25**, 733–742.

13. D. Zhao, Z. Fei, T. J. Geldbach, R. Scopelliti and P. J. Dyson, *J. Am. Chem. Soc.*, 2004, **126**, 15876–15882.

14. J. C. Xiao, B. Twamley and J. M. Shreeve, *Org. Lett.*, 2004, **6**, 3845–3847.

15. N. Audic, H. Clavier, M. Mauduit and J. C. Guillemin, *J. Am. Chem. Soc.*, 2003, **125**, 9248–9249.

16. Q. Yao and Y. Zhang, *Angew. Chem., Int. Ed. Engl.*, 2003, **42**, 3395–3398.

17. D. Rix, F. Caijo, I. Laurent, L. Gulajski, K. Grela and M. Mauduit, *Chem. Commun.*, 2007, 3771–3773.

18. C. S. Consorti, G. L. P. Aydos, G. Ebeling and J. Dupont, *Org. Lett.*, 2008, **10**, 237–240.

19. C. S. Consorti, G. L. P. Aydos, G. Ebeling and J. Dupont, *Organometallics*, 2009, **28**, 4527–4533.

20. S.-W. Chen, J. H. Kim, K. Y. Ryu, W.-W. Lee, J. Hong and S.-g. Lee, *Tetrahedron*, 2009, **65**, 3397–3403.

21. Y. Peng, Y. Cai, G. Song and J. Chen, *Synlett*, 2005, 2147–2150.

22. Y. Liu, H.-J. Zhang, Y. Lu, Y.-Q. Cai and X.-L. Liu, *Green Chem.*, 2007, **9**, 1114–1119.

23. C. E. Song, *Chem. Commun.*, 2004, 1033–1043.

24. V. I. Parvulescu and C. Hardacre, *Chem. Rev.*, 2007, **107**, 2615–2665.

25. S. V. Malhotra, V. Kumar and V. S. Parmar, *Curr. Org. Synth.*, 2007, **4**, 370–380.

26. C. Baudequin, J. Baudoux, J. Levillain, D. Cahard, A.-C. Gaumont and J.-C. Plaquevent, *Tetrahedron: Asymmetry*, 2003, **14**, 3081–3093.

27. P. Wasserscheid and P. Schulz, in *Ionic Liquids in Synthesis*, eds. P. Wasserscheid and T. Welton, Wiley-VCH, Weinheim, 2008, vol. 2, pp. 369–463.

28. S. Guernik, A. Wolfson, M. Herskowitz, N. Greenspoon and S. Geresh, *Chem. Commun.*, 2001, 2314–2315.

29. S.-g. Lee, Y. J. Zhang, J. Y. Piao, H. Yoon, C. E. Song, J. H. Choi and J. Hong, *Chem. Commun.*, 2003, 2624–2625.

30. H. U. Blaser, C. Malan, B. Pugin, F. Spindler, H. Steiner and M. Studer, *Adv. Synth. Catal.*, 2003, **345**, 103–151.

31. H. U. Blaser, W. Brieden, B. Pugin, F. Spindler, M. Studer and A. Togni, *Top. Catal.*, 2002, **19**, 3–16.

32. B. Pugin, M. Studer, E. Kuesters, G. Sedelmeier and X. Feng, *Adv. Synth. Catal.*, 2004, **346**, 1481–1486.

33. X. Feng, B. Pugin, E. Küsters, G. Sedelmeier and H.-U. Blaser, *Adv. Synth. Catal.*, 2007, **349**, 1803–1807.
34. C. Wang, X. Wu and J. Xiao, *Chem. Asian J.*, 2008, **3**, 1750–1770.
35. T. J. Geldbach and P. J. Dyson, *J. Am. Chem. Soc.*, 2004, **126**, 8114–8115.
36. I. Kawasaki, K. Tsunoda, T. Tsuji, T. Yamaguchi, H. Shibuta, N. Uchida, M. Yamashita and S. Ohta, *Chem. Commun.*, 2005, 2134–2136.
37. A. Berger, R. F. de Souza, M. R. Delgado and J. Dupont, *Tetrahedron: Asymmetry*, 2001, **12**, 1825–1828.
38. H. L. Ngo, A. Hu and W. Lin, *Chem. Commun.*, 2003, 1912–1913.
39. A. Hu, H. L. Ngo and W. Lin, *Angew. Chem., Int. Ed. Engl.*, 2004, **43**, 2501–2504.
40. M. Berthod, J.-M. Joerger, G. Mignani, M. Vaultier and M. Lemaire, *Tetrahedron: Asymmetry*, 2004, **15**, 2219–2221.
41. K. H. Lam, L. Xu, L. Feng, J. Ruan, Q. Fan and A. S. C. Chan, *Can. J. Chem.*, 2005, **83**, 903–908.
42. C. Baleizao, B. Gigante, H. Garcia and A. Corma, *Tetrahedron*, 2004, **60**, 10461–10468.
43. B. Gadenne, P. Hesemann and J. J. E. Moreau, *Tetrahedron Lett.*, 2004, **45**, 8157–8160.
44. B. Gadenne, P. Hesemann and J. J. E. Moreau, *Tetrahedron: Asymmetry*, 2005, **16**, 2001–2006.
45. M. Lombardo, M. Chiarucci and C. Trombini, *Chem. Eur. J.*, 2008, **14**, 11288–11291.
46. Š. Toma, B. Gotov, I. Kmentová and E. Solčániová, *Green Chem.*, 2000, **2**, 149–151.
47. R. Šebesta, M. Mečiarová, V. Poláčková, E. Veverková, I. Kmentová, E. Gajdošiková, J. Cvengroš, R. Buffa and V. Gajda, *Coll. Czech. Chem. Commun.*, 2007, **72**, 1057–1068.
48. R. Šebesta and F. Bilčík, *Tetrahedron: Asymmetry*, 2009, **20**, 1892–1896.
49. S. Doherty, P. Goodrich, C. Hardacre, J. G. Knight, M. T. Nguyen, V. I. Pârvulescu and C. Paun, *Adv. Synth. Catal.*, 2007, **349**, 951–963.
50. S. Doherty, P. Goodrich, C. Hardacre, V. Pârvulescu and C. Paun, *Adv. Synth. Catal.*, 2008, **350**, 295–302.
51. C. Vallée, Y. Chauvin, J.-M. Basset, C. C. Santini and J.-C. Galland, *Adv. Synth. Catal.*, 2005, **347**, 1835–1847.
52. K. N. Gavrilov, S. E. Lyubimov, O. G. Bondarev, M. G. Maksimova, S. V. Zheglov, P. V. Petrovskii, V. A. Davankov and M. T. Reetz, *Adv. Synth. Catal.*, 2007, **349**, 609–616.

CHAPTER 2

Catalysis with Supported Organocatalysts

ŠTEFAN TOMA

Department of Organic Chemistry, Faculty of Natural Sciences, Comenius University in Bratislava, Mlynska dolina CH-2, SK-84215 Bratislava, Slovakia

2.1 Introduction

The first application of small chiral organocatalysts was published nearly a century ago. The addition of HCN to benzaldehyde was catalyzed by quinine and quinidine,[1] but the reaction gave product with just 10% ee. The next paper[2] was published nearly a half century later and described the addition of methanol to methyl phenyl ketene. O-Acetylquinine (1 mol%) was used as the catalyst and product was isolated with 74% ee. Ten years later (S)-proline was used as the catalyst for intramolecular aldol cyclodehydration of Wieland–Miescher ketone, which is an important intermediate in steroid synthesis.[3,4] Reactions were going with good stereoselectivity, as product with 71–93% ee was isolated. Another half century elapsed till the real boom of organocatalysis started in 2000.[5–7]

The terms 'organocatalyst' and 'organocatalysis' were coined by MacMillan.[7] Organocatalyst, according to him, is a small chiral organic molecule formed from C, H, O and N atoms, which catalyzes stereoselective reactions (organocatalytic reactions). Organocatalysts are usually accessible in both enantiomeric forms and are not sensitive to the moisture or air, which is the main advantage over transition metal catalyzed reactions. The term 'organocatalyst' is applied

RSC Green Chemistry No. 15
Enantioselective Homogeneous Supported Catalysis
Edited by Radovan Šebesta
© Royal Society of Chemistry 2012
Published by the Royal Society of Chemistry, www.rsc.org

now in a broader sense and all small organic molecules having some catalytic activity, even phosphines, are declared to be organocatalysts, but in this chapter organocatalysts are recognized as just those that catalyze stereoselective reactions.

A large amount of organocatalysts were usually used at organocatalytic reactions and work-up procedures were sometime complicated by difficult organocatalyst separation. It is not surprising that early attempts have been made to solve this problem by organocatalyst immobilization either on cross-linked polymers, polyethylene glycol, dendrimers etc. and especially appended organocatalyst on an ionic liquid molecule (organocatalysts with an ionic tag).

2.2 Catalysis with Organocatalysts with an Ionic Tag

Organocatalysts with an ionic tag are sometimes named as chiral ionic liquids, which is not strictly correct because term 'ionic liquid' is used when such molecules are used as solvents. We can find therefore organocatalysts with ionic tags in reviews on chiral ionic liquids.[8–12] The term 'organocatalysts with ionic tags' is also used in some reviews.[13–16] There is just one review describing proline derivatives also immobilized on other supports.[17] Supported organocatalysts are also elaborated in two monographs.[18,19] Catalysts with an ionic tag are salts that are soluble in water, ionic liquids and very polar organic solvents, but insoluble in non-polar solvents. These properties allow one to perform reactions in homogeneous medium and at the same time ensuring easy recyclability of the organocatalyst. Among the above mentioned reviews a special attention should be paid to the review of Lombardo and Trombini,[15] as the authors are trying to explain a special role of the ionic tag on the activity of the appended catalyst. They suppose that an ionic tag can interact with developing charges in a polar transition state which result in decreasing of the free-energy barrier of the reaction. For such interplay of coulombic and steric interaction of the ionic tag with transition state they use term 'electrosteric activation'.

2.2.1 Aldol Reactions

The first aldol reaction with a supported catalyst did not in fact use a catalyst with an ionic tag, but (*S*)-proline was immobilized in a thin layer of an ionic liquid on the surface of silica which was modified by an immobilized ionic liquid moiety.[20] Reaction of acetone with benzaldehyde was carried out in homogeneous medium (in [bmim]BF$_4$ ionic liquid) and the catalyst was, after extraction of the product, recovered by filtration and could be reused after a short drying under reduced pressure. The catalytic system was stable and 51% of the aldol product with 64% ee was isolated even after the 3rd cycle. Unfortunately, reactions were carried out with large excess of acetone (1 mL of acetone to 0.5 mmol of aldehyde). Achieved yields as well as enantioselectivity were slightly lower than in the true homogeneous reaction in the same ionic liquid. In the 'Notes' of this paper, the authors have mentioned that the same supported catalyst can also be successfully used in the three component (benzaldehyde, 4-methoxyaniline and acetone) Mannich reaction. The same authors later described[21] that this catalytic system can be reused with reproducible enantioselectivity 13 times.

The first organocatalyst with an ionic tag, catalyst **1**, described by Miao and Chan,[22] was used in aldol reactions of acetone or butanone (large excess) with different aromatic aldehydes (Scheme 2.1).

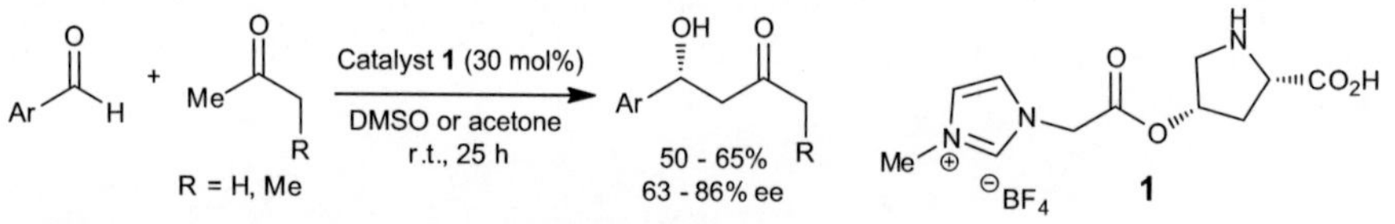

Scheme 2.1

Both chemical yields as well as enantioselectivity are a function of the aldehyde used. The same reactions were performed also with unsupported (*S*)-proline for comparison, and it was found that reactions with ionic tag catalyst are going slightly faster than with (*S*)-proline, but enantioselectivity is practically the same. The advantage of catalyst **1** is that it is insoluble in dichloromethane and can be, after extraction of the product into dichloromethane, re-used at least four times without losing its activity. However, no reactions with smaller catalyst loading were tested.

Luo *et al.*[23] have prepared a series of ionic liquid tag organocatalysts derived from pyrrolidine; some of them are depicted on Figure 2.1.

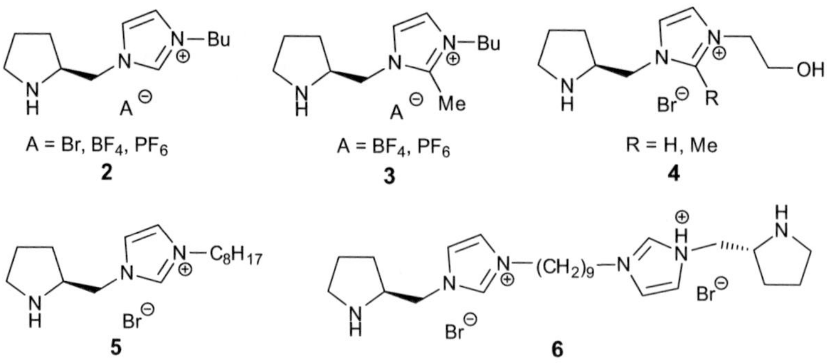

Figure 2.1 Ionic liquid tags organocatalysts derived from pyrrolidine.

Catalysts were tested on reaction of acetone with *p*-nitrobenzaldehyde and as the best ones were found to be **2** (with BF_4^- anion), **4** and **6**. Reactions were performed in water with acetic acid as a co-catalyst. Full conversion of the aldehyde was observed, but the reaction was complicated by formation of the Claisen–Schmidt product (water elimination). Reactions of several acyclic ketones with aromatic aldehydes were also studied (Scheme 2.2).

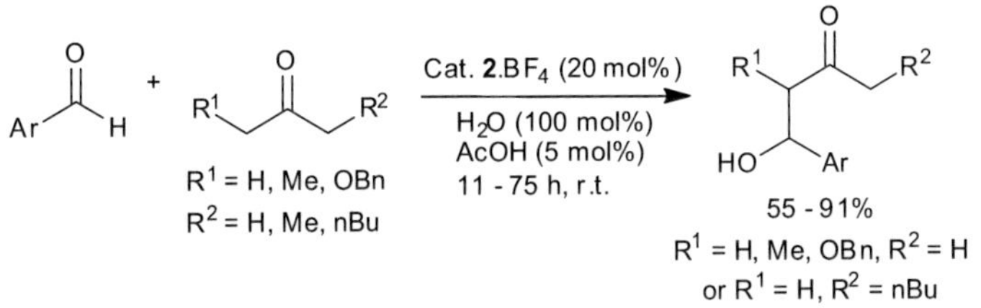

Scheme 2.2

It is of interest that regioselectivity of the reaction is substituent dependent and that reaction is not going on the methylene group bearing the *n*-butyl group. Diastereoselectivity of the reaction is small and enantioselectivity was not described. Aldol reactions of 4-nitrobenzaldehyde with cyclic ketones were performed under the same conditions. The reaction time to full conversion was dependent on the cycle size. In the case of cyclopentanone it was 6 h, with cyclohexanone 12 h but with cycloheptanone and cyclooctanone even after 105 h the product was isolated only in 50% or 35% yields. Diastereoselectivity was rather low, going from *syn/anti* 2.8:1 in the case of the five-membered ring to 1:1.2 in the case of the seven-membered ring. Enantioselectivity was small and also ring dependent (from 32% to 5% ee). The positive result was that the catalyst can be recycled six times, while to reach 92% yield it was necessary to prolong reaction time up to 17 h. Diastereo- and enantioselectivity was the same also in the last cycle.

Lombardo *et al.*[24–26] have used imidazolium tagged catalysts prepared from *trans*-4-hydroxy-proline (Figure 2.2) for aldol reaction of cyclohexanone as well as acetone and hydroxyacetone with aromatic aldehydes. In the first paper,[24] 10 mol% of catalyst **7** was used in the reaction of cyclic ketones in water. Four equivalents of ketones were used and high yields of aldol products with an *anti/ syn* ratio 95:5 and ee up to 99% were obtained.

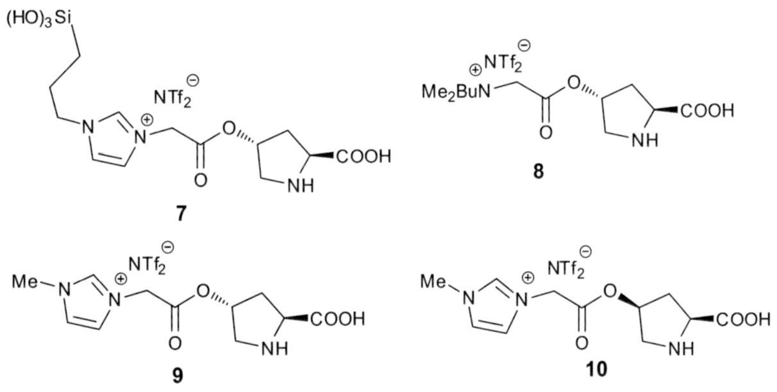

Figure 2.2   Imidazolium tagged catalysts prepared from *trans*-4-hydroxyproline.

Good yields were achieved even with 5 mol% of the catalyst **8** in [bmim]NTf$_2$ (75%, 85% ee). Lowering the catalyst amount to 1 mol% resulted in 25% yield of the product, but ee was still high (83%). Catalyst **9** was just a little less active.[25] In the next paper they compared reactions catalyzed with **9** in water and ionic liquid [bmim]NTf$_2$ and proved that reactions in water gave better results, especially with respect to diastereo- and enantioselectivity.[26] The best results were achieved with cyclohexanone (98%, *anti/syn* up to 98:2 and 99% ee), while with cyclopentanone reaction resulted in *anti/syn* 58:42 and 80% ee. It should be mentioned that reactions used 5 to 10 equivalents of ketones. In the next paper[27] they used a catalyst prepared from *cis*-4-hydroxyproline (catalyst **10**) and found that catalyst **10** is, in reaction of cyclohexanone with different aromatic aldehydes, much more active and stereoselective than its *trans* analogue and that the catalyst can be recycled four times without losing its activity (Scheme 2.3).

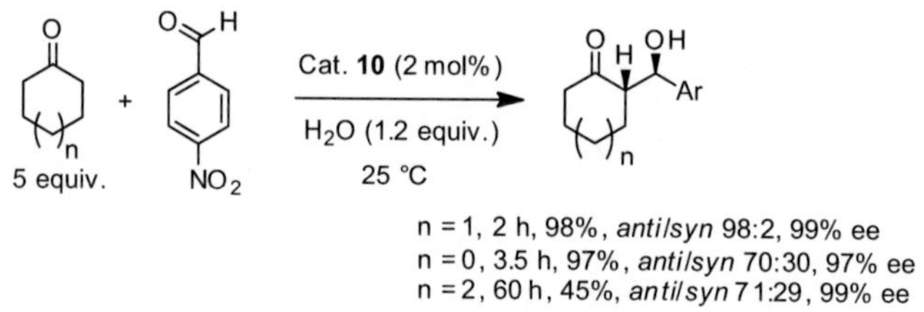

n = 1, 2 h, 98%, *anti/syn* 98:2, 99% ee
n = 0, 3.5 h, 97%, *anti/syn* 70:30, 97% ee
n = 2, 60 h, 45%, *anti/syn* 71:29, 99% ee

Scheme 2.3

Authors are explaining this interesting observation[15] by better electrosteric activation because the charged ionic tag is on the same side of the plane as the carboxyl group, which activates the aldehyde group *via* a hydrogen bond.

Zhang *et al.*[28] described a pyrrolidine-derived catalyst similar to catalyst **3**, using anions of α-amino- or α-azido carboxylic as well as mono anions of several dicarboxylic acids, with the hope that the second chiral element will improve the enantioselectivity of the reaction. Catalysts were tested on the aldol reaction of cyclohexanone (10 equivalents) with 4-nitrobenzaldehyde without additional solvent. Unfortunately their hopes were not realized as the best *anti/syn* ratio (70:30) was achieved with the catalyst in which the anion was the achiral monoanion of phthalic acid. Enantioselectivity was also just medium (55 or 73% ee respectively).

Much better results were achieved with 1,2,3-triazolium-tagged prolines described by Shah *et al.*,[29] which are depicted on Figure 2.3.

Catalysts were tested on the reaction of cylohexanone with several substituted benzaldehydes. All three catalysts were nearly of the same activity giving in 24–48 h product yields of 91–98% and *anti/syn* ratios up to 98:2 and ee of the *anti* isomer up to 96%. Reactions were performed with 20 mol% of the catalyst either under solvent-free conditions (excess of cyclohexane) or in quanidinium or imidazolium ionic liquids where just equimolar quantities of

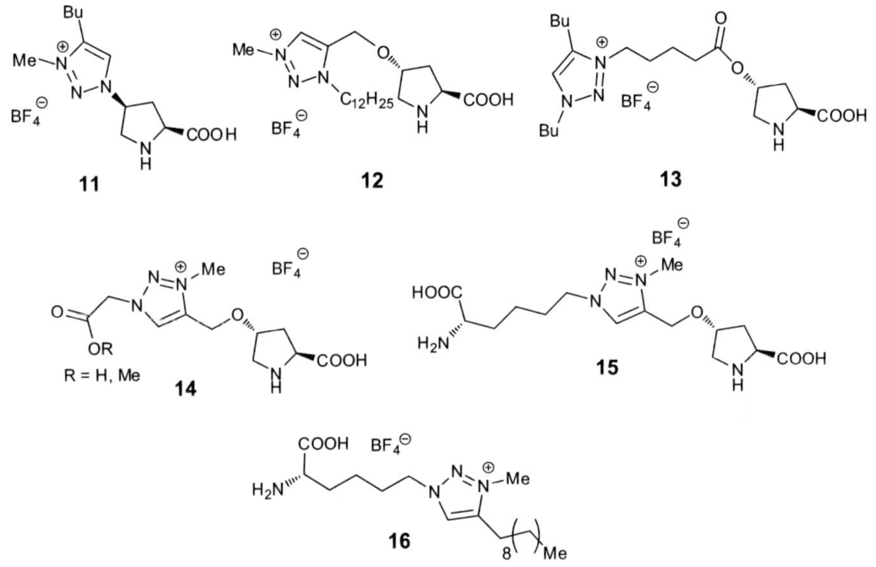

Figure 2.3 1,2,3-triazolium-tagged proline-derived organocatalysts.

reactants were used. Products were isolated by extraction either with diethyl ether or cyclohexanone and catalyst (oil in solvent-free reactions) or its solution in ionic liquid was re-used in subsequent experiments. Five recycles were possible without lowering diastereoselectivity, but ee of the product was lowered after 4th recycle. The same group prepared new types of triazole-derived organocatalysts,[30] which are shown on Figure 2.3.

Catalysts were tested in the aldol reaction of 4-nitrobenzaldehyde with 5 molar excess of cyclohexanone. The highest *anti/syn* ratio (up to 98:2), as well as enantioselectivity (98% ee) were achieved with 20 mol% of catalyst **16**. High yields as well as high enantioselectivity were achieved also after fourth recycle, but diastereoselectivity dropped down to 80:20. Lower enantioselectivity was observed in reaction with cyclopentanone (92% ee) as well as pentan-3-one (82%).

Siyutkin *et al.* have prepared several *trans*-4-hydroxy-(*S*)-proline catalysts with ionic tags and have used them in aldol reactions in the presence of water.[31–33] The best of the prepared catalysts are depicted on Figure 2.4.

In the first paper,[31] an amphiphilic catalyst **17** was used, with the results shown on Scheme 2.4.

The catalyst **17** was recycled five times without lowering its activity. In the next paper[32] they used catalyst **18**, which proved to be more active as well as more selective. Using just 15 mol% of the catalyst gave, in 15 h with cyclohexanone, 98% of the product with ratio *anti/syn* 98:2 and 99% ee. The same results were achieved also in the 8th cycle. In the most recent paper[33] they tested catalyst **19**, which was even better because just 1 mol% of the catalyst must be used to achieve practically same results as described above.

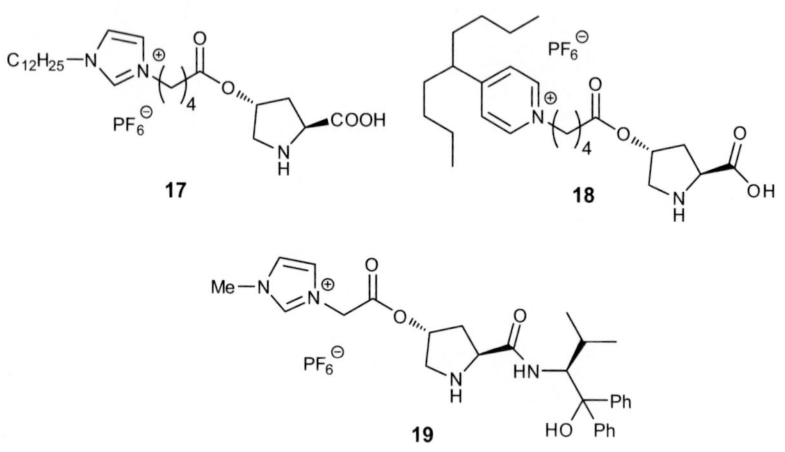

$C_{12}H_{25}$
PF_6
COOH
17
PF_6
OH
18
Me
PF_6
HN
Ph
HO
Ph
19

Figure 2.4 *trans* 4-hydroxy-(*S*)-proline catalysts with ionic tags.

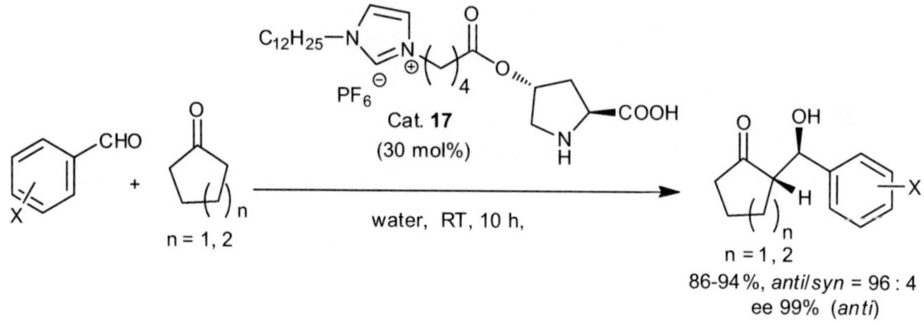

$C_{12}H_{25}$
PF_6
COOH
Cat. 17
(30 mol%)
CHO
X
n = 1, 2
water, RT, 10 h,
OH
X
n = 1, 2
86-94%, *anti*/*syn* = 96 : 4
ee 99% (*anti*)

Scheme 2.4

Yang *et al.*[34] prepared the prolinamide-derived ionic tag catalyst **20** and examined its activity in aldol reactions of aromatic as well as aliphatic aldehydes with cyclic ketones under solvent-free conditions. To their surprise no aldol product was isolated, but high yields of Claisen–Schmidt products were isolated (Scheme 2.5).

2.2.2 Michael Additions

The first paper describing Michael addition of cyclohexanone to β-nitrostyrenes was published by Xu *et al.*[35] (Scheme 2.6).

They used catalysts **2** (Figure 2.1) in aldol reaction. Ionic liquids or polar organic solvents were used as reaction media. The best results were achieved in ionic liquid [bmim]PF_6 and catalyst was easily recovered by addition of diethyl ether. Ionic liquid containing catalyst can be reused four times, but the reaction time had to be prolonged to 100 h. Chemical yields did not change, but

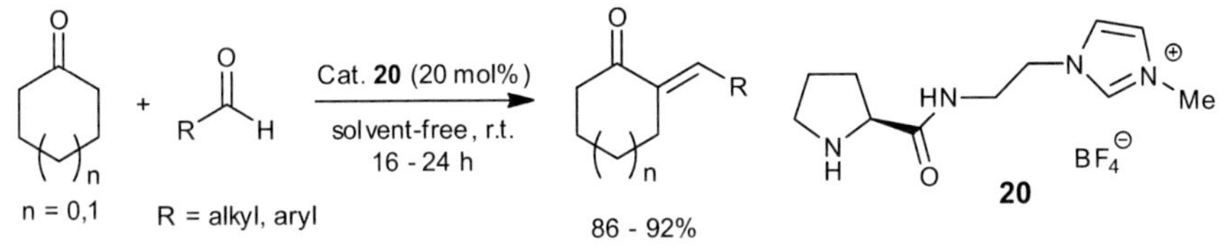

Scheme 2.5

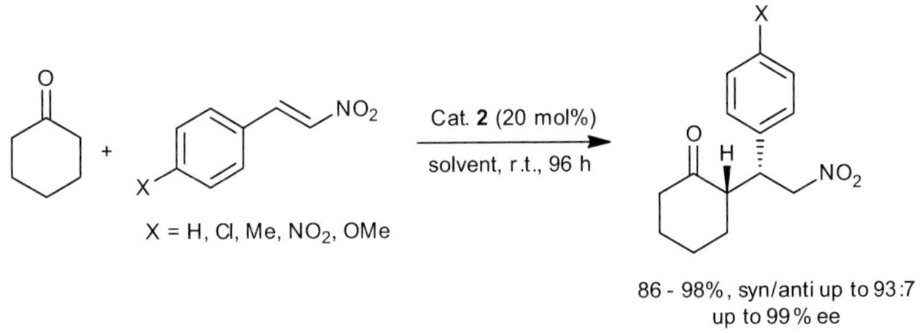

X = H, Cl, Me, NO₂, OMe

86 - 98%, syn/anti up to 93:7
up to 99% ee

Scheme 2.6

diastereoselectivity lowered to 90:10. This indicates that some catalyst was lost during the work-up procedure. Reactions in DMSO, DMF or *i*-PrOH proceeded with much lower yields (53–45%) and lower diastereoselectivity. Michael additions of alicyclic ketones and aliphatic aldehydes were also described. In the same year a study of the same reaction by Luo *et al.*[36] was published. They used 15 mol% of catalysts 2–5 and 5 mol% of trifluoroacetic acid as co-catalyst and excess of cyclohexanone was used as the solvent. Reactions were relatively fast (8–18 h) under such conditions. Excellent results were achieved, especially with catalysts 2 and 3 with BF₄ anions (100% yield, *syn/anti* up to 99:1 and 98% ee). Addition of diethyl ether to the reaction mixture resulted in precipitation of the catalyst which was separated by filtration and used in the next experiment. Four recycles of the catalyst were possible without affecting its selectivity. The reaction time should be prolonged to 48 h at the fourth recycle, which indicates again that some catalyst was lost during the work-up procedure. Addition of trifluoroacetic acid as co-catalyst was necessary as just 33% of the addition product was isolated after 40 h in the experiment without TFA. Catalyst 2 was more active than analogous catalyst in which imidazole ring was not alkylated. The authors examined the possibility of using other ketones or aliphatic aldehydes as donors as well. Additions of acetone as well as 3-methylbutanal went smoothly, but the reaction with cyclopentanone was much slower and less selective. After 60 h product was isolated in 87% yield with *syn/anti* ratio 63:37 and 80% ee. Same authors later studied desymmetrization of prochiral ketones in Michael addition to β-nitrostyrene.[37] They used several organocatalysts with ionic tags, such as **2–6**, but the best results were achieved with a new catalyst **21** (Scheme 2.7).

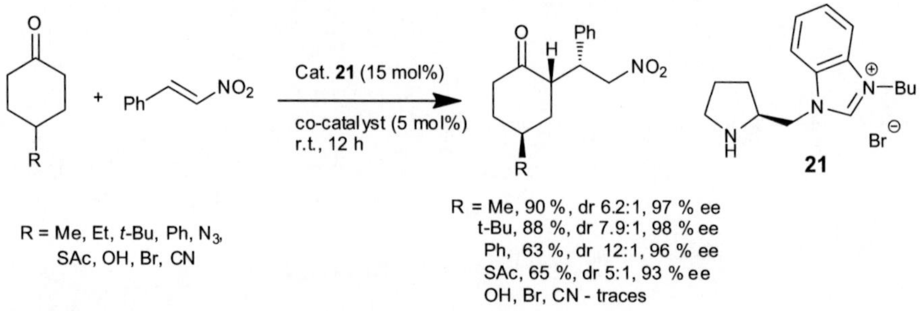
Cat. 21 (15 mol%)
co-catalyst (5 mol%)
r.t., 12 h
21
R = Me, Et, t-Bu, Ph, N3, SAc, OH, Br, CN
R = Me, 90 %, dr 6.2:1, 97 % ee
t-Bu, 88 %, dr 7.9:1, 98 % ee
Ph, 63 %, dr 12:1, 96 % ee
SAc, 65 %, dr 5:1, 93 % ee
OH, Br, CN - traces

Scheme 2.7

Trifluoroacetic acid and salicylic acid were tested as co-catalysts. The reaction with salicylic acid resulted in a little bit higher diastereoselectivity. From the data given on Scheme 2.7 it is clear that higher diastereoselectivity was achieved with more bulky substituents on cyclohexanone.

Ni *et al.* have prepared several new organocatalysts with an ionic tag,[38–41] which are depicted on Figure 2.5.

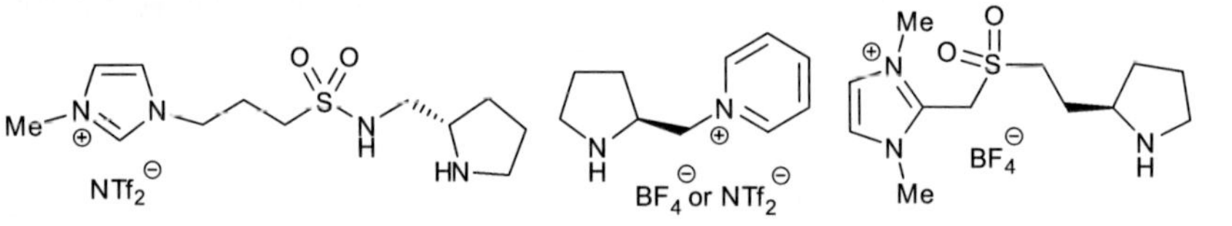

22 **23** **24**

Figure 2.5 Pyrrolidine-derived organocatalysts with an ionic tag.

Catalyst **22** was used in the Michael addition of aliphatic aldehydes to β-nitrostyrene[38] and its derivatives. Screening of 11 different solvents was performed in the addition of iso-butyraldehyde (Scheme 2.8).

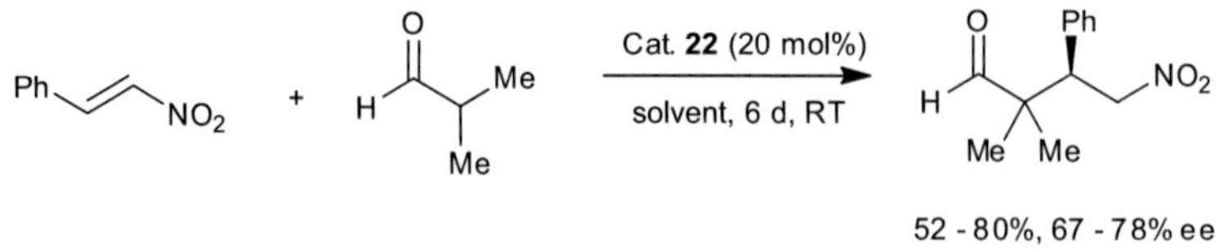
Cat. 22 (20 mol%)
solvent, 6 d, RT
52 - 80%, 67 - 78% ee

Scheme 2.8

The best yields and the lowest ee was observed at reaction in methanol (80%, 67% ee). In diethyl ether yields were worse, but enantioselectivity higher

(52%, 78% ee). Selectivity of the reaction rose to 82% ee when the reaction temperature was 4 °C. After addition of diethyl ether to the reaction mixture it was possible to recover solid catalyst, which was used in two subsequent reactions without losing activity or selectivity. It is interesting to note that reactions with non-branched aldehydes were going with high diastereoselectivity (*syn/anti* 97:3). In the next paper,[39] addition of cyclohexanone to β-nitrostyrene and its derivatives was studied. Catalyst **18** (15 mol%) was used and reactions were run in excess of cyclohexanone as the solvent. Addition of 5 mol% TFA dissolved in methanol was necessary to reach 91% of practically pure *syn* isomer (*syn/anti* 99:1) with 99% ee. Such results were achieved using catalyst with BF_4^- anion. It is of interest that the addition of cyclopentanone or acetone was a better catalyst with NTf_2 anion. In the next paper,[40] the authors studied the same reaction, and just 10 mol% of catalyst **24** was used and reactions were run in propan-2-ol. Small excess of cyclohexanone (3–5 equiv.) was used and no co-catalyst was necessary to get similar diastereo- and enantioselectivity as was achieved with catalyst **23** and TFA as co-catalyst. This can be explained by hypothesis that the amide NH of the catalyst **24** is acidic enough to take over a co-catalyst function *via* formation of strong hydrogen bond with the ketone group. The catalyst was easy to recover and was used in five subsequent experiments without losing either enantio- or diastereoselectivity, but the reaction time had to be prolonged from 12 h to 40 h to reach similar yields, which were 92% in the first and 80% in the fifth experiment with the same catalyst. This indicates that some catalyst must be lost during the work-up procedure.

A series of ionic tag catalysts in which a triazole ring is inserted between the catalytic and ionic moieties were prepared by Wu *et al.*,[42] Miao *et al.*[43] and Yakob *et al.*[29,44] These catalysts are depicted on Figure 2.6.

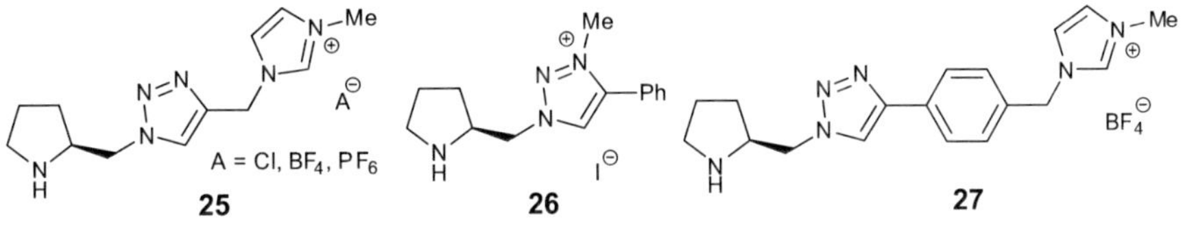

Figure 2.6 2-triazolylmethylpyrrolidine organocatalysts with ionic tag.

Wu *et al.*[42] have studied the Michael addition of cyclohexanone with a series of substituted β-nitrostyrenes. They found that 15 mol% of catalyst **27** and 5 mol% of TFA as co-catalyst is an excellent catalytic system for this Michael addition, which can be recycled four times and still give 95% of the products with *syn/anti* ratio up to 97:3 in the case of substituents in 3- or 4- position on the styrene ring and *syn/anti* ratio up to 49:1 when the substituent is in position 2. Products always have ee higher than 94%. It is of interest that reaction with cyclopentanone was much slower and gave just 63% yield of the product with *syn/anti* ratio 5:3 and 81% ee.

For the same reaction, Miao *et al.*[42] have used a combination of 10 mol% of the catalyst **25** and 5 mol% of TFA as co-catalyst. Reactions were carried out either in ethanol, chloroform or water. The best catalyst was that with chloride anion when reactions were carried in ethanol. Chemical yields were up to 90%, *syn/anti* ratio up to 99:1 and ee 96%. The reaction without a co-catalyst resulted in slightly lower yield as well as lower enantioselectivity.

Yakob *et al.*[29,44] have used catalyst **11** (Figure 2.3), and its analogue having phenyl instead of *n*-butyl group. They used 10 mol% of the catalyst and 2 mol% of TFA as co-catalyst. The best results were achieved in reactions in ethanol. A large amount of cyclohexanone was also used as the solvent. Chemical yields as well as diastereo- and enantioselectivity were high (up to 99% yield, *syn/anti* ratio up to 95:5 and 99% ee) and catalyst was recycled three times without losing its activity. It is interesting that using just 5 mol% of the catalyst needed 40 h to reach similar yields. This catalytic system was used also in addition of acetone and aliphatic ketones, but reactions were much less selective.

Xu *et al.*[45] have prepared several ionic organocatalysts derived by alkylation of azaheterocycles, which are depicted on Figure 2.7.

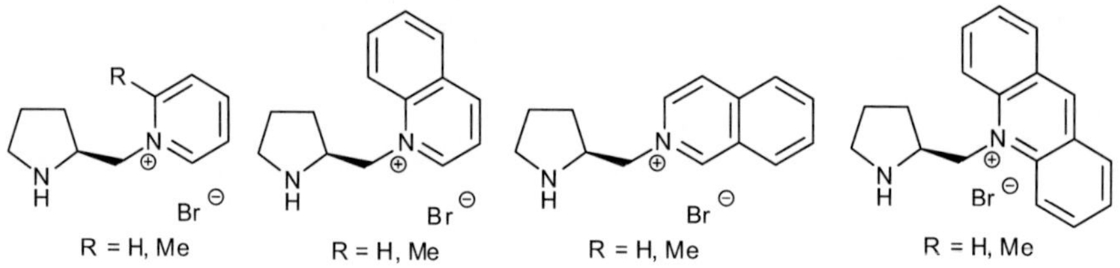

Figure 2.7   Ionic organocatalysts derived by alkylation of azaheterocycles.

Again, a Michael addition of cyclohexanone to β-nitrostyrene was studied and 20 mol% of the catalyst was used. Ionic liquids were used as solvents, the best one was [bmim]BF$_4$, and the best catalyst was **30**. Ionic liquid containing catalyst was recycled four times giving, after 24 h, 90% yield of the product with *syn/anti* ratio 96:4 and ee 96%.

Lombardo *et al.* have published[46] the first organocatalyst in which the ionic tag was bound to the oxygen of diphenylprolinol **32**. This catalyst was very active in catalysis of Michael addition of aliphatic aldehydes to a broad range of aromatic as well as aliphatic nitroethenes (Scheme 2.9).

This catalyst can be used in a range of solvents, from ionic liquids and water to hexane, as well as in solvent-free conditions and aldehyde is used in small excess. Reaction time could be shortened to 15–90 min if 5 mol% of catalyst is used. Unfortunately authors did not perform experiments with recovered catalyst. Wang *et al.* have prepared prolineamide-derived organocatalyst **33** having ionic tag bond to amide nitrogen (Figure 2.8).[47] This catalyst was effective for Michael addition of aldehydes to β-nitrostyrene and its analogues.

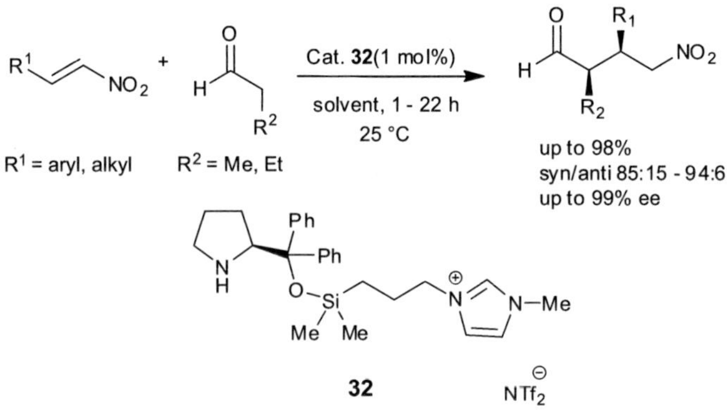

Scheme 2.9

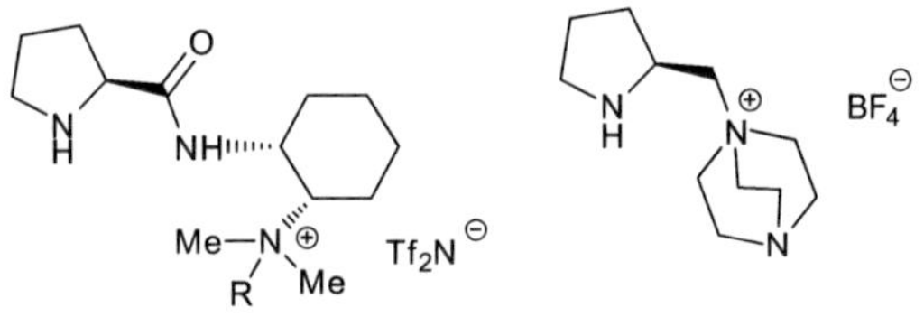

Figure 2.8 Prolineamide and DABCO-derived organocatalysts.

Reactions were carried out either in dichloromethane or THF and tri-fluoroacetic acid was used as co-catalyst. Products were isolated in high yields, but diastereoselectivity and enantioselectivity were just mediocre (*syn/anti* 70:30 – 80:20 and 80–86% ee). Recycling experiments showed that the *syn/anti* ratio as well as ee of the products are the same even after the 3rd cycle, but the reaction time must be drastically prolonged. Xu *et al.*[48] described proline catalyst with DABCO ionic tail **34** (Figure 2.8), which was successfully applied in Michael addition of cyclic ketones to β-nitrostyrene derivatives and its heterocyclic analogues. The best results were achieved when ionic liquids were used as solvents. Over 90% yield of the product was isolated even after 6th cycle, and, what is even more interesting, diastereoselectivity as well as enantioselectivity were very high (*syn/anti* 95:5 and 91% ee respectively).

Maltsev *et al.* have published[49] very good organocatalyst **35** for Michael addition of dimethyl malonate to substituted cinnamic aldehyde. Reactions were performed either in ethanol or in water (Scheme 2.10). The catalyst was easy to recover, after the reaction was over, by addition of diethyl ether to the reaction mixture. Precipitated catalyst was filtered off and reused in the next experiment. Five recycles were possible without catalyst deactivation. In the sixth cycle it was necessary to prolong reaction time from 1 d to 4 d to reach 70% yield and 98% ee. The same catalyst as that derived from *cis*-4-hydroxyproline was successfully used also at addition of nitroalkanes to substituted *trans*-cinnamic aldehyde[50] (Scheme 2.11).

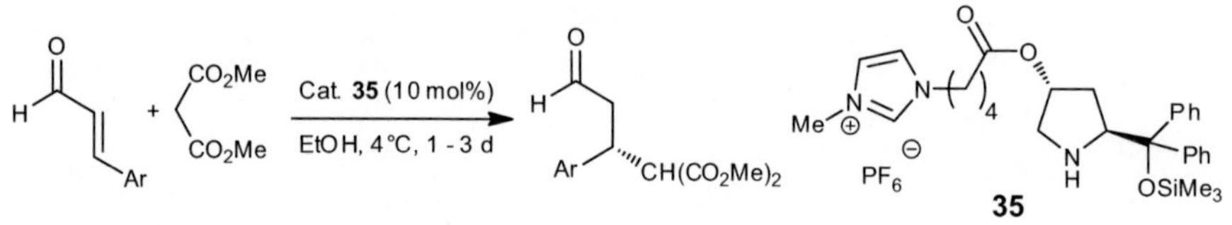

Scheme 2.10

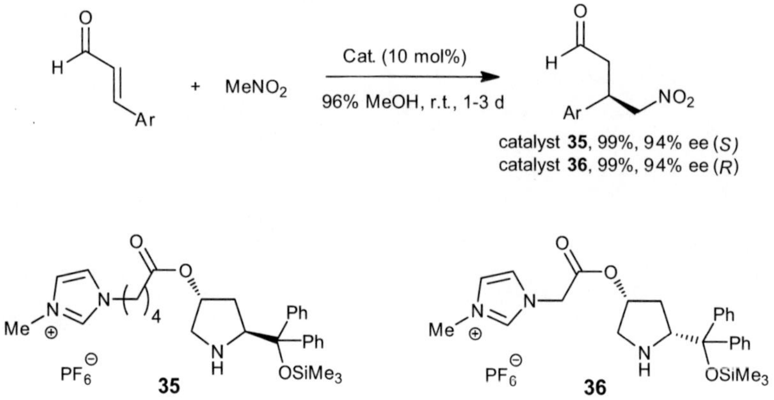

catalyst **35**, 99%, 94% ee (*S*)
catalyst **36**, 99%, 94% ee (*R*)

Scheme 2.11

By this method, both enantiomers of important intermediates for medicaments Phenibut, Baclofen and Rolipram were prepared. The same group has recently published a series of new organocatalysts with ionic tags, which are depicted on Figure 2.9, and used them in domino reactions of substituted cinnamaldehydes with *N*-protected hydroxylamines[51] (Scheme 2.12).

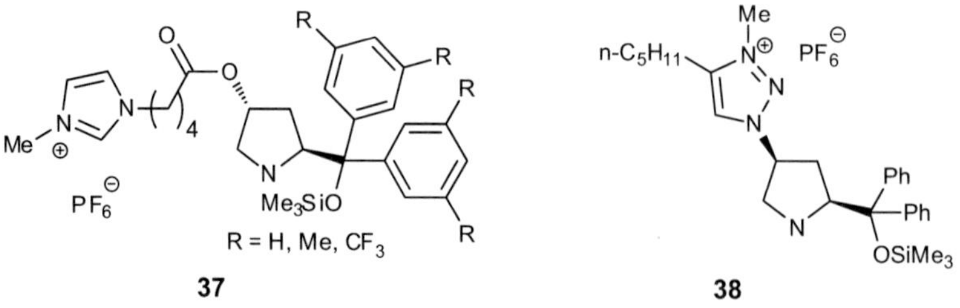

R = H, Me, CF$_3$

37   **38**

Figure 2.9 Hayashi–Jørgensen catalysts with ionic tag.

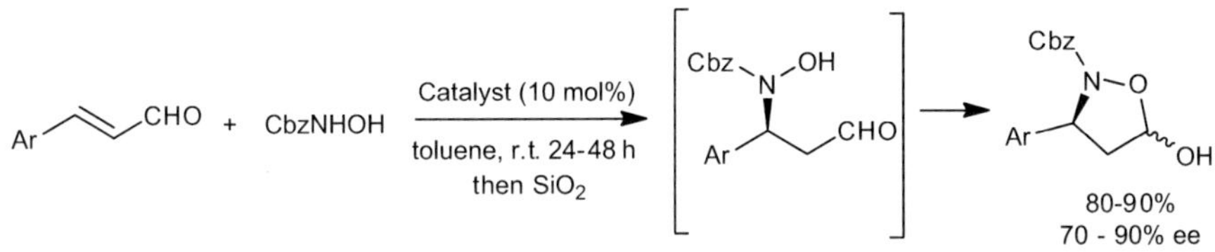

Scheme 2.12

Several solvents have been tested, but the best results were achieved in toluene. Crude 5-hydroxyisoxazoline was single diastereoisomer, but epimerization occurred during purification on silica. The same results were achieved also after the 3rd recycle of the catalyst, which was performed by evaporation of the solvent from the reaction mixture and extraction of the product into diethyl ether. Catalyst **38** was less active than catalysts **37**. Vo-Thanh has prepared[52] organocatalysts based on ephedrine and used them at Michael addition of diethyl-2-acetamidomalonate to chalcones under solvent-free conditions (Scheme 2.13). The best results were achieved with catalyst **40** (70% ee) and the worst with catalyst **41** (40% ee).

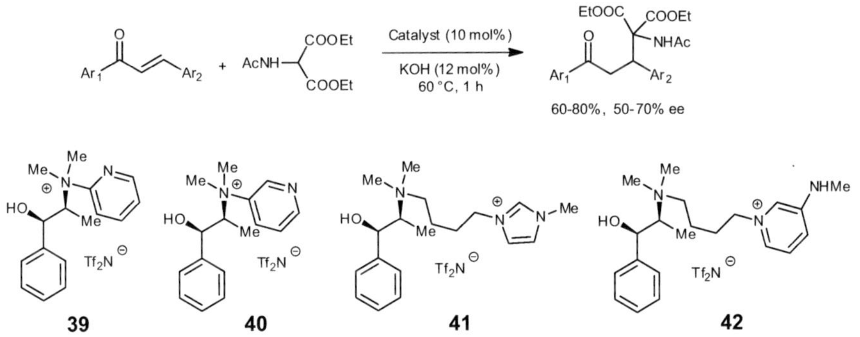

Scheme 2.13

2.2.3 Miscellaneous

Organocatalyst with ionic tag **43** was successfully applied by Ding *et al.*[53] at α-aminoxylation of aldehydes and ketones and ionic liquid was used as solvent (Scheme 2.14).

Reactions with cyclic ketones needed a longer reaction time and gave slightly lower product yields. The resulting α-aminoxyketones did not racemize quickly and therefore it was not necessary to reduce them to alcohols. The same group used the same catalyst also at additions of aldehydes to dibenzyl azodicarboxylate.[54] Reactions went very fast (20 min) in ionic liquid [bmim]BF$_4$

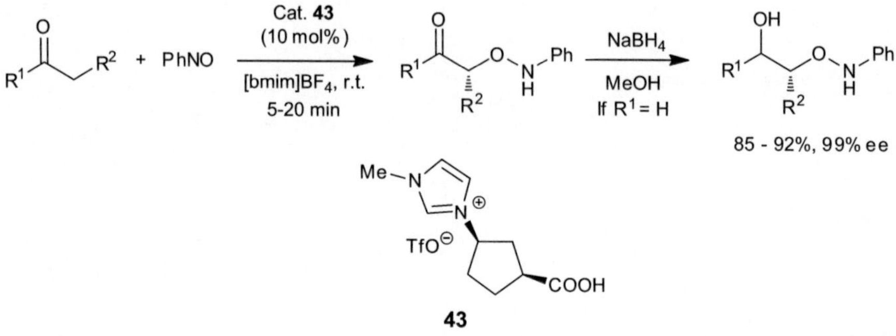

Scheme 2.14

at 0 °C. Formed β-aminoaldehydes were immediately reduced by NaBH$_4$. Resulting β-amino alcohols were isolated in 90–96% yields with 92–98% ee.

Yang et al.[55] attempted aldol reactions of different aromatic as well as aliphatic aldehydes with cyclic ketones under solvent-free conditions using 20 mol% of catalyst **20**. Water was eliminated under such conditions and Claisen–Schmidt products were isolated in high yields (81–94%). Yang et al.[56] prepared prolinol sulfonamide catalyst **44** and used it at BH$_3$·SMe$_2$ reduction of substituted acetophenones. Toluene was used as the solvent. After quenching the reaction mixture with water, products were extracted into diethyl ether. From the water solution of the catalyst, water was evaporated under reduced pressure, and the resulting catalyst was used in next reaction. This process was repeated thrce times with good yields, but a decrease of enantioselectivity was observed in the case of acetophenone from 73% to 65% ee (Scheme 2.15).

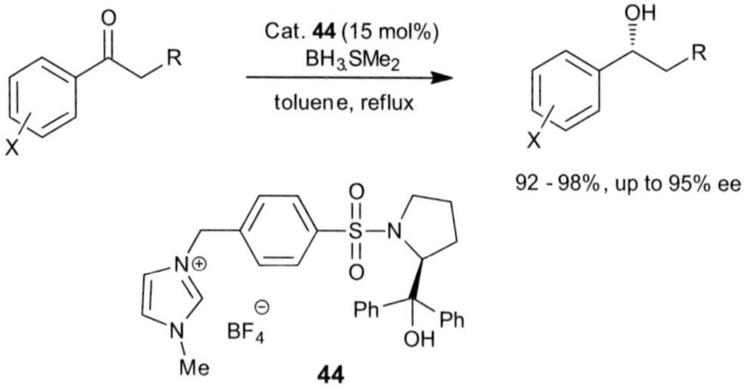

Scheme 2.15

It is interesting that reductions were also going with very high yields, as well as high enantioselectivity in the case of α-haloacetophenones (R = Cl, Br).

Lombardo *et al.*[57] prepared the diphenylprolinol-derived catalyst (discussed in more detail in chapter 1) and used it at diethylzinc addition to aromatic aldehydes (Scheme 2.15).

Zhang *et al.*[58] described asymmetric S_N1 α-alkylation of cyclic ketones, the method invented by Cozzi *et al.*,[59] but they used different organocatalysts and also used organocatalysts with ionic tags (Scheme 2.16).

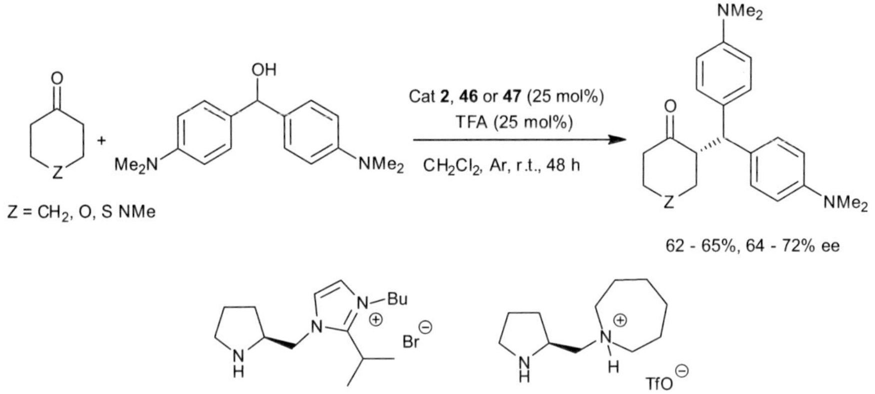

Scheme 2.16

The same authors have published[60] a thorough study of alkylation of a large series of cyclic as well as acyclic ketones. The best results were achieved using catalyst **21** and phthalic acid as a co-catalyst. Up to 80% of the desired product with 75–82% ee was isolated when dichloromethane, 1,2-dichloroethane and chloroform were used as solvents.

Special cases are papers in which chiral ionic liquids are used as catalysts. Hu and co-workers[61] prepared a proline-derived chiral ionic liquid, (2-hydro-xyethyl)trimethylammonium (*S*)-2-pyrrolidine carboxylic acid salt ([Choline]-[Pro]) (**45**), and used it as the catalyst (co-solvent) at aldol reaction of aromatic aldehydes with excess of acetone and cyclic ketones in water. Both mono- and bis-aldol products were isolated. The best results are depicted on Scheme 2.17.

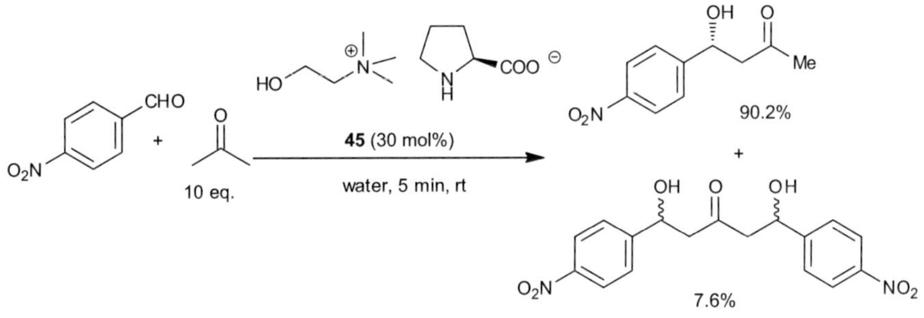

Scheme 2.17

Enantioselectivity of the reaction was, unfortunately, very low (less than 10% ee was obtained in all experiments). A similar chiral ionic liquid, 1-ethyl-3-methylimidazolium prolinate [emim][Pro], was described as a competent catalyst for a one-pot three-component Mannich reaction.[62] It was found that reaction of cyclohexanone, aniline and 4-nitrobenzaldehyde in [bmim]PF$_6$ proceeded reasonably well (32% yield, *syn/anti* 56:43, 82% ee), but better yields were achieved in DMF and DMSO. Qian *et al.*[63] have used chiral ionic liquid **46** (Figure 2.10) as an effective catalyst for aldol reactions of substituted benzaldehydes with cyclohexanone and other cyclic ketones. The ionic liquid [bmim]BF$_4$ was used as the solvent. Reactions went smoothly, but with low diastereoselectivity and medium enantioselectivity. Luo *et al.*[64] have used chiral ionic liquids of the type **47** (Figure 2.10), where the cation as well as the anion of the ionic liquid is chiral, as catalysts in tandem oxa-Michael-aldol reaction (Scheme 2.18). Several chiral acids were tested, but the best results were achieved with Mosher's acid.

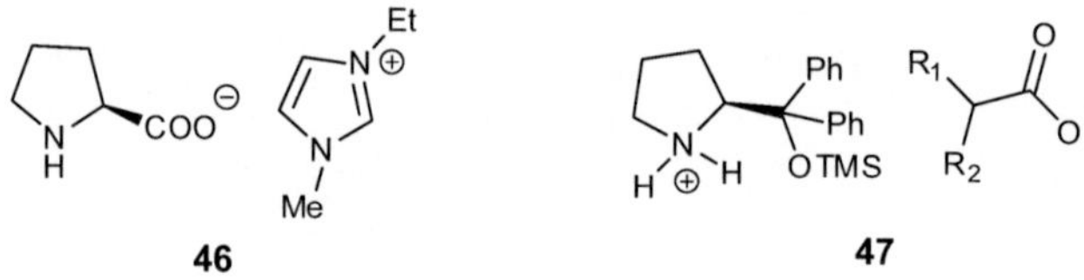

Figure 2.10   Chiral ionic liquid organocatalysts.

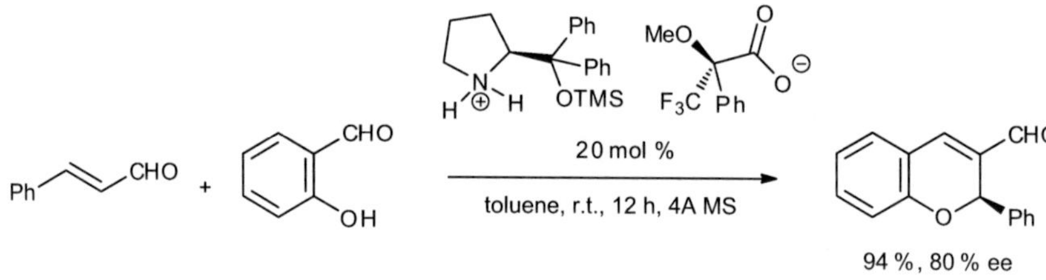

Scheme 2.18

Other cases are protonated amino acids or its derivatives, which were used as the catalysts for aldol reactions in ionic liquids (Figure 2.11).

Xu *et al.*[65] have studied Michael addition of cyclic ketones to β-nitrostyrene and used the protonated proline-imidazole catalyst **48**. The best results were obtained in the mixture DMSO/polyethylene glycol (PEG 720 and higher). Polyethylene glycol can form a special pocket, which can solvate the cation of the catalyst (proved by ESI MS) and thus improve its catalytic activity. Practically full conversion of starting materials was achieved after 12 h and high diastereoselectivity (*syn/anti* ratio 97:3) as well as enantioselectivity (97% ee) were observed. Lombardo *et al.*[66] have used protonated arginine or protonated

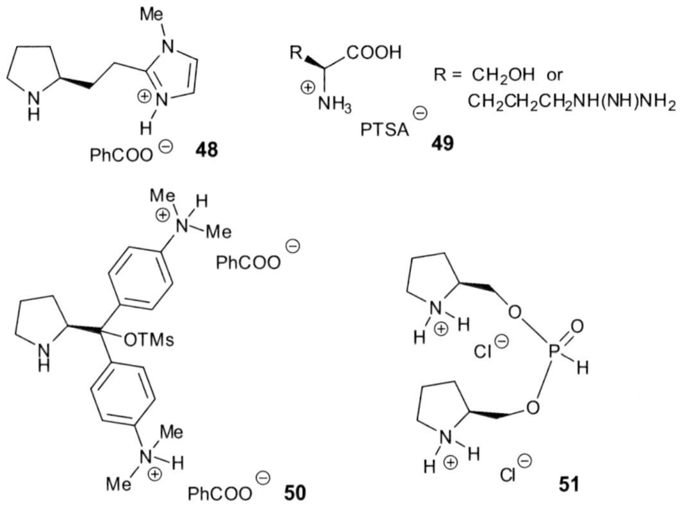

Figure 2.11 Ionic organocatalysts derived by protonation of the basic precursors.

lysine **49** as catalysts for aldol reaction of benzaldehyde with cyclohexanone. Two ionic liquids, [bmim]N(CN)$_2$ and [bmpy]OTf, as well as DMSO were tested as the solvents. The best results were achieved using arginine-PTSA in [bmpy]OTf – 95% yield, *anti/syn* 70:30 and 94% ee. These results were comparable or even better than in DMSO/water 20:1 mixture. The reaction in [bmim]N(CN)$_2$ gave lower yield as well as ee, but better diastereoselectivity (90:10). Ghosh *et al.*[67] described Michael addition of nitromethane to substituted β-nitrostyrenes. Reactions were carried out in water/propan-2-ol 3:1 mixture and catalyst was *in situ* formed by protonation of Hayashi–Jørgensen catalyst with benzoic acid **50**. Reactions have been run at 4 °C for 45–68 h and products were isolated in 62–82% yields having 84–94% ee. Xu *et al.*[68] described Michael addition of cyclic ketones to substituted β-nitrostyrenes using catalyst **51** and NaHCO$_3$ as an additive. Several solvents have been tested and the best results (99% yields, *syn/anti* ratio up to 98:2 and ee up to 91%) were achieved in [bmim]BF$_4$.

There are several papers describing reactions which could be specified as organocatalytic reaction only into some extent. Mi *et al.* described[69,70] quinuclidine derivatives with ionic tags **52**, **53** which were successfully used in a Morita–Baylis–Hillman reaction. Several solvents have been tested but the best results were achieved in methanol (84% yield after 8 h and 74% yield in the third cycle). Catalyst **53** was more active than catalyst **52** (Figure 2.12).

Wu *et al.*[71] described TEMPO immobilized on an ionic tag **54** and used it as the catalyst for oxidations of different primary and secondary alcohols to aldehydes or ketones. Sodium hypochlorite (NaClO) was used as oxidant (1.24 equiv. and 10 mol% of KBr). Reactions were performed in the mixture of [bmim]PF$_6$ and water at 0 °C. Just 1 mol% of the catalyst was necessary to reach 90–95% yield in 5–10 min. Miao *et al.*[72] found very good conditions for

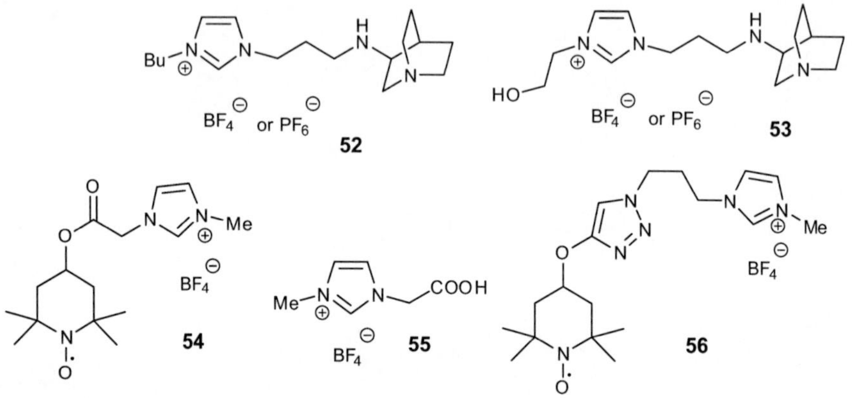

Figure 2.12 Catalysts with ionic tag for Morita-Baylis-Hillman reaction and alcohol oxidations.

aerobic oxidation of alcohols to aldehydes or ketones. They used a combination of 5 mol% of catalyst **54** with 10 mol% of catalyst **55** and 5 mol% of NaNO$_2$. Reactions were performed in water at 60 °C and 1 atm of air. At such conditions full conversion of alcohols was reached in 30 min. Zhu, Ji and Wei[73] have used catalyst **54** for oxidation of primary as well as secondary alcohols with tetra(*n*-butyl)ammonium hydrogenpersulfate (*n*-Bu$_4$NHSO$_5$). Reactions went smoothly in [bmim]PF$_6$ and higher than 90% yield of carbonyl compounds were formed in 1 h in the case of benzyl alcohols and 3 h in the case of aliphatic alcohols. Catalyst **54** did not lose activity even after the 5th cycle, while catalytic activity of simple TEMPO derivative decreased sharply. Fall *et al.*[74] have used for the same reaction catalyst **56**, the oxidant being BAIB (bis-acetoxyiodoniumbenzene). Oxidations were performed at room temperature in dichloromethane and higher than 90% yield of products were formed in 30 min. Reactions in ionic liquids were slower (60 min). Recyclation experiments in [hmim]BF$_4$ proved that practically no loss of activity was observed up to the 3rd cycle.

2.3 Catalysis with Soluble Supported Organocatalysts

There are many papers describing organocatalysts anchored on insoluble supports such as cross-linked polystyrene or silica gel. Such organocatalysts are well described in several reviews.[18,19,76,77] Much less attention has been paid to date to organocatalysts anchored on soluble supports other than ionic liquids. This subject is partially covered in literature.[17,78] In this part of our chapter we shall discuss reactions which are catalyzed by such anchored organocatalysts which allow reactions under homogeneous or quasi-homogeneous conditions.

The research group of Wang has described organocatalysts with perfluoroalkyl tags.[79,80] They studied Michael addition of cyclic ketones to β-nitrostyrenes and used catalyst **57**. Reactions were performed at room temperature in water with 10 mol% of the catalyst. Reactions gave up to 98% of the product with diastereomeric ratio 50:1 and ee up to 95%. The catalyst was used five times without losing its activity or selectivity[79] (Figure 2.13).

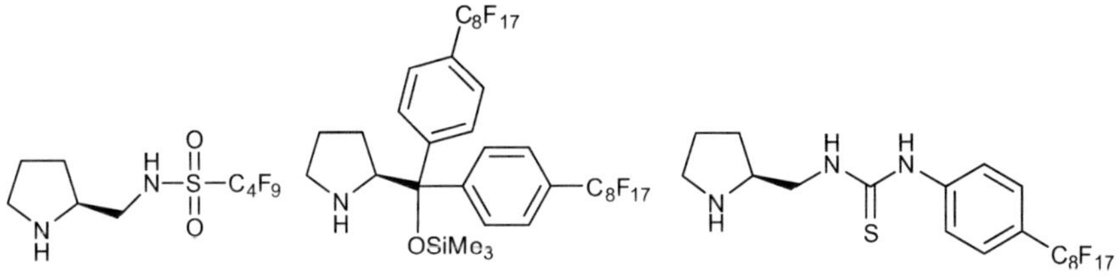

Figure 2.13 Organocatalysts with an perfluoroalkyl tags.

In the next paper[80] they used catalyst **58** in Michael addition of aldehydes to β-nitrostyrene. They used 20 mol% of the catalyst and reactions were run at room temperature in trifluoromethylbenzene. Reactions gave up to 89% yield of the product having diastereomeric ratio 24:1 and 99% ee. The catalyst could be recycled up to six times but reaction time should be prolonged from 4 h to 36 h. Wang *et al.*[81] have prepared (*S*)-proline-thiourea catalyst with perfluoroalkyl group **59** and used it at chlorination of aldehydes with NCS (Scheme 2.19). The best solvent was dichloromethane and catalyst can be recovered by solid phase extraction and reused.

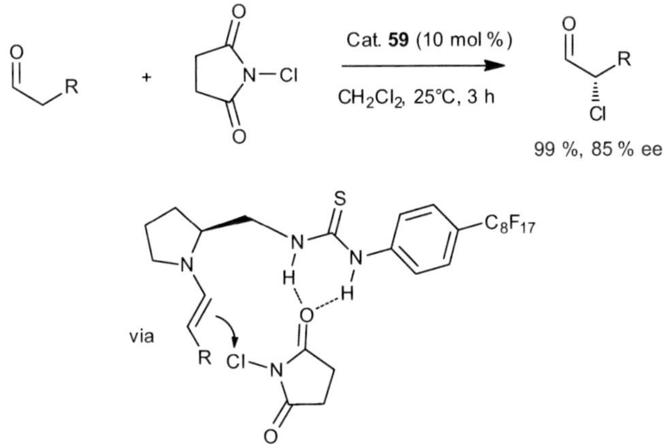

Scheme 2.19

There are several types of polymers on which the organocatalyst could be anchored. The most frequently used ones are re-crosslinked polystyrene polymers, but these are insoluble. Soluble polymers, to which organocatalyst could be anchored *via* a spacer, are polyethylene glycols or its methyl ethers (PEG and MPEG respectively), linear polystyrene (LPS) and recently organocatalysts anchored on polymethacrylate have been invented. Peptides can be considered as special type of polymers too. Organocatalysts on dendrimers could be also soluble. There exist also several types of soluble organocatalysts, which are immobilized to carriers without spacers or chemical bonds.

The first organocatalyst bound to MPEG were prepared by Benaglia *et al.*[82,83] Catalysts **60** and **61** (Figure 2.14) were prepared by anchoring *trans*-4-hydroxy-(*S*)-proline to mono methyl ether of PEG$_{5000}$ *via* a succinate spacer. These catalysts were successfully used in aldol reactions of different aldehydes with acetone in DMF as the solvent. Reactions resulted, after 40–60 h, in up to 80% yields of aldol product with 96% ee. Catalyst was recovered by simple addition of diethyl ether to the reaction mixture and separated by filtration. Recovered catalyst was re-used without losing its activity. Similar results were obtained with unsupported *trans*-4-hydroxy-(*S*)-proline. Similar results were achieved also with hydroxyacetone. The same catalysts were used at Mannich reaction both with pre-formed imines or in a three-component version of this reaction and *syn*-β-amino ketones were isolated in 80% yields and 97% ee.

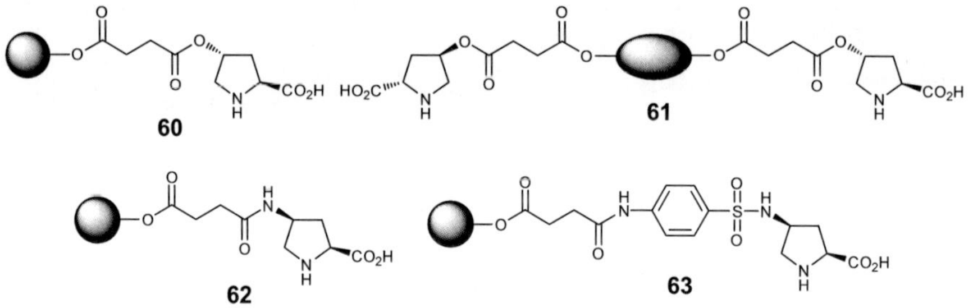

Figure 2.14 Organocatalyst bound to soluble polyethylene glycol type polymers.

Catalyst **60** was tested also at Michael additions of cyclic ketones to β-nitrostyrene in methanol as well as addition of 2-nitropropane to cyclohexenone in 2-propanol with NaOH as co-catalyst.[84] Products were isolated with reasonable 40–60% yields but having just 20–40% ee.

Gu *et al.* have prepared analogous catalysts **62** and **63** (Figure 2.14) from *cis*-4-amino-(*S*)-proline.[85] Catalysts (5 mol%) were tested at Michael addition of cyclohexanone to β-nitrostyrene in methanol DMF, DMSO and THF as well as dichloromethane. The best chemical yields were achieved in dichloromethane and THF (90% and 88% respectively), but ee of the product was just 48% and 34% respectively. Enantioselectivity was a bit better in MeOH (50% ee). It is of interest that reactions were going in all solvents with excellent diastereoselectivity (*syn/anti* 98:2). Similar results were achieved also with model acetamide or sulfonamide catalysts which were not anchored to MPEG.

Benaglia *et al.* have also immobilized MacMillan's imidazolidin-4-one catalyst on MPEG$_{5000}$ and used this catalyst (**64**) at Diels–Alder reaction of acrolein to 1,3-cyclohexadiene and 2,3-dimethyl-1,4-butadiene (Scheme 2.20).[86] The catalyst was recycled four times without losing its activity. The same catalyst, just with HBF$_4$ instead of TFA, was used also in 1,3-dipolar cycloaddition of nitrones to unsaturated aldehydes (Scheme 2.21).[87]

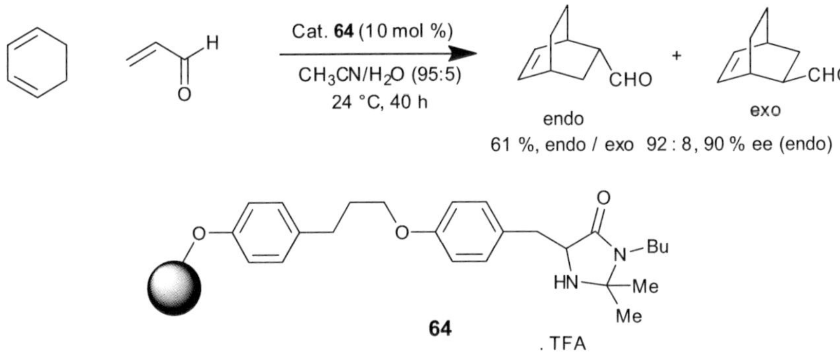

endo

61 %, endo / exo 92 : 8, 90 % ee (endo)

64

. TFA

Scheme 2.20

73 %, trans/cis 86 : 14, 90 % ee (trans)

64

HBF$_4$

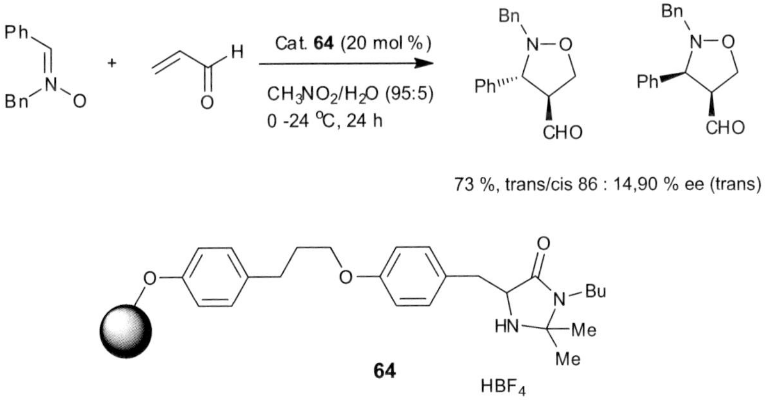

Scheme 2.21

Recycled catalyst has smaller catalytic efficiency, but practically the same selectivity. Thierry *et al.* have immobilized cinchona alkaloids on MPEG$_{5000}$ and used them as phase-transfer catalysts at the synthesis of α-amino acids[88] (Figure 2.15, Scheme 2.22).

Several solvents were tested and better enantioselectivity was observed at reactions in non-polar solvents such as toluene or trifluomethylbenzene or carbon tetrachloride. Just 1–6% ee was observed in reactions in

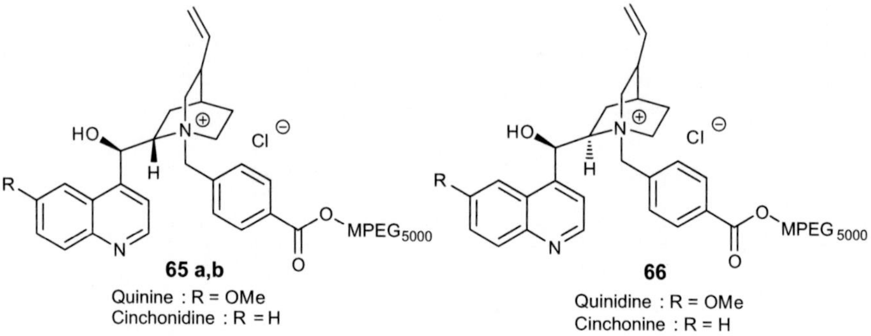

65 a,b

Quinine : R = OMe
Cinchonidine : R = H

66

Quinidine : R = OMe
Cinchonine : R = H

Figure 2.15 MPEG immobilized cinchona alkaloids.

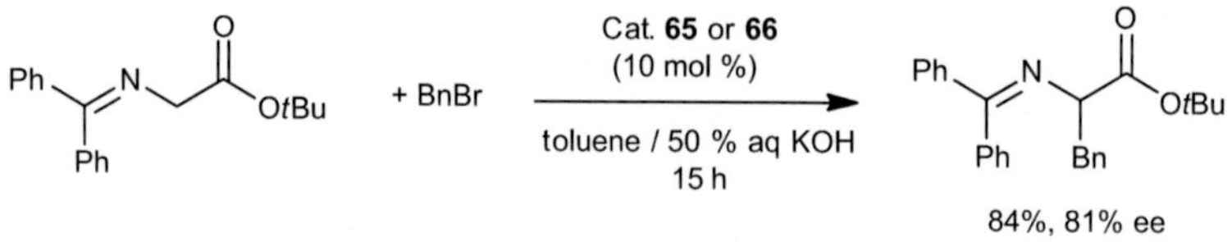

Scheme 2.22

dichloromethane or THF. The catalyst with the cinchonine moiety gave the (*R*)-enantiomer and that with the cinchonidine moiety gave the (*S*)-enantiomer. Attempts on re-use of recovered catalyst failed. The catalyst was partially destroyed in the alkaline medium.

There are just few papers describing organocatalysts anchored on soluble polystyrene. Grutadauria *et al.*[89,90] have prepared catalysts **67** and **68** (Figure 2.16) and used them at aldol reaction of cyclohexanone with substituted benzaldehydes. Excess of cyclohexanone was used as the solvent and under such conditions the authors observed 71–98% conversion and high diastereo- as well as enantioselectivity (*anti/syn* 96:4, 93–98% ee). The catalyst was recycled four times without decrease of stereoselectivity.

Solvent screening proved that reactions can be carried out well in water or in DMSO or DMF in presence of water. This was explained by a model in which hydrophilic proline lies in the polymer/water interface. There is an inner hydrophobic core in which reaction take place. The catalyst was recycled five times without losing its activity.[90] The same catalysts (30 mol% of **67** or 5 mol% of **68**) were used also for selenylation of aliphatic aldehydes. Reactions were carried out in dichloromethane and gave 40–96% of the selenation product. Catalyst **68** was better at recyclation experiments.

Liu *et al.* prepared[91,92] organocatalysts anchored on linear polystyrene with $M_w = $ ca. 5000. Catalysts **69** and **70** were derived from *N*-Cbz-protected *cis*-4-amino-(*S*)-proline methyl ester and linear aminomethyl-polystyrene while

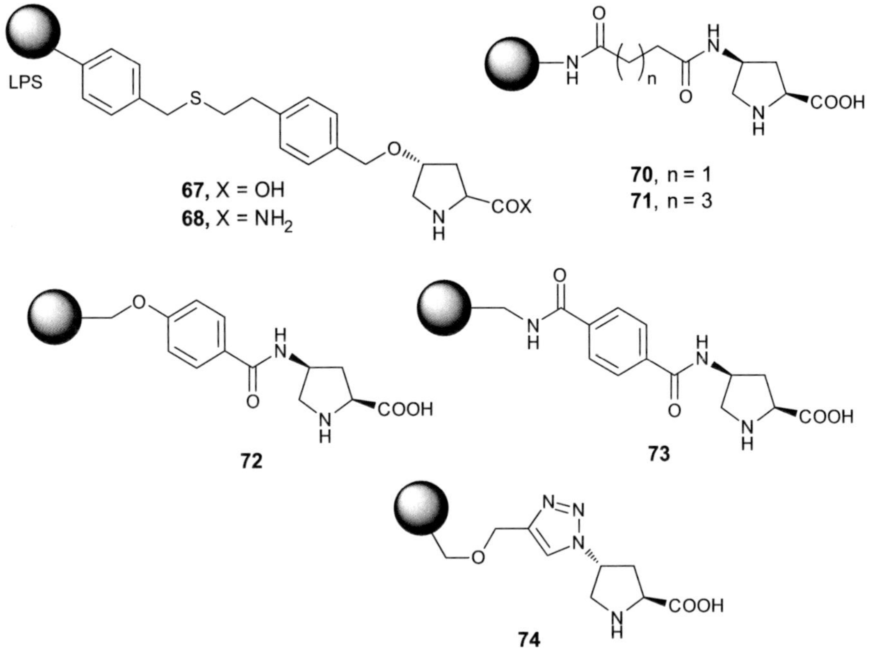

Figure 2.16 Organocatalysts anchored on soluble polystyrene.

succinic anhydride or hexanoic acid were used as the spacer. In the case of catalysts **71** and **72**, the spacer was made from 4-hydroxybenzoic or ter-ephthalic acid respectively (Figure 2.16).

Catalysts **70** and **71** (5 mol%) was used at aldol reaction of cyclohexanone with substituted benzaldehydes in DMF/H$_2$O 15:1 mixture. Reactions pro-ceeded with 46–94% yields having diastereoselectivity ratio 83:17–95:5 and ee up to 94%. The catalyst was recovered by addition of diethyl ether to the reaction mixture and filtered off. No decrease of diastereoselectivity ratio or ee was observed even after four cycles, but chemical yields were decreasing.[91] The same paper also described reactions with acetone and 3-pentanone, but very large excesses of these ketones (15–36 equiv.) were used. Similar results were also achieved[92] with catalysts **72** and **73**.

Font *et al.*[93] described organocatalyst **74** in which proline was anchored on polystyrene *via* a 1,2,3-triazole fragment. This catalyst (10 mol%) was used for catalysis of aldol reaction of benzaldehyde with cyclohexanone in water. In this medium the catalyst was swollen and gave 74% of the product, with *anti/syn* ratio 96:4 and 98% ee. The catalyst can be used also in dichloromethane/water mixture, but the reaction is going in a swollen catalyst, which contains 36 water molecules per one proline unit. The same catalyst was used as an insoluble catalyst at α-aminoxylations of aldehydes and ketones in DMF[94] and with very good results at Michael addition of cyclohexanone to β-nitrostyrene at solvent-free conditions, but TFA has to be used as co-catalyst[95] (80%, dr 94:6, 67% ee). Reactions in water or DMF gave much lower product yields (55 or 31%).

Li *et al.*[96] described a catalyst **75**, which is an organocatalyst bound to a Merrifield resin *via* an ionic liquid spacer. This catalyst was successfully used at Michael additions of cyclohexanone and its analogues to substituted nitrostyrenes (Figure 2.17). Reactions were catalyzed with 10 mol% of the catalyst in different solvents. Very good results were achieved in H_2O, EtOH, DMF, toluene, hexane (80–95%, dr 98:2, ee 98%) but even better results were achieved under solvent–free conditions. It is necessary to note that just 2 equivalents of cyclohexanone were used. The catalyst was reused eight times without losing its activity.

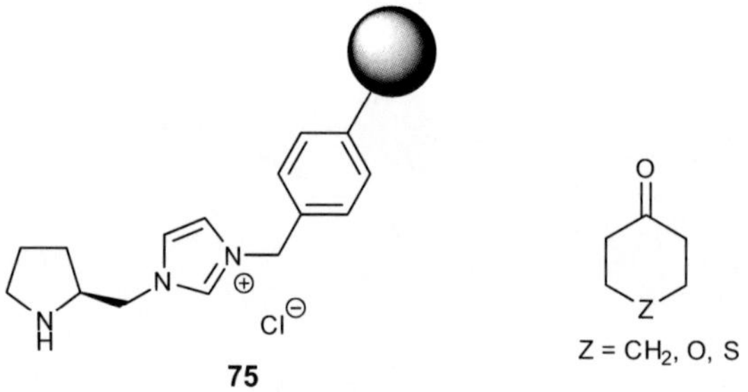

Figure 2.17   Organocatalyst bound to a Merrifield resin *via* an ionic liquid spacer.

Doyagüez *et al.*[97] have prepared methacrylate polybetains **76** and **77** (Figure 2.18). The synthesis started by preparation of *trans*-4-acryoyloxy-(S)-proline *via* reaction of *trans*-4-hydroxy-(S)-proline with acryloyl chloride. Prepared acryloyl ester was then polymerized with AIBN to get catalyst **76**. Catalyst **77** was prepared in similar manner. These polybetains were used as catalysts for

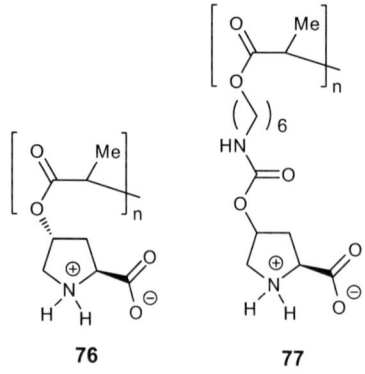

Figure 2.18   Methacrylate polybetains organocatalysts.

aldol reaction of 4-nitrobenzaldehyde with 2,2-dimethyl-1,3-dioxan-5-one in DMF or water. High conversion of aldehyde was observed in 48 h, but reactions went with rather small diastereo- as well as enantioselectivity (dr 5:1, ee 58–78%).

A rather interesting paper describing the preparation of a polymer-supported organocatalyst *via* acrylic copolymerization was published by Kristensen *et al.*[98] Successful synthesis of such an organocatalyst was based on several transformations described by the authors. They found these interesting reactions:

1. A very simple and high yielding transformation of *trans*-4-hydroxy-(*S*)-proline to *cis*-4-hydroxy-(*S*)-proline, just by acylation of *trans*-4-hydroxy-(*S*)-proline by acetic anhydride followed by HCl-catalyzed hydrolysis of the reaction product
2. Esterification of the 4-hydroxy group without protection of the NH group of *cis*-4-hydroxy-(*S*)-proline *via* its acylation with acyl chlorides in TFA.
3. One-pot preparation of trans-4-hydroxydiphenylprolinol by reaction of protonated ethyl ester of *trans*-4-hydroxy-(*S*)-proline with phenylmagnesium bromide in Et₂O followed by work up with TFA/water in MeOH.

The authors have prepared intermediates **78**–**82** (Figure 2.19).

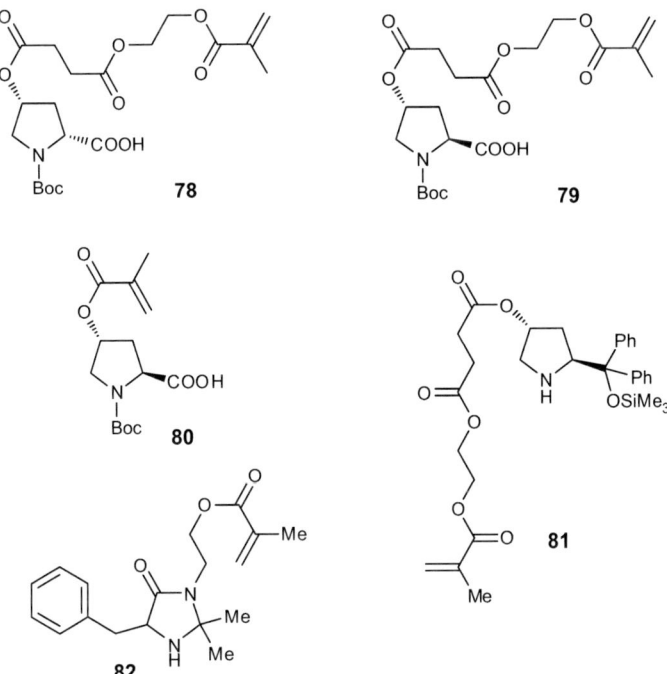

Figure 2.19 Intermediates for methacrylate polybetaine organocatalysts.

From **78** and **79**, final compounds **83** and **84** (Figure 2.20) were prepared by standard peptide coupling with diphenyl phenylglycinol intermediates **80** and **81**.

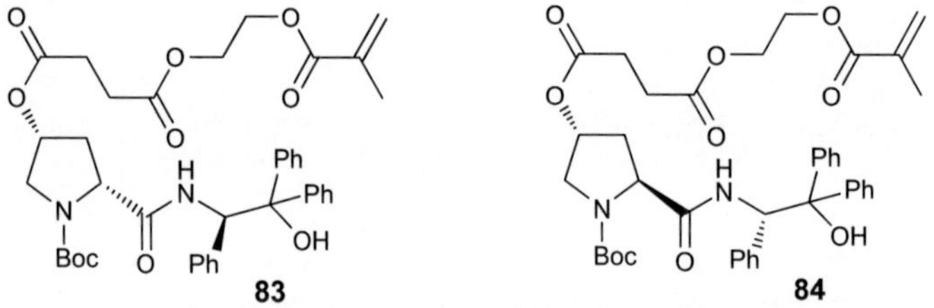

Figure 2.20   Diphenyl phenylglycinol intermediates.

By dispersion copolymerization with benzyl methacrylate and 2 mol% of 1,2-di-methacryloyoxy ethane (diester of ethylene glycol with methacrylic acid) or styrene with 2 mol% of 1,4-divinylbenzene, polymeric organocatalysts were prepared which were used in aldol reactions of substituted benzaldehydes with cyclohexanone (catalysts derived from **79** and **80**) in water or substituted benzaldehydes with acetone in water (catalysts derived from **83** and **84**). The authors[98] have used immobilized Hayashi-Jørgensen catalyst for Enders's type[99] cascade reaction (Scheme 2.23).

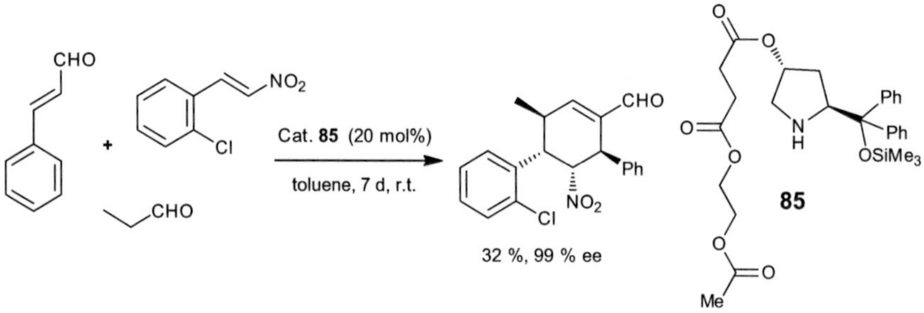

Scheme 2.23

They used also MacMillan's immobilized catalyst for Diels–Alder reaction of 4-nitrocinnamaldehyde with cyclopentadiene (Scheme 2.24). Chemical yields of the products were excellent and ee of the products were also good, just the *exo/endo* selectivity was low. Unfortunately, the catalyst was losing its activity after the third reuse. Similar results were described also with monomeric catalyst.

Peptides were also immobilized on a polymer support. The first paper was published by Berkessel *et al.*[100] who used such catalyst in Julia–Colonna epoxidation of chalcone (Scheme 2.25). The authors found that yields are especially very dependent on the peptide length and the best results were achieved with peptides having chain length of 15–20. Very similar results were

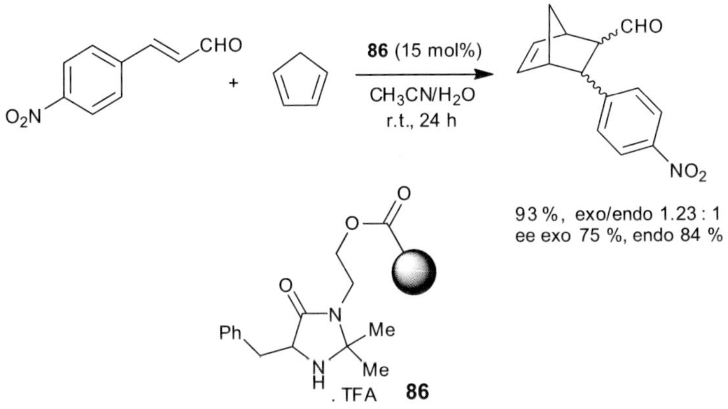

Scheme 2.24

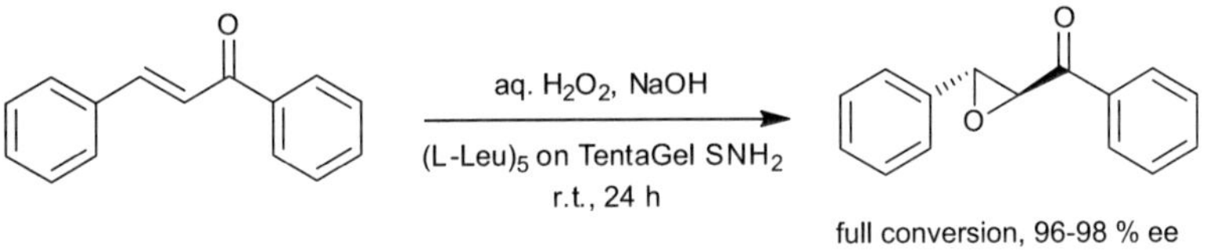

Scheme 2.25

achieved also by Flood *et al.*[101] with polyleucine bound to polyethylene glycol and by Qui *et al.*[102] with polyleucine bound to ionic liquid moiety.

Dipeptides Pro-Ser and Pro-Thr were immobilized on aminomethyl polystyrene resin **87** and successfully tested in an aldol reaction.[103] Several solvents were tested and the best results were achieved in acetone at –25 °C. A variety of tri- and tetrapeptides was immobilized onto polyethylene glycol grafted on cross-linked polystyrene **88**, but in this case the catalyst was insoluble.[104] Nevertheless aldol reaction of aliphatic aldehydes with acetone catalyzed with 20 mol% of the catalyst was possible in 1:1:1 mixture of acetone/water/THF and afforded up to 99% yield and 64–73% ee, but ZnCl$_2$ must be used as co-catalyst.

Tripeptide H-Pro-Pro-Asp-NH$_2$ was anchored on polystyrene (PS), polyacrylamide (SPAR), TentaGel and polyethylene glycol polystyrene (PEGA)[105] **89** (Figure 2.21). These insoluble catalysts were evaluated in aldol reaction of acetone with 4-nitrobenzaldehyde. TentaGel-derived catalyst was the best and 1 mol% of the catalyst was enough to reach good results (up to 93% and 80% ee), but NMM must be used as the base.

Carpenter *et al.*[106] have prepared peptide organocatalysts **90** and **91** from 4-*t*-butoxyproline and *t*-butoxythreonine and bound it on TentaGel, as a relatively inert and hydrophilic polymer. These catalysts were excellent at aldol reaction of acetophenone with substituted benzaldehydes, Michael addition of

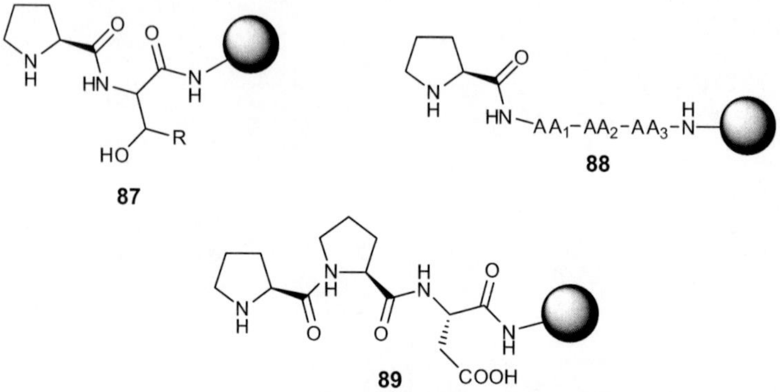

Figure 2.21 Di- and tri-peptides immobilized on aminomethyl polystyrene resin.

acetophenone to *t*-butyl cinnamate, Robinson annulation of acetaldehyde with but-3-en-2-one as well as Mannich reaction of propiophenone, formaldehyde and dimethylamine, but found special applications in one-pot preparations of chromanones (Scheme 2.26).

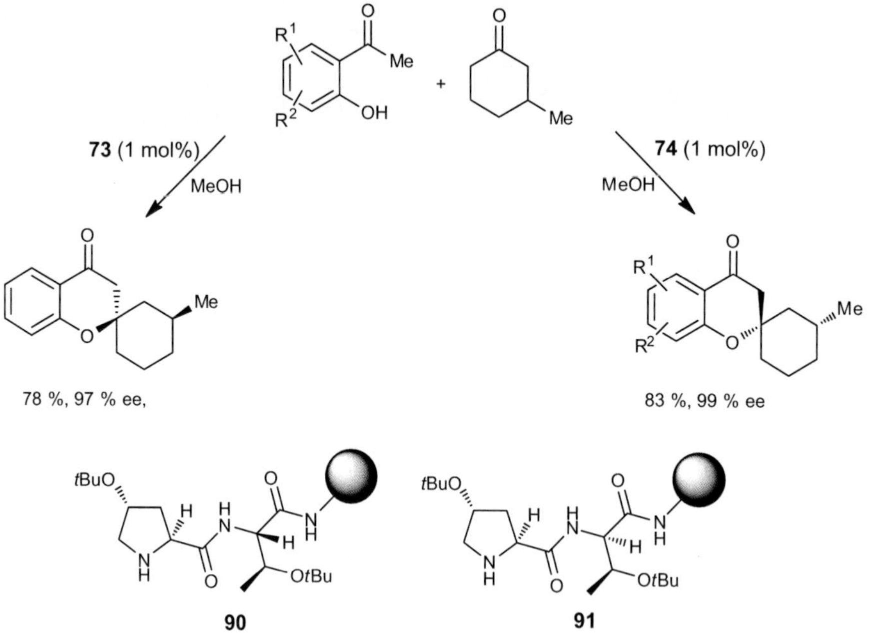

Scheme 2.26

It is of interest to note that such good enantioselectivity was observed only when reactants and the catalyst was dissolved in methanol and solution was stirred for 1 h at room temperature followed by microwave irradiation at

110 °C for 11 min. Enantioselectivity was substantially reduced when the reaction mixture was heated immediately.

Akagawa *et al.*[107] found that resin-supported polyleucine **92** is a good catalyst for hydrogen transfer reduction of unsaturated aldehydes in aqueous media (Scheme 2.27).

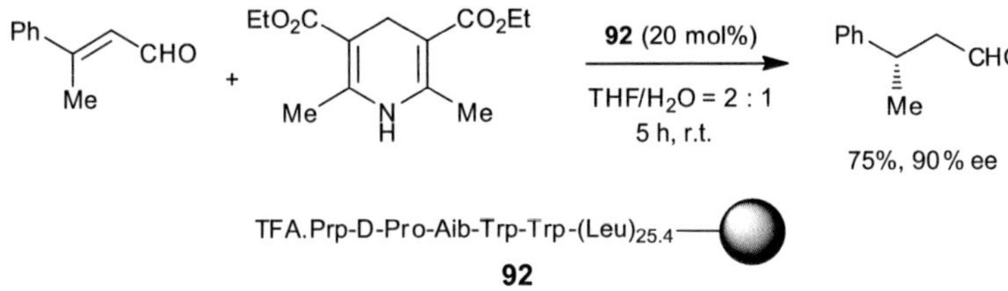

Scheme 2.27

Bellis and Kokotos[108] immobilized proline on poly(propyleneimine) dendrimers **93** and used it as catalyst in an aldol reaction of acetone with 4-nitrobenzaldehyde. This dendrimeric organocatalyst was soluble in DMF and therefore the reaction was running in homogeneous medium. Reactions were much faster than those catalyzed by (*S*)-proline and just 6.5 mol% of dendrimeric catalyst was necessary to reach 75% yield with 65% ee (Figure 2.22).

Wu *et al.*[109] prepared dendritic catalyst **94** with *N*-prolylsulfonamide organocatalyst. This dendritic catalyst is soluble in water and aldol reactions of cyclohexanone with aromatic aldehydes and 10 mol% of the catalyst afforded excellent results, *e.g.* 99% yield, *anti/syn* 99:1 and up to 99% ee. The catalyst was easily recovered, just by addition of appropriate solvent (*n*-hexane/ethyl acetate) and filtration. The catalyst had the same activity after five reuses.

Li *et al.*[110] described soluble dendritic catalyst **95** (Figure 2.22) and used it in Michael addition of aliphatic aldehydes to β-nitrostyrene and its analogues. Additions with 10 mol% of the catalyst in CCl$_4$ at room temperature for 5 days afforded 75–81% yields of the products with *syn/anti* ratio around 90:10 and 99% ee. The catalyst was reused five times without losing selectivity, just with slightly lowered yields (65%).

Another dendrimeric catalyst was described by Lv *et al.*,[111] and was tested in the same reaction, but cyclohexanone was used in 5–10 molar excess. Just 5 mol% of the catalyst with 2.5 mol% of TFA resulted in 96–98% yields of the product with 94–96% ee. Catalyst was reused six times without losing activity. Wang *et al.*[112] have used catalyst **78**, but without protected –OH groups, at reductions of ketones with BH · Me$_2$S in THF. Using 5 mol% of the catalyst resulted in 95–98% yields of chiral alcohols with 96% ee. The catalyst was reused five times without losing activity.

Figlus *et al.*[113] have prepared dendrimeric catalyst **96** and used it at reductions of PMP imines with Cl$_3$SiH. Reductions went with high yields as well as high enantioselectivity (up to 90% yields and 94% ee). Liu and Shi prepared[114] dendrimeric catalyst **97** and used it at asymmetric aza–Morita–Baylis–Hillman

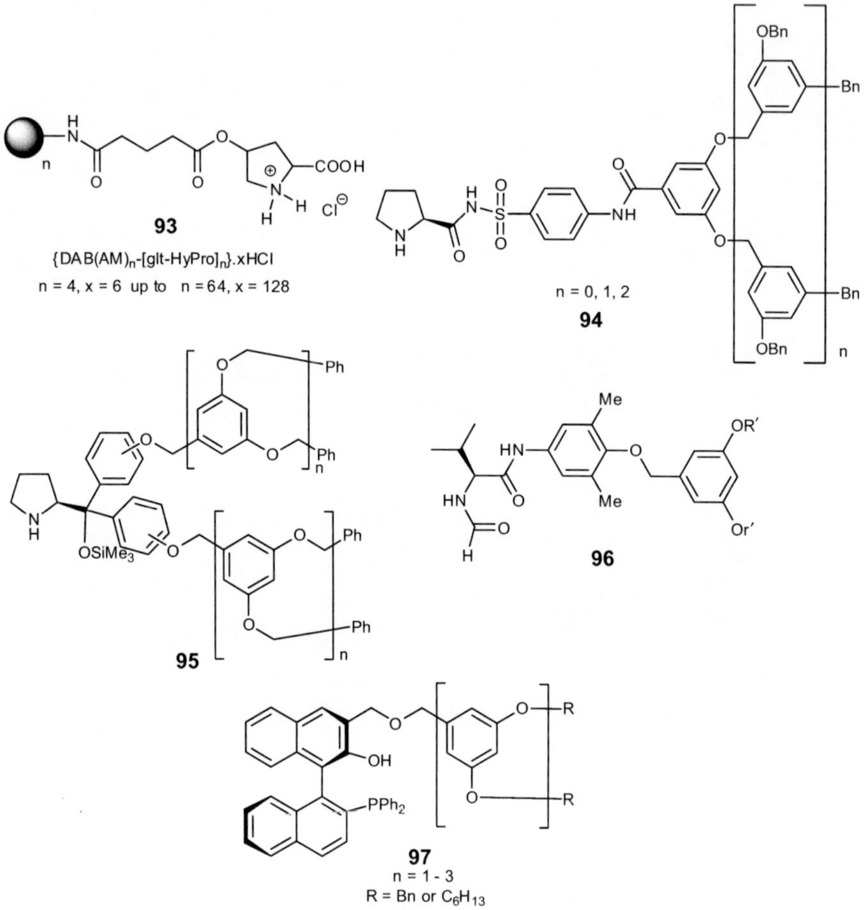

Figure 2.22   Organocatalyst immobilized on dendrimers.

reaction of *N*-sulfonated imines with activated alkenes. Reactions were carried out in different solvents at 25 °C and gave 99% yields with 87–97% ee (Figure 2.22).

Mager and Zeitler prepared[115] a Hayashi–Jørgensen catalyst immobilized on methoxy polyethylene glycol *via* triazole ring **98** and have used it at Michael addition of nitromethane to substituted cinnamic aldehydes (Scheme 2.28). The catalyst did not lose activity even after four reuses.

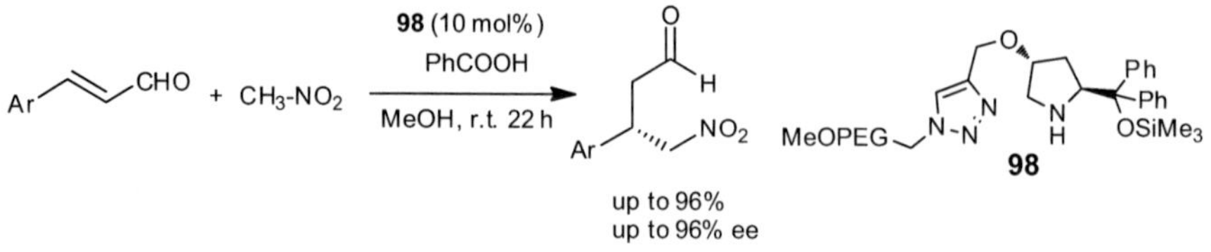

Scheme 2.28

2.4 Organocatalytic Reactions with Organocatalysts Immobilized in Microenvironment

There are many examples of immobilization of organocatalysts in ionic liquids or water by structural modification of an organocatalyst.[116,117] A special case is immobilization of the organocatalyst in organic as well as inorganic micro-environment without the formation of a chemical bond. Reactions are carried out in such microenvironments under homogeneous conditions. Luo *et al.*[118] described partially sulfonated linear or slightly cross-linked (1% of *p*-divinylbenzene) polystyrene (M_w 25 000), which can protonate chiral diamine organocatalyst and in such a way immobilize the organocatalyst in the polymer cavities. Such catalysts were used in aldol reactions of different ketones with substituted benzaldehydes in dichloromethane at room temperature. With cyclohexanone and cyclopentanone the best results were achieved with catalyst **99** (97%, *anti/syn* 91:9, 97 and 91% ee, *anti/syn* 68:32, 94% ee). The reaction was much slower with acetone (72 h versus 24 h) and the best catalyst was **100** (88%, 83% ee) (Figure 2.23).

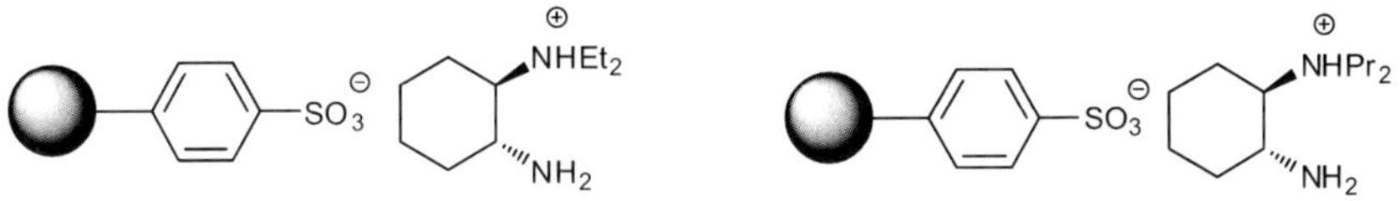

99   **100**

Figure 2.23 Protonated chiral diamine organocatalyst in polystyrene ($M_W = 25\,000$) microenvironment.

An interesting paper was published by Chi *et al.*[119] They prepared soluble star polymers with highly branched non-interpenetrating catalytic cores. Some inside pores of such polymers were sulfonated, and can protonate and incorporate MacMillan's organocatalyst as described above. The rest of the inside pores were occupied by methoxydiphenyl prolinol. These two kinds of micropores cannot penetrate each other. Such star polymers together with the ethyl ester of 3,4-dihydrobenzoic acid, as H-bond catalyst, were used in a cascade reaction shown on Scheme 2.29. The course of this reaction is on Scheme 2.30.

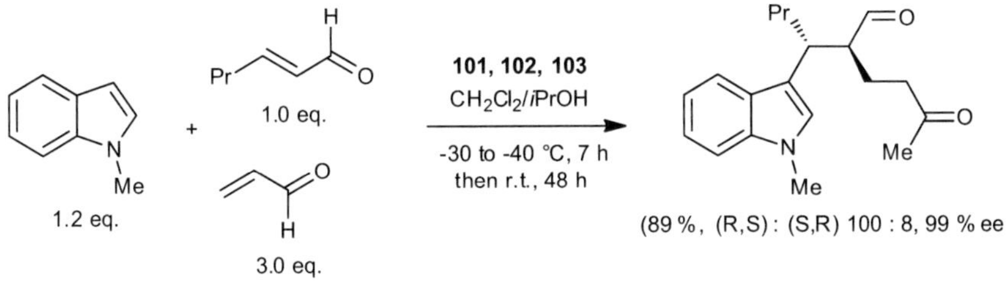

Scheme 2.29

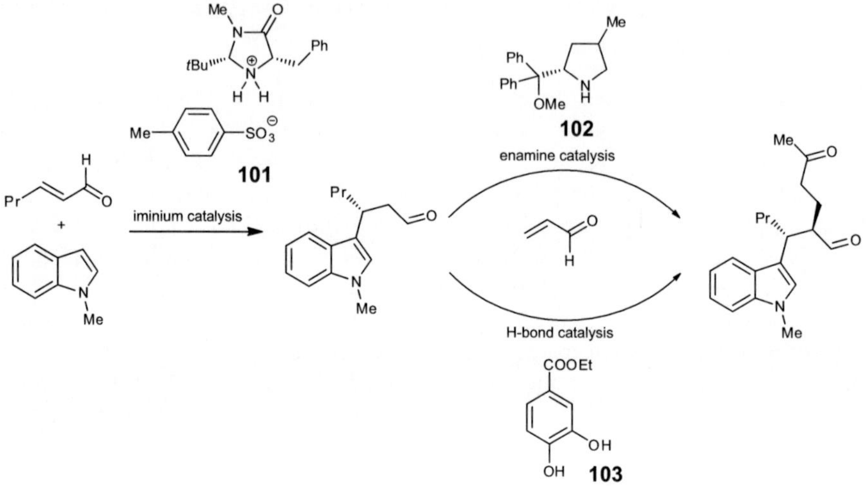

Scheme 2.30

It is of interest to note that the reaction without star polymer catalysts, that is just with monomeric catalysts, did not occur.

Organocatalysts can form complexes with β-cyclodextrin. Shen *et al.*[120] described such a complex with *cis*-4-phenoxy-(*S*)-proline and used it as a catalyst of an aldol reaction with a great excess of acetone with substituted benzaldehydes. With *o*-nitrobenzaldehyde, using 10 mol% of such catalyst, they achieved 90% yields with 83% ee in 16 h. The same results were obtained also in the fourth run with the same catalyst. Huang *et al.*[121] prepared similar complexes with *trans*-4-(4-*t*-butylphenoxy-(*S*)-proline as well as with *cis*-4-(4-*t*-butylphenoxy-(*S*)-proline and used them in aldol reactions of cyclohexanone with benzaldehyde. Reactions were performed in water and after 48 h resulted in 84% yields of the product with *anti/syn* 90:10 and 93% ee. Li *et al.*[122] described *cis*-4-hydroxy-(*S*)-proline bonded *via* its –OH group on calix[4]arene scaffold and used it at aldol reaction of cyclohexanone with substituted benzaldehydes in water. Using 2 mol% of the catalyst they achieved the product within 48 h in 50–73% yields of the product with *anti/syn* 55:45 up to 90:10 and 63–93% ee.

Organocatalysts can be incorporated also into cavities or interlayer space of inorganic carriers such as silica, montmorillonite, etc. Mitsumode *et al.*[123] exchanged sodium cation in the montmorillonite by protonated MacMillan's catalyst and obtained in this way a heterogeneous catalyst in which reaction took place in the interlayer space of such montmorillonite, which can be filled either by reactant (under solvent-free conditions) or by their solution in the solvent. Reactions are therefore carried out under homogeneous conditions, but the catalyst can be easily recycled by simple filtration. This catalyst (15 mol%) was used in Diels–Alder reaction of cyclohexa-1,3-diene as well as cyclopentadiene with acrolein. Reactions were performed in CH_3CN/H_2O mixture at room temperature. The reaction with cyclohexa-1,3-diene went very well (82%) and with high *endo/exo* selectivity (96:4) as well as enantioselectivity (92% ee). The reaction with cyclopentadiene went with lower *endo/exo* selectivity. Reactions can be accelerated by addition of trimethyl acetic acid as co-catalyst and catalyst was recycled four times without lowering its activity. Hagiwara *et al.*[124] dissolved salts of MacMillan's catalyst with such acids as HCl, TFA and NTf$_2$ in ionic liquid [bmim]PF$_6$ and after addition of SiO$_2$ obtained solid-supported organocatalysts called Mac-SILC, which were used as catalyst of Diels–Alder reaction of cyclopentadiene with cinnamaldehyde. Reactions were carried out with high yields as well as high enantioselectivity (up to 96% ee), but the *endo/exo* ratio of the product was just nearly 1:1. Reactions were performed in EtOH or *t*-amylOH and catalyst did not lose activity even after the 5th reuse.

Srivastava *et al.*[125] intercalated (*S*)-proline into montmorillonite and used such catalyst in aldol reaction of 2,4-dinitrobenzaldehyde with acetone. Several solvents were tested, DMSO being the best (78%, 70% ee). Both yields as well as selectivity were improved by addition of 1 equiv. of trimethylbutylammonium bis-trifluormethylimide. Hara *et al.*[126] have intercalated *N*-(2-thiphenesulfonyl)prolinamide **104** into the interlayer spaces of montmorillonite and used it at aldol reaction of substituted isatin derivatives with acetaldehyde or acetone. TFA was used as co-catalyst and reactions went smoothly in water, giving up to 99% of the product with 92–96% ee.

Wang *et al.*[127] have prepared a Hayashi–Jørgensen catalyst immobilized on superparamagnetic nanoparticles **105** (Figure 2.24). This catalyst was effectively used at Michael addition of nitroethane to β-nitrochalcone. Reaction was carried out in different solvents, the best being either pure water or EtOH/H$_2$O 1:1 mixture; up to 85% yields of the product with dr 99:1 having 84% ee was isolated. The advantage of this catalyst was that it can be easily recovered by an external magnet and used in subsequent experiments.

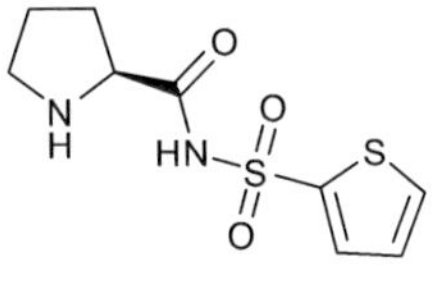

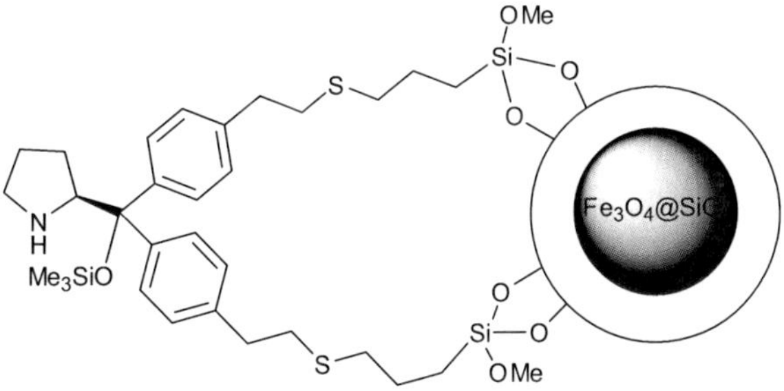

Figure 2.24 Hayashi–Jørgensen catalyst immobilized on superparamagnetic nanoparticles.

Li *et al.*[128] utilized the fact that polyoxometalate hydrides are acidic and can therefore protonate chiral diamine organocatalysts. Such catalysts were used in aldol reaction of 4-nitrobenzaldehyde with acetone or cyclohexanone.

Reactions were carried out either in excess of ketone or water. Vijaikumar *et al.*[129] anchored (*S*)-proline on hydrotalcite clays and the catalyst prepared was used at Michael addition of acetone to β-nitrostyrene or nitromethane to benzylideneacetone. Good conversions, but just medium enantioselectivity, were observed.

Massi *et al.*[130] have prepared *trans*-4-hydroxyproline derivative immobilized on silica **126** and have used it as the bed in a micro-reactor for aldol reaction of cyclohexanone with 4-nitrobenzaldehyde or α-amination of *n*-propanal with dibenzyl azodicarboxylate. Liu *et al.*[131] incorporated the earlier described[132] TEMPO-IL catalyst **127** together with CuCl$_2$ into mesoporous silica. Such catalyst (TEMPO-IL/CuCl$_2$/silica gel) was used with preference at oxidation of different primary and secondary aldehydes to carbonyl derivatives using oxygen as the oxidant. The best results were achieved when *n*-octane was used as the solvent, with over 90% yield after the 5th reuse (Figure 2.25).

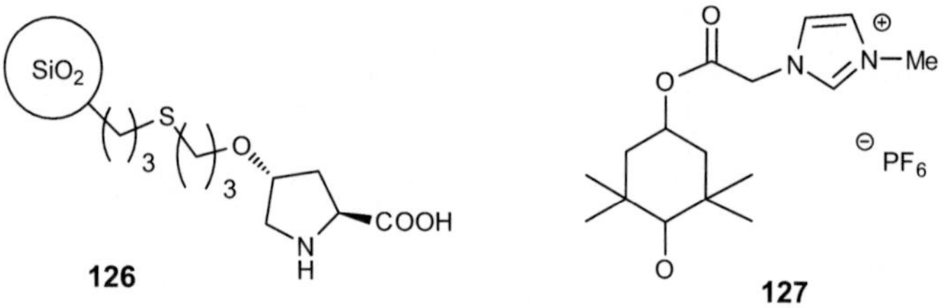

Figure 2.25   Catalysts anchored on hydrotalcite clay or mesoporous silica.

2.5   Conclusions

Immobilization of organocatalysts either on ionic liquids or on a soluble polymer was proved to be a very effective and green methodology to make organocatalytic reactions greener and more effective. Immobilization of organocatalysts on an ionic liquid moiety has especially very good perspective, because there are many structurally different ionic liquids and therefore really tailored organocatalysts can be prepared.

References

1. G. Bredig and W. F. Fiske, *Biochem. Z.*, 1912, 7.
2. H. Pracejus, *Justus Liebigs Ann. Chem.*, 1960, **634**, 9–22.
3. Z. G. Hajos and D. R. Parish, *J. Org. Chem.*, 1974, **39**, 1615–1621.
4. U. Eder, G. Sauer and R. Wiechert, *Angew. Chem., Int. Ed. Engl.*, 1971, **10**, 496–497.
5. M. S. Sigman and E. N. Jacobsen, *J. Am. Chem. Soc.*, 1998, **120**, 4901–4902.

6. B. List, R. A. Lerner and C. F. Barbas III, *J. Am. Chem. Soc.*, 2000, **122**, 2395–2396.
7. K. A. Ahrendt, C. J. Borths and D. W. C. MacMillan, *J. Am. Chem. Soc.*, 2000, **122**, 4243–4248.
8. X. Chen, X. Li, A. Hu and F. Wang, *Tetrahedron: Asymmetry*, 2008, **19**, 1–14.
9. A. Winkel, P. Vasu, G. Reddy and R. Wilhelm, *Synthesis*, 2008, 999–1016.
10. K. Bica and P. Gaertner, *Eur. J. Org. Chem.*, 2008, 3235–3250.
11. S. Luo, L. Zhang and J. P. Cheng, *Chem. Asian J.*, 2009, **4**, 1184–1195.
12. P. D. de Maria, *Angew. Chem., Int. Ed. Engl.*, 2008, **47**, 6960–6968.
13. R. Šebesta, I. Kmentová and Š. Toma, *Green Chem.*, 2008, **10**, 484–496.
14. S. G. Zlotin, G. V. Kryshtal, G. M. Zhdankina, A. S. Kucherenko, A. V. Bogolyubov and D. E. Siyutkin, *Pure Appl. Chem.*, 2009, **81**, 2059–2068.
15. M. Lombardo and C. Trombini, *ChemCatChem*, 2010, **2**, 135–145.
16. C. Huo and T. H. Chan, *Chem. Soc. Rev.*, 2010, **39**, 2977–3006.
17. M. Gruttadauria, F. Giacalone and R. Noto, *Chem. Soc. Rev.*, 2008, **37**, 1666–1688.
18. K. Ding and Y. Uozumi (eds.), *Handbook of Asymmetric Catalysis*, J. Wiley, 2008.
19. M. Benaglia (ed.), *Recoverable and Reusable Catalysts*, J. Wiley, 2009.
20. M. Gruttadauria, S. Riela, P. Lo Meo, F. D'Anna and R. Noto, *Tetrahedron Lett.*, 2004, **45**, 6113–6116.
21. M. Gruttadauria, S. Riela, C. Aprile, P. Lo Meo, F. D'Anna and R. Noto, *Adv. Synth. Catal.*, 2006, **348**, 82–92.
22. W. Miao and T. K. Chan, *Adv. Synth. Catal.*, 2006, **348**, 1711–1718.
23. S. Luo, X. Mi, L. Zhang, S. Lu, H. Xu and J. P. Cheng, *Tetrahedron*, 2007, **63**, 1923–1930.
24. M. Lombardo, S. Easwar, A. De Marco, F. Pasi and C. Trombini, *Org. Biomol. Chem.*, 2006, **6**, 4224.
25. M. Lombardo, F. Pasi, S. Easwar and C. Trombini, *Adv. Synth. Catal.*, 2007, **349**, 2061–2065.
26. M. Lombardo, F. Pasi, S. Easwar and C. Trombini, *Synlett*, 2008, 2471–2474.
27. M. Lombardo, S. Easwar, F. Pasi and C. Trombini, *Adv. Synth. Catal.*, 2009, **351**, 276–282.
28. L. Zhang, S. Luo, X. Mi, S. Liu, Y. Qiao, H. Xu and J. P. Cheng, *Org. Biomol. Chem.*, 2008, **6**, 567–576.
29. J. Shah, S. Khan, H. Blumenthal and J. Liebscher, *Synthesis*, 2009, 3975–3982.
30. S. Khan, J. Shah and J. Liebscher, *Tetrahedron*, 2010, **66**, 5082–5088.
31. D. Siyutkin, A. S. Kucherenko, M. I. Struchkova and S. G. Zlotin, *Tetrahedron Lett.*, 2008, **49**, 1212–1216.
32. D. Siyutkin, A. S. Kucherenko and S. G. Zlotin, *Tetrahedron*, 2009, **65**, 1366–1372.

33. D. Siyutkin, A. S. Kucherenko and S. G. Zlotin, *Tetrahedron*, 2010, **66**, 513–518.
34. S. D. Yang, L. Y. Wu, Z. Y. Yan, Z. L. Pan and M. Liang, *J. Mol. Cat. A: Chem.*, 2007, **268**, 107–111.
35. D. Xu, S. Luo, H. Yue, L. Wang, Y. Liu and Z. Xu, *Synlett*, 2006, 2569–2572.
36. S. Luo, X. Mi, L. Zhang, S. Liu, H. Xu and J. P. Cheng, *Angew. Chem., Int. Ed. Engl.*, 2006, **45**, 3093–3097.
37. S. Luo, L. Zhang, X. Mi, L. Qiao and J. P. Cheng, *J. Org. Chem.*, 2007, **72**, 9350–9352.
38. B. Ni, Q. Zhang and A. D. Headley, *Green Chem.*, 2007, **9**, 737–739.
39. B. Ni, Q. Zhang and A. D. Headley, *Tetrahedron Lett.*, 2008, **49**, 1249–1252.
40. Q. Zhang, B. Ni and A. D. Headley, *Tetrahedron*, 2008, **64**, 5091–5097.
41. B. Ni, Q. Zhang, K. Dhungana and A. D. Headley, *Org. Lett.*, 2009, **11**, 1037–1040.
42. L. Y. Wu, Z. Y. Yan, Y. X. Xie, Y. N. Niu and Y. M. Liang, *Tetrahedron: Asymmetry*, 2007, **18**, 2086–2090.
43. T. Miao, L. Wang, P. Li and J. Yan, *Synthesis*, 2008, 3828–3834.
44. Z. Yakob, J. Shah, J. Leistner and J. Liebscher, *Synlett*, 2007, 2342–2344.
45. D. Q. Xu, B. T. Wang, S. P. Luo, H. D. Yue, L. P. Wang and Z. Y. Xu, *Tetrahedron: Asymmetry*, 2007, **18**, 1788–1794.
46. M. Lombardo, M. Chiarucci, A. Quintavalla and C. Trombini, *Adv. Synth. Catal.*, 2009, **351**, 2801–2806.
47. W. H. Wang, X. B. Wang, K. Kodama, T. Hirose and G. Y. Zhang, *Tetrahedron*, 2010, **66**, 4970–4976.
48. D. Z. Xu, Y. Liu and Y. Wang, *Tetrahedron: Asymmetry*, 2010, **21**, 2530–2534.
49. O. V. Maltsev, A. S. Kucherenko and S. G. Zlotin, *Eur. J. Org. Chem.*, 2009, 5134–5137.
50. O. V. Maltsev, A. S. Kucherenko, I. P. Beletskaya, V. A. Tartakovsky and S. G. Zlotin, *Eur. J. Org. Chem.*, 2010, 2927–2933.
51. O. V. Maltsev, A. S. Kucherenko, A. L. Chimishkyan and S. G. Zlotin, *Tetrahedron: Asymmetry*, 2010, **21**, 2659–2670.
52. T. K. T. Truong and G. Vo-Thanh, *Tetrahedron*, 2010, **66**, 5277–5282.
53. X. Ding, W. Tang, C. J. Zhu and Y. X. Cheng, *Adv. Synth. Catal.*, 2010, **352**, 108–112.
54. X. Ding, H. L. Jiang, C. J. Zhu and Y. X. Cheng, *Tetrahedron Lett.*, 2010, **51**, 6105–6107.
55. S. D. Yang, L. Y. Wu, Z. Y. Yan, Z. L. Pan and Y. M. Liang, *J. Mol. Catal. A: Chem.*, 2007, **268**, 107–111.
56. S. D. Yang, Y. Shi, Z. H. Sun, Y. B. Zhao and Y. M. Liang, *Tetrahedron: Asymmetry*, 2006, **17**, 1895–1900.
57. M. Lombardo, M. Chiarucci and C. Trombini, *Chem. Eur. J.*, 2008, **14**, 11288–11291.

58. L. Zhang, L. Cui, X. Li, J. Li, S. Luo and J. P. Cheng, *Chem. Eur. J.*, 2010, **16**, 2045–2049.
59. P. G. Cozzi, F. Benfanti and L. Zoli, *Angew. Chem., Int. Ed. Engl.*, 2009, **48**, 1313–1316.
60. L. Zhang, L. Cui, X. Li, J. Li, S. Luo and J. P. Cheng, *Eur. J. Org. Chem.*, 2010, 4876–4885.
61. S. Hu, T. Jiang, Z. Zhang, A. Zhu, B. Han, J. Song, Y. Xie and W. Li, *Tetrahedron Lett.*, 2007, **48**, 5613–5617.
62. X. Zheng, Y. B. Qian and Y. Wang, *Eur. J. Org. Chem.*, 2010, 515–522.
63. Y. Qian, X. Zheng and Y. Wang, *Eur. J. Org. Chem.*, 2010, 3672–3677.
64. S. P. Luo, Z. B. Li, L. P. Wang, Y. Guo, A. B. Xia and D. Q. Xu, *Org. Biomol. Chem.*, 2009, **7**, 4539–4546.
65. D. Q. Xu, S. P. Luo, Y. F. Wang, A. B. Xia, H. D. Yue, L. P. Wan and Z. Y. Xu, *Chem. Commun.*, 2007, 4393–4395.
66. M. Lombardo, S. Easwar, F. Pasi, C. Trombini and D. D. Dhavale, *Tetrahedron*, 2008, **64**, 9203–9207.
67. S. K. Ghosh, Z. Zheng and B. Ni, *Adv. Synth. Catal.*, 2010, **352**, 2378–2382.
68. D. Z. Xu, Y. Liu, H. Li and Y. Wang, *Tetrahedron*, 2010, **66**, 8899–8903.
69. X. Mi, S. Luo and J. P. Cheng, *J. Org. Chem.*, 2005, **70**, 2338–2341.
70. X. Mi, S. Luo, H. Xu, L. Zhang and J. P. Cheng, *Tetrahedron*, 2006, **62**, 2537–2544.
71. X. E. Wu, L. Ma, M. X. Ding and L. X. Gao, *Synlett*, 2005, 607–610.
72. C. X. Miao, L .N. He, J. Q. Wang and J. L. Wang, *Adv. Synth. Catal.*, 2009, **351**, 2209–2216.
73. C. Zhu, L. Ji and Y. Wei, *Catal. Commun.*, 2010, **11**, 1017–1020.
74. A. Fall, M. Sene, M. Gaye, G. Gómez and Y. Fall, *Tetrahedron Lett.*, 2010, **51**, 4501–4504.
75. A. Fall, M. Sene, M. Gay, G. Gomez and Y. Fall, *Tetrahedron Lett.*, 2010, **51**, 4501–4504.
76. M. Benaglia, A. Puglisi and F. Cozzi, *Chem. Rev.*, 2003, **103**, 3401–3429.
77. M. Benaglia, *New J. Chem.*, 2006, **30**, 1525–1533.
78. T. E. Kristensen and T. Hansen, *Eur. J. Org. Chem.*, 2010, 3179–3204.
79. L. Zu, J. Wang, H. Li and W. Wang, *Org. Lett.*, 2006, **8**, 3077–3079.
80. L. Zu, H. Li, J. Wang, X. Zu and W. Wang, *Tetrahedron Lett.*, 2006, **47**, 5131–5134.
81. L. Wang, C. Cai, D. P. Curran and W. Zhang, *Synlett.*, 2010, 433–436.
82. M. Benaglia, G. Celentano and F. Cozzi, *Adv. Synth. Catal.*, 2001, **343**, 171–173.
83. M. Benaglia, M. Cinquini, F. Cozzi, A. Puglisi and G. Celentano, *Adv. Synth. Catal.*, 2002, **344**, 533–542.
84. M. Benaglia, M. Cinquini, F. Cozzi, A. Puglisi and G. Celentano, *J. Mol. Catal. A: Chem.*, 2003, **204–205**, 157–163.
85. L. Gu, Y. Wu, Y. Zhang and G. Zhao, *J. Mol. Catal. A: Chem.*, 2007, **263**, 186–194.

86. M. Benaglia, G. Celentano, M. Cinquini, A. Puglisi and F. Cozzi, *Adv. Synth. Catal.*, 2002, **344**, 149–152.

87. A. Puglisi, M. Benaglia, M. Cinquini, F. Cozzi and G. Celentano, *Eur. J. Org. Chem.*, 2004, 567–573.

88. B. Thierry, J. C. Plaquevent and D. Cahard, *Tetrahedron: Asymmetry*, 2003, **14**, 1671–1677.

89. F. Giacalone, M. Gruttadauria, A. Mossuto Marculescu and R. Noto, *Tetrahedron Lett.*, 2007, **48**, 255–259.

90. M. Gruttadauria, F. Giacalone, A. Mossuto Marculescu, S. Riela and R. Noto, *Eur. J. Org. Chem.*, 2007, 4688–4698.

91. Y. X. Liu, Y. N. Sun, H. H. Tan, W. Liu and J. C. Tao, *Tetrahedron: Asymmetry*, 2007, **18**, 2649–2656.

92. Y. X. Liu, Y. N. Sun, H. H. Tan and J. C. Tao, *Catal. Lett.*, 2008, **120**, 281–287.

93. D. Font, S. Sayalero, A. Bastero, C. Jimeno and M. A. Pericas, *Org. Lett.*, 2008, **10**, 337–340.

94. D. Font, A. Bastero, S. Sayalero, C. Jimeno and M. A. Pericas, *Org. Lett.*, 2007, **9**, 1943–1946.

95. E. Alza, X. C. Cambeiro, C. Jimeno and M. A. Pericas, *Org. Lett.*, 2007, **9**, 3717–3720.

96. P. Li, L. Wang, M. Wang and Y. Zhang, *Eur. J. Org. Chem.*, 2008, 1157–1160.

97. E. G. Doyagüez, F. Parra, G. Corrales, A. Fernandez-Mayoralas and A. Gallardo, *Polymer*, 2009, **50**, 4438–4446.

98. T. E. Kristensen, K. Vestli, M. G. Jakobsen, F. K. Hansen and T. Hansen, *J. Org. Chem.*, 2010, **75**, 1620–1629.

99. D. Enders, M. R. M. Hüttl, G. Raabe and J. W. Bats, *Adv. Synth. Catal.*, 2008, **350**, 267.

100. A. Berkessel, N. Gasch, K. Glaubitz and C. Koch, *Org. Lett.*, 2001, **3**, 3839–3842.

101. R. W. Flood, T. P. Geller, S. A. Petty, S. M. Roberts, J. Skidmore and M. Volk, *Org. Lett.*, 2001, **3**, 683–686.

102. W. Qui, L. He, Q. Chen, W. Luo, Z. Yu, F. Yang and J. Tang, *Tetrahedron Lett.*, 2009, **50**, 5225–5227.

103. M. R. M. Andreae and A. P. Davis, *Tetrahedron: Asymmetry*, 2005, **16**, 2487–2492.

104. K. Akagawa, S. Sakamoto and K. Kudo, *Tetrahedron Lett.*, 2005, **46**, 8185–8187.

105. J. D. Revell, D. Gantenbein, P. Krattinger and H. Wennemers, *Biopolymers*, 2006, **84**, 105–113.

106. R. D. Carpenter, J. C. Fettinger, K. S. Lam and M. J. Kurth, *Angew. Chem., Int. Ed. Engl.*, 2008, **47**, 6407–6410.

107. K. Akagawa, H. Akabane, S. Sakamoto and K. Kudo, *Org. Lett.*, 2008, **10**, 2035–2037.

108. E. Bellis and G. Kokotos, *J. Mol. Catal. A: Chem.*, 2005, **241**, 166–174.

109. Y. Wu, Y. Zhang, M. Yu, G. Zhao and S. Wang, *Org. Lett.*, 2006, **8**, 4417–4420.
110. Y. Li, X. Y. Liu and G. Zhao, *Tetrahedron: Asymmetry*, 2006, **17**, 2034–2039.
111. G. Lv, R. Jin, W. Mai and L. Gao, *Tetrahedron: Asymmetry*, 2008, **19**, 2568–2572.
112. G. Y. Wang and X. Y. Liu, G. Zhao, *Synlett*, 2006, 1150–1154.
113. M. Figlus, S. T. Caldwell, D. Walas, G. Yesibag, G. Cooke, P. Kočovský, A. V. Malkov and A. Sanyal, *Org. Biomol. Chem.*, 2010, **8**, 137–141.
114. Y. H. Liu and M. Shi, *Adv. Synth. Catal.*, 2008, **350**, 122–128.
115. I. Mager and K. Zeitler, *Org. Lett.*, 2010, **12**, 1480–1483.
116. C. E. Song, D. H. Kim and D. S. Choi, *Eur. J. Inorg. Chem.*, 2006, 2927–2935.
117. Š. Toma, R. Šebesta and M. Mečiarová, *Curr. Org. Chem.*, 2011, **15**, 2257–2281.
118. S. Luo, J. Li, L. Zhang, H. Xu and J. P. Cheng, *Chem. Eur. J.*, 2008, 1273–1281.
119. Y. Chi, S. T. Scroggins and J. N. J. Frechet, *J. Am. Chem. Soc.*, 2008, **130**, 6322–6323.
120. Z. Shen, J. Ma, Y. Liu, C. Jiao, M. Li and Y. Zhang, *Chirality*, 2005, **17**, 556–558.
121. J. Huang, X. Zhang and D. W. Armstrong, *Angew. Chem., Int. Ed. Engl.*, 2007, **46**, 9073–9077.
122. Z. Y. Li, J. W. Chen, L. Wang and Y. Pan, *Synlett*, 2009, 2356–2360.
123. T. Mitsudome, K. Nose, T. Mizugaki, K. Jitsukawa and K. Kaneda, *Tetrahedron Lett.*, 2008, **49**, 5464–5466.
124. H. Hagiwara, T. Kuroda, T. Hoshi and T. Suzuki, *Adv. Synth. Catal.*, 2010, **352**, 909–916.
125. V. Srivastava, K. Gaubert, M. Pucheault and M. Vaultier, *ChemCatChem*, 2009, **1**, 94–98.
126. N. Hara, S. Nakamura, N. Shibata and T. Toru, *Adv. Synth. Catal.*, 2010, **352**, 1621–1624.
127. B. G. Wang, B. C. Ma, Q. Wang and W. Wang, *Adv. Synth. Catal.*, 2010, **352**, 2923–2928.
128. J. Li, S. Hu, S. Luo and J. P. Cheng, *Eur. J. Org. Chem.*, 2009, 132–140.
129. S. Vijaikumar, A. Dhakshinamoorthy and K. Pitchumani, *Appl. Cat. A*, 2008, **340**, 25–38.
130. A. Massi, A. Cavazzini, L. Del Zoppo, O. Pandoli, V. Costa, L. Pasti and P. P. Giovanni, *Tetrahedron Lett.*, 2011, **52**, 619–622.
131. L. Liu, J. Ma, J. Xia, L. Li, C. Li, X. Zhang, J. Gong and Z. Tong, *Catal. Commun.*, 2011, **12**, 323–326.
132. L. Lin, J. Ma, J. Liuyan and W. Yunyang, *J. Mol. Cat. A: Chem.*, 2008, **291**, 1–4.

Asymmetric Catalysis in Ionic Liquids with 'Unmodified' Catalysts

PETER GOODRICH, CRISTINA PAUN AND
CHRISTOPHER HARDACRE

School of Chemistry and Chemical Engineering/QUILL, Queen's University,
Stranmillis Road, Belfast, Northern Ireland BT9 5AG, UK

3.1 Introduction

In the last 25 years ionic liquids (ILs) have been thoroughly investigated as solvents or additives in a number of reactions. The first report of the use of an IL as a catalyst was in a Friedel–Crafts acylation, reported in 1986.[1] However, it is only in the past 15 years that there has been an exponential increase in their use in a number of physical and chemical applications.

As well as many other properties, ILs provide very different solvent–solute interactions cases which can give rise to distinct chemistries compared with molecular solvent systems.[2] This has had the effect that ILs have been shown to stabilize catalytic complexes or even act as modifiers that accelerate the reaction that leads to an enhancement in selectivity/enantioselectivity.[3] A number of good reviews cover the area of ILs in general.[4–8] Although they are often considered as alternatives to volatile organic solvents, their expense (bulk material cost, synthetic cost or purification) can significantly affect the economics of any process. Impurities in ILs such as water, halides and organic

RSC Green Chemistry No. 15
Enantioselective Homogeneous Supported Catalysis
Edited by Radovan Šebesta

solvents not only have a profound effect on their physical properties[9] but have also resulted in significant changes in reaction chemistry.[10] Within both these areas significant steps have been made in the creation of cleaner and more economic routes to the preparation of ILs;[11,12] however, purification is often a costly process and, recently, the use of ionic liquids as additives/modifiers in small quantities rather than as a bulk solvent has been examined.[13]

In terms of catalysis, the immobilization of homogeneous catalysts with the aid of ILs is a field of growing interest due to the advantages of the ionic layer such as catalyst stabilization and immiscibility with a number of common organic solvents. This has resulted in the concept of biphasic IL catalysis.[14,15] Such systems consist of two phases with the catalyst residing in the IL phase and the substrate/product in the other phase. The technique is particularly applicable to reactions where there is a significant difference in polarity between the starting material and products. The disadvantages of these systems include the leaching of either the IL or catalyst into the alternate phase, and reduced activity due to mass transfer limitations arising from the inherent high viscosity of the IL phase. Furthermore, significant difficulties can be envisaged when considering scale up of either monophasic and biphasic systems, both from a financial and an engineering perspective.

Although numerous reports indicate that covalent attachment of IL salts to achiral or chiral catalysts increased the preferential solubility of the catalysts resulting in successful recycle,[16–18] this generally requires lengthy synthetic procedures and will not be reviewed in this chapter. The most frequently employed strategy to overcome these problems is to prepare an IL-containing heterogeneous catalyst. The immobilization of the catalyst complex in a thin layer of IL on a solid support is known as supported ionic liquid phase (SILP) catalysis.[19–21] This concept emerged as a follow-up to supported aqueous phase catalysis. Moreover, when utilizing SILP catalysts the expense of the IL is negated, as only a small volume is required compared to that when the IL is employed as the bulk solvent.

The aim of this chapter is to provide a review of the use of ILs in metal-catalyzed enantioselective reactions. Where possible examples will be selected to highlight the potential for process optimization of systems originally based on the use of the IL as a bulk solvent through to biphasic liquid–liquid systems, and yet further on to SILP systems.

Although this chapter is entitled 'Asymmetric Catalysis in Ionic Liquids with Unmodified Catalysts', there will be some examples in the text where more recent understanding and analysis of the catalysts show that even so-called 'benign' ILs are now not considered chemically passive, and are capable of modifying the catalyst.[22] This is unsurprising given the numerous publications in which ILs give different chemistries, rates and selectivities compared to molecular solvents.

Within this chapter, 1-alkyl-3-methyl imidazolium cations are denoted as $[C_nC_1im]^+$, 1-alkyl-2,3-dimethyl imidazolium cations are denoted as $[C_ndC_1im]^+$, 1-alkyl-3-alkyl' imidazolium cations are denoted as $[C_nC_nim]^+$, *N*-alkylpyridinium cations are denoted as $[C_npyr]^+$, *N*-alkyl, *N*-methyl pyrrolidinium cations are denoted as $[C_nC_1pyrr]^+$, tetraalkylphosphonium cations are denoted as $[P_{nn'n''n'''}]^+$, and tetraalkylammonium cations are denoted as $[N_{nn'n''n'''}]^+$ where n represents the alkyl chain length attached to the N or P centre. The anion bis(trifluoromethylsulfonyl)imide, i.e. $[(CF3-SO_2)_2N]^-$, is denoted as $[NTf_2]^-$ and triflate is denoted as $[OTf]^-$. BINAP = 2,2'-bis(diphenylphosphino)-1,1'-binaphthyl, BIPHEP = 2,2'-bis(diphenyl-phosphino)-1,1'-biphenyl, QUINAPHOS = 2-(1-naphthyl)-8-diphenylpho-sphino-1-[3,5-dioxa-4-phospha-cyclohepta[2,1-a;3,4-a']dinaphthalen-4-yl]-1, 2-dihydroquinoline, TPPTS = trisodium salt of tri-(m-sulfophenyl)-phos-phine, $P(m\text{-}C_6H_4SO_3Na)_3$, BINAPS = 2,2-bis(diphenylphosphino)-1,1-binaphthyl), DPEN = 1,2-diphenylethylenediamine, BFFPDEA = N-methyl-N-[bis(hydroxyethyl)methy1l-][1',2-bis-(diphenylphosphino)ferrocenyl]ethyl-amine; Phospherox = 2-[2-(diphenylphosphino)ferrocenyl]-4-(1-methylethyl) oxazoline, PHOX = 2-[2-(diphenylphosphino)phenyl]-4-isopropyl-4,5-dihy-dro-1,3-oxazole, XYLIPHOS = 1{2-[bisdiphenyl)phosphino]ferrocenyl}ethyldi-3,5-xylylphosphine.

3.2 Supported Catalysis using Ionic Liquids

Economically and environmentally, SILP catalysis is more advantageous than homogeneous and biphasic catalysis in ILs, due to the reduced amount of ionic liquid, ease of separation and the possibility of use in fixed bed reactors. As previously mentioned, there are a number of ways for the immobilization of homogeneous complexes including SILP systems, for example cross-linking with organic[23–25] fixation on solid supports (oxides,[26,27] polymers,[28] clays, zcolites) or fixation in a liquid phase.[29,30] However, the covalent attachment of a homogeneous catalyst onto a support can be expensive and often leads to changes in its catalytic performance. These changes in the catalyst performance have often been associated with changes in the structure of the catalyst and/or in the conformational preferences of the ligand, but the nature of these changes are not fully understood. One method of preparation consists of dissolving a catalyst complex in small amount IL with a help of volatile organic solvents. To this solution is added the support material (i.e. mesoporous silica materials, zeolites, clays, alumina and titania); organic polymer materials; membranes; carbon nanofibres and sintered metal fibres (SILP type 1). Other methods might employ the use of a covalently bonded ionic liquid to the support material (SILP type 2) to which further portions of ionic liquid can be added (SILP type 3). The resulting materials are solids, with the active catalytic species solvated by the IL and still acting as a homogeneous catalyst, preserving in the majority of the cases their high activity and selectivity and in some cases even resulting in an improved catalytic activity (Figure 3.1).

SILP catalysis was originally developed and is particularly suited for gas-phase reactions as several ILs show high thermal stability[31] and low vapour pressure, and, consequently, they are excellent media for reactions that require high elevated temperatures at which conventional solvents would evaporate or

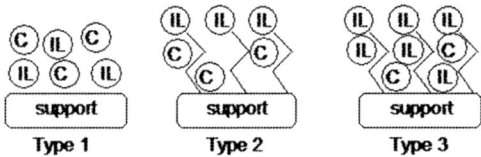

Figure 3.1 Types of supported ionic liquid phase (SILP) catalysts.

decompose. However, if the catalyst, substrates or products are thermally unstable then liquid phase catalysis must be considered. When running a SILP system in liquid-phase conditions, there are a series of factors that have to be taken into consideration, such as the polarity of all the reaction constituents in order to prevent catalyst and or IL leaching, choosing an IL that has a low solubility in the liquid reactant/product mixture and prevention of the physical removal of the IL film by mechanical forces. For this purpose, non-polar solvents such as alkanes (hexane, heptanes, dodecane) are generally used, although functionalized solvents such as toluene and diethyl ether are suitable under specific conditions. Furthermore, water, ethanol or propan-2-ol can be used in combination with hydrophobic ionic liquids, *e.g.* containing $[PF_6]^-$ or $[NTf_2]^-$ anions, although even in this case, limited extraction of the IL and the catalyst in the solvent is difficult to avoid. Most liquid phase SILP systems reported require an additional (organic) solvent as reaction medium, but there are examples when solvent-free conditions can be applied, when the reagents and the products form a separate organic layer. Recently, an alternative methodology has been proposed making use of supercritical carbon dioxide ($scCO_2$). Although $scCO_2$ shows good solubility in some ILs, the reverse is not the case, with no detectable IL solubilization in the $scCO_2$ phase.[32] The $scCO_2$ reduces the viscosity of the IL and thus improves the solubility and mass transport of permanent gases. Moreover, due to high diffusion rates, mass transfer effects are reduced.[33] The main disadvantages of $scCO_2$ are the high investment and operating costs and the limited solvating ability of $scSO_2$ in comparison with classical organic solvents.

3.3   Catalyst and Ionic Liquid Characterization

Investigation of the IL surface and interface characteristics is a field of growing interest due to the importance of the interface of ionic liquid and its environment (solid, liquid or gaseous).[34] Some of the important surface properties of ILs include surface tension and wettability, heat and mass transport characteristics as well as surface order and reactivity.[35] For SILPs, as in any heterogeneous catalyst, surface interfaces play a key role and in order to explain fundamental macroscopic surface properties it is important to understand the liquid–vapour (gas phase) interface at molecular level. Studies using SFG spectroscopy (surface-specific sum frequency generation) vibrational, surface tension measurements and variation of the IL structure show how the balance between surface charge and dispersion forces are manifested. For a range of [Rmim][R-OSO$_3$] ILs (R = C$_1$–C$_4$) the longer alkyl chains dominate shifting the polar moieties shift away from the surface, resulting in a weaker polarity resulting in a decrease in the surface tension.[36] Therefore, the chemical composition of the near-surface region and the molecular arrangement at the surface can be different to that in the bulk.

Initially, characterization of the surface properties of ILs was done using techniques such as sum frequency generation,[37] X-ray and neutron reflectivity,[38] surface tension measurements,[36] grazing incidence X-ray diffraction[39] and simulations.[40] Normally, surface investigations at atomic level precision require ultra high vacuum (UHV, pressures in the 10^{-9} mbar range or below), in order to ensure surface cleanliness and to avoid interactions of the used probes with gas-phase molecules. Once it was realized that due to their very low vapour pressure ILs are stable under ultra high vacuum conditions and do not contaminate the sensitive UHV systems, techniques such as X-ray photoelectron spectroscopy (XPS), UV photoelectron spectroscopy (UPS), inverse photoelectron spectroscopy (IPES), near edge X-ray absorption spectroscopy (NEXAFS), metastable ion spectroscopy (MIES), direct recoil spectroscopy (DRS), high resolution electron energy loss spectroscopy (HREELS), low energy ion scattering (LEIS), time-of-flight secondary mass spectroscopy (TOF-SIMS), soft X-ray emission spectroscopy (SXES) and Rutherford backscattering (RBS) have successfully been utilized to study IL surfaces. All these techniques have helped to open the new field of ionic liquid surface science.[41]

As previously discussed, in the SILP concept a catalyst (homogeneous or heterogeneous) is immobilized in the IL, and so the reactants have to pass the surface, diffuse to the active complex, react, diffuse back to and pass through the surface again. If the catalyst is preferentially enriched in the near-surface region, the diffusion pathways are reduced and the total process is more efficient. Maier *et al.*[42] used angle-resolved XPS to demonstrate that for a Pt catalyst with a nominal concentration below 0.1 mol% an enrichment in the surface region by a factor of 200 was attained. However, until now only one system with such a pronounced enrichment effect has been observed. For future researchers the next challenge would be to understand the physical and chemical properties to an extent which will allow to selectively tune the IL and/or catalyst properties to achieve the desired enrichment (or in other cases depletion). Furthermore, it would also be desirable to tune the solubility of reactants and products in the IL such that the reagent was soluble and the product insoluble. Also, when dealing with ILs supported on solids, surface versus bulk composition and surface orientation are another series of factors that have to be taken into consideration. For example, using angle-resolved XPS it has been demonstrated that for imidazolium-based ILs containing alkyl chains a preferential enrichment of the alkyl chains at the surface occurs and the degree of enrichment increases with chain length and decreases with the size of the anion.[43] This is an important finding since it could provide further information on the way the homogeneous catalyst is orientated/positioned in the IL layer as well as how reagents/products will be attracted or repelled from the surface.

3.4 Hydrogenation Reactions

Hydrogenation is one of the most studied catalytic transformations in the presence of ILs.[44] While the vast majority of studies indicate lower reaction rates compared with molecular solvents due low hydrogen solubility[45] when ILs are used as bulk solvents, tailoring the reaction parameters such as the use of a co-solvent or high substrate concentration can increase the overall hydrogen solubility in the reaction mixture has been shown to lead to similar activities using ILs. Other processes to reduce the impact of mass transfer limitations involve the use of biphasic process liquid–liquid,[45] solid–liquid systems and the use of transfer hydrogenations.

3.4.1 Purity of Ionic Liquids

The importance of purity of many ILs for hydrogenation reactions has been highlighted by Pugin *et al.*[46] For the Ru-catalyzed asymmetric hydrogenation reaction of α-acetamidocinnamic acid (R = Ph, Scheme 3.1) conducted in ILs containing significant amounts of halide, detrimental catalyst performances were observed. The role of water also had a significant effect in catalyst activity/selectivity. IL and water combinations resulted in superior enantioselectivities (ee) compared with conventional organic solvents and biphasic IL/organic solvent systems. In addition, although lower ee values were observed at higher pressure in all solvents, the use of wet ILs enabled a reduction in the sensitivity towards pressure. This effect was thought to be caused by different solvation properties of the IL/water system.

The role of water has also been shown to be critical in enabling high asymmetric induction in the hydrogenation of acetophenone catalyzed by $[RuCl_2(TPPTS)]_2$-dpends/KOH catalyst in $[C_nC_1im][OTs]$ ILs.[47] The addition of appropriate amounts of water (V_{IL}/V_{H2O} 2:1) was found to increase the ee from 45% to 80%. This increase was attributed to the enhancement of the solubility of the water-soluble catalyst in the IL. However, a lower activity and ee (36%) was observed when only water is used as a solvent. Moreover, an increase in alkyl chain length of the imidazolium cation decreases the catalyst solubility resulting in lower activity and enantioselectivity. A similar phenomenon

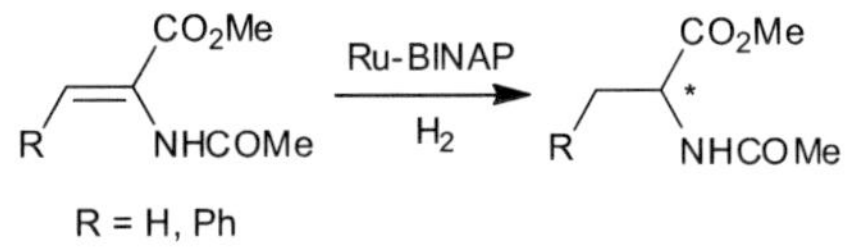

R = H, Ph

Scheme 3.1 Enantioselective hydrogenation of enamides.

was also observed for the same catalyst in the asymmetric hydrogenation of benzalacetone under similar conditions.[48]

3.4.2 Homogeneous Enantioselective Hydrogenation

Mass transfer limitations in ILs have arguably had the most impact in asymmetric hydrogenation reactions. In 2001, Dupont was the first to study the effect of the concentration of hydrogen in an IL on the product ee. Significant changes in ee, 81% in $[C_4C_1im][PF_6]$ and 93% in $[C_4C_1im][BF_4]$ were observed for the hydrogenation of α-acetamidocinnamic acid (R = Ph) using a Ru-(S)-BINAP catalyst (see Scheme 3.1). This was attributed to the higher hydrogen solubility in the $[BF_4]^-$ based IL compared with the IL based on $[PF_6]^-$, under the same pressure, indicating that mass-transfer limitations played an important role in product selectivity.

Jessop *et al.*[49] studied this further and compared the IL- and molecular solvent-mediated hydrogenation of tiglic and atropic acid using Ru-(S)-BINAP catalysts (Scheme 3.2). The hydrogenation of tiglic acid with high ee is known be independent of hydrogen concentration and $[C_2C_1im][NTf_2]$ was shown to have an optimized ee of 95% compared with only 88% in methanol. The use of 'wet' $[C_4C_1im][PF_6]$ was also reported to give higher ee values (up to 92%) compared with dry $[C_4C_1im][PF_6]$ (88% ee) or i-PrOH/$[C_4C_1im][PF_6]$ (up to 40% ee). This beneficial effect was attributed to a lower hydrogen concentration in ILs than in the organic solvent. In contrast, in the hydrogenation of atropic acid, the ee is hydrogen-pressure dependent. Using the same catalyst, the optimized ee for atropic acid was 87% in the $[C_4C_1im][PF_6]$–methanol mix but 92% in pure methanol and extremely low ee values in the pure ILs.

The enantioselective hydrogenation of trimethylindolenine using an Ir-(R,S) XYLIPHOS catalyst was also subjected to mass transfer limitations even under 40 bar H_2 pressure.[50] This was in part due to the high viscosity of the $[C_{10}C_1im][NTf_2]$ used in the reaction. To reduce the viscosity effect, elevated temperatures were used which resulted in comparable reactions times to that encountered with toluene. In addition, an improvement in the ee from 65% to 72% was noted with increasing temperature in IL which was comparable to the 80% ee observed with toluene.

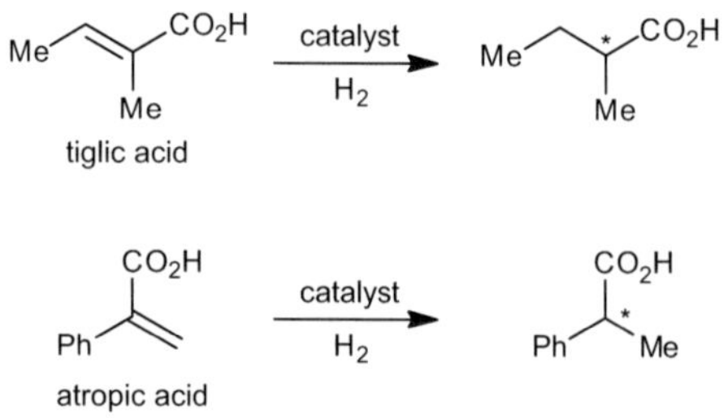

Scheme 3.2 Hydrogenation of tiglic and atropic acid.

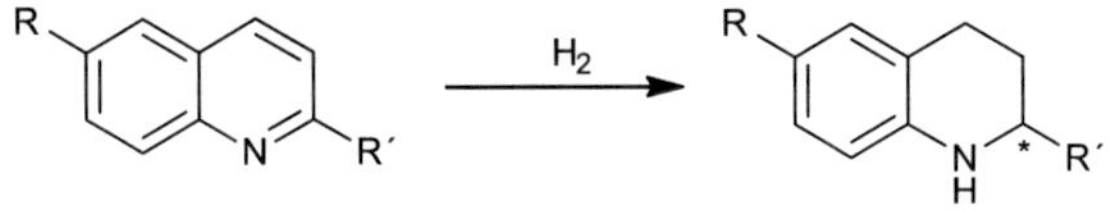

Scheme 3.3   Enantioselective hydrogenation of quinolines.

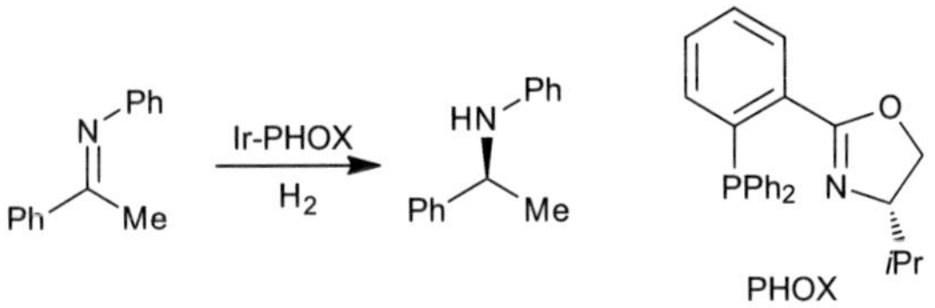

Scheme 3.4   Enantioselective hydrogenation of *N*-(1-phenylethylidene)aniline.

For substrates that require high hydrogen pressures, lower viscosity ILs can be successfully employed as bulk solvents without offsetting the ee. For example, see the asymmetric hydrogenation of substituted quinolines at 50 bar H_2 catalyzed by a [η^6-*p*-cymene Ru-Ts-DPEN] complex in [C_4C_1im][PF_6] to the corresponding tetrahydroquinolines.[51] For a range of substituted quinolines (R = H, Me, OMe, F and R' = Me, Et, Pr, Bu) high asymmetric induction 94–99% ee has been observed (Scheme 3.3). The increased air stability of the phosphine-free catalyst and its immobilization in an IL facilitated a highly effective recyclable catalyst–IL system.

In an attempt to overcome the problems of H_2 solubility in ILs the Ir-catalyzed hydrogenation of *N*-(1-phenylethylidene)aniline was used as a chemical probe to evaluate the potential of biphasic ionic liquid/supercritical CO_2 reaction media.[52] The cationic Ir-PHOX complex showed almost complete inactivity in [C_2C_1im][NTf_2] under 30 bar H_2; however, at 100 bar, 97% conversion with 58% ee was obtained (Scheme 3.4).

Under IL/scCO_2, quantitative formation of the product with a similar ee was observed under 30 bar hydrogen. NMR studies indicated that the H_2 concentration in the IL increased from 0.01 M to 0.14 M in the IL/CO_2 system. This increase in hydrogen solubility was also coupled with a decrease in the viscosity of the IL, helping the poor mass transfer and gas availability that is present in the monophasic IL reaction. Moreover, the use of scCO_2 was seen to prevent catalyst leaching from the IL phase.

3.4.3   Supported Enantioselective Hydrogenations

While the use of ILs as a catalyst immobilization medium has been shown to be effective for transition metals and their complexes, for asymmetric metal catalysts; retention, stability and recycle is a necessity due to the added expense of chiral ligands. In 2003, Wolfson *et al.*[53] developed a SILP type 2 catalyst based

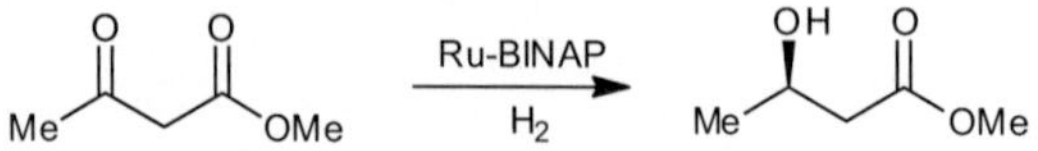

Scheme 3.5 Hydrogenation of methyl acetoacetate.

on a Ru-(*S*)-BINAP catalyst in [C$_4$C$_1$im][PF$_6$] and a poly-(diallyldimethylammonium chloride) polymeric support for the liquid phase hydrogenation of methyl acetoacetate (Scheme 3.5). The SILP reaction in isopropanol (TOF 29 h^{-1}, 97% ee) showed higher activity than the biphasic reaction (TOF 16 h^{-1}, 97% ee); however, comparable performance of the homogeneous reaction in an IPA-MeOH mix (TOF 74–103 h^{-1}, 99% ee) could not be reached (TOF = turnover frequency). The main advantage of the heterogeneous system was the catalyst reusability without significant loss in activity, chemoselectivity and ee.

In a quest to understand the nature of the catalytically active species in the supported ionic phase catalysts, Fow *et al.*[54] used Rh and Ru complexes with (*R*)-BINAP immobilized on silica using thin films of phosphonium based ILs as heterogeneous catalysts for the hydrogenation of acetophenone. Compared with the analogous homogeneous reactions, which showed 0% ee, 1-cyclohexylethanol was formed with ee values up to 74% using the solid supported catalyst. Nitrogen adsorption analysis showed that the surface area of the pure silica only decreased slightly with the addition of the IL; however, the pores with diameters ≤3 nm diameter were filled entirely with IL. The remainder of the IL formed an even film (thickness ~1.1 nm) inside the larger pores. ^{31}P MAS NMR of the supported catalysts showed a shift downfield of 0.4–0.6 ppm caused by the different dielectric constant of the IL, while in the ^{1}H NMR strong line broadening of 134 Hz was observed in comparison with the line width in the neat ILs of 12–21 Hz. Sievers *et al.*[55] associated the line broadening to reduced mobility of the IL molecules within the liquid film. It was speculated that the IL gave rise to so called 'solvent cages', which consisted of supramolecular aggregates of one complex molecule and 16–33 ion pairs of the ionic liquid. In order to analyze the catalytic behaviour as a function of the structure of the catalyst, the homogeneous (THF) and the heterogeneous catalysts systems were compared. The heterogeneous Rh-(*R*)-BINAP complexes yielded good ee values with high conversions. In comparison, the corresponding homogeneous catalysts showed slower reaction rates with no asymmetric induction, clearly indicating a different binding mode in the IL in contrast with the molecular solvent. A binding model was proposed, as shown in Figure 3.2, where acetophenone was found to bind preferentially *via* both the carbonyl group and phenyl ring to the metal centre. The resultant η^2/η^2-coordinated acetophenone fitted into the pocket formed by the phenyl rings of the chiral BINAP ligand, resulting in high stereoselection.

Lou *et al.*[56] immobilized a chiral ruthenium complex, Ru-(*S,S*)-DPEN, in an imidazolium-based IL phase confined on the surface a series of mesoporous materials MCM-48, MCM-41, SBA-15 and amorphous SiO$_2$ all of which were

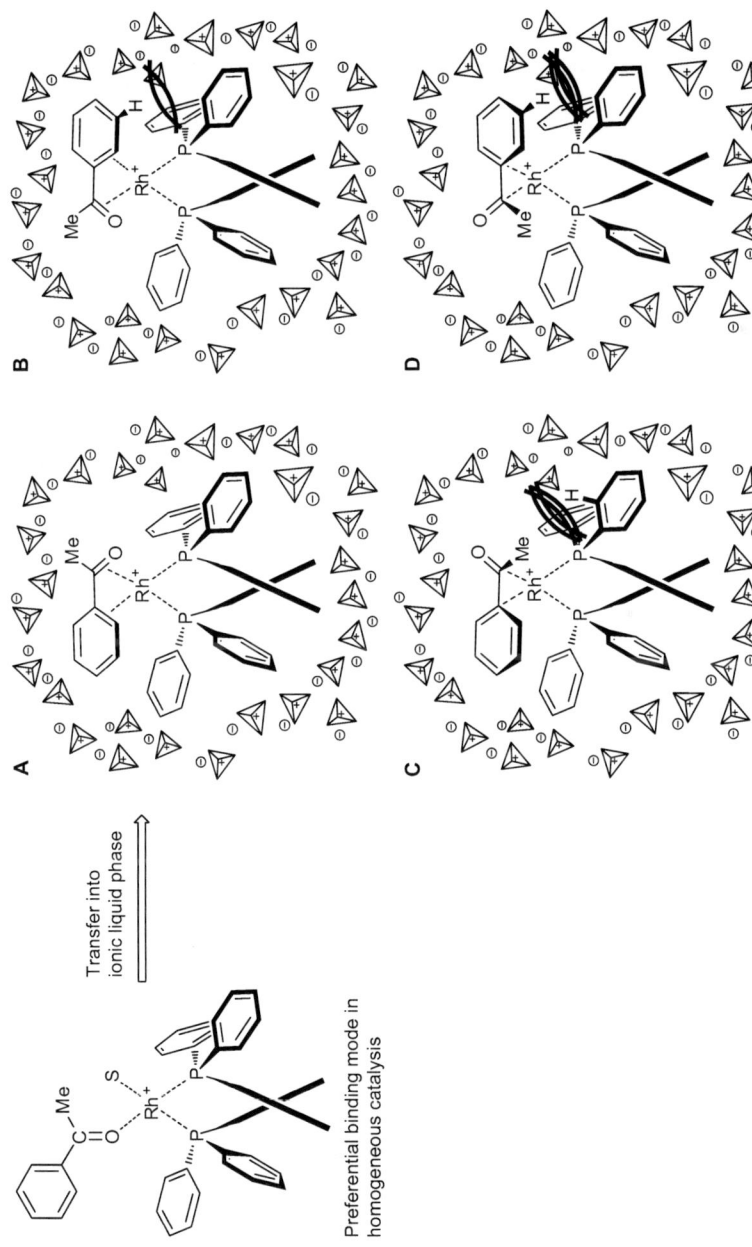

Figure 3.2 Preferential coordination mode in homogenous catalysis and schematic representation of the geometry of the diastereomeric [Rh((*R*)-BINAP)(acetophenone)]⁺ catalyst-substrate adduct in supported ionic liquids. A is the preferred coordination mode. (Reprinted from K. L. Fow, S. Jaenicke, T. E. Müller and C. Sievers, *J. Mol. Catal. A: Chem.*, 2008, **279**, 239, with permission from Elsevier.)

modified with an covalently grafted IL in a SILP type 3 system. The corresponding heterogeneous catalysts were evaluated in the asymmetric hydrogenation of acetophenone in propan-2-ol. All of the catalysts exhibited increased activity with TONs of *ca.* 1000 compared with *ca.* 200 for the analogous homogeneous systems. For the heterogeneous system up to four recycles of the catalyst were possible. In the absence of either the chemisorbed (SILP type 2) or physisorbed IL (SILP type 1), poor conversions were noted upon recycle indicating a synergistic effect of both ILs. The modified SILP type 3 MCM48 Ru-IL complex showed slight increase in ee compared to other systems SILP type 3 systems. This was attributed to the confinement effect of the 3D channel of MCM-48 resulting in the higher ee compared with the 1D channel of MCM-41 and SBA-15.

A highly efficient continuous-flow asymmetric catalysis has been achieved by combining a SILP type 1 catalyst with scCO$_2$ as the mobile phase, for the hydrogenation of dimethyl itaconate (Scheme 3.6) in the presence of a Rh-(*S*,*R*)-QUINAPHOS type complex.[57] Initial reactions in a biphasic system ([C$_2$C$_1$im][NTf$_2$]/scCO$_2$) at 40 °C and 120 bar total pressure (including 20 bar H$_2$) gave turnover frequencies (TOFs) as high as 10 500 h^{-1} with 100% ee. In contrast, no reaction was observed when the Rh catalyst was supported on SiO$_2$ without the IL, confirming the homogeneous nature of the active catalytic species formed in the SILP matrix was critical under scCO$_2$ flow conditions. A representative run of over 65 h on-stream showed the total turnover number reached a remarkable 115 000 moles of substrate per mole of Rh, corresponding to a high space-time yield of 0.3 kg h^{-1} l^{-1}. However, despite the high level of activity, the ee was not maintained with time on-stream. After the first 10 h on-stream the ee decreased to 70–75%. The decrease has been ascribed to a partial decomposition of the active catalyst forming unselective Rh hydrogenation catalyst. This was supported by discoloration of the used SILP catalyst at the front end of the fixed bed after 65 h reaction. These results are significant as it

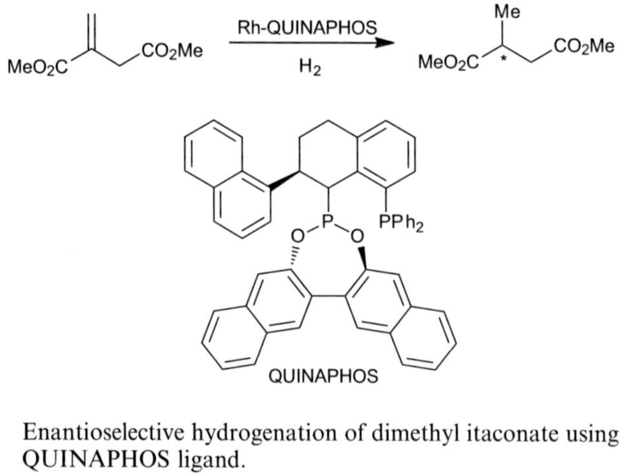

Scheme 3.6 Enantioselective hydrogenation of dimethyl itaconate using the modified QUINAPHOS ligand.

was proved that the combination of the SILP concept with supercritical flow systems opens very promising approaches to continuous enantioselective catalysis with chiral organometallic catalysts. Moreover, in direct comparison to other immobilization techniques,[58] such as the concept of self-supported catalysts, approximately one order of magnitude higher space-time yields are achieved with the SILP-scCO$_2$ system.

3.4.4 Transfer Hydrogenation

Due to the safety problems associated with gaseous hydrogen, transfer hydrogenation has also been employed in the presence of ILs. Asymmetric hydrogenation has been limited to use of [Ru(amine)(arene)] Noyori-type catalysts. The asymmetric reduction of acetophenone using a triethylamine-formic acid (TEAF) azeotrope was successfully applied resulting in 96% conversion and 93% ee for the product 2-phenylethanol.[59] Similar Noyori-type Ru catalysts were also used for the hydrogenation of acetophenone using TEAF.[60] In a range of imidazolium, ammonium and phosphonium ILs similar ee values and TOFs were noted compared to dichloromethane.

3.4.5 Chiral Ionic Liquids as Promoters for Asymmetric Hydrogenations

The first successful example of a solvent promoting chirality was first reported by Seebach and Oei,[61] which was the synthesis of 2,3-diphenyl-2,3-butanediol (23% ee) from acetophenone in the presence of the amino ether (*S*,*S*)-(+)-1,4-bisdimethylamino-2,3-dimethoxybutane. Menthol has also been employed in the asymmetric oxidation of sulfides and alkenes with very low ee values (10%).[62] The hydroboration of acetophenone in the chiral solvent (*S*)-methyl lactate exhibited ee values of up to 60%.[63]

Recently, a chiral ionic liquid based on (*S*)-proline (Figure 3.3) has been successfully used in the enantioselective hydrogenation of dimethylitaconate and methyl *N*-acetamidoacrylate.[64] In the case of dimethylitaconate, when the IL was used as an additive or as a solvent, identical product ee values could be obtained using a racemic Rh-BINAP catalyst compared with the enantiopure Rh-(*S*)-BINAP catalyst. Kinetic and spectroscopic studies showed chiral poisoning as a principal mechanism for the differentiation of two enantiomeric catalytically active species, with the IL preferentially complexing to the Rh-(*S*)-BINAP. This was further evidenced by a change in product configuration in the hydrogenation of methyl acetamidoacrylate from 25% (*R*) in dichloromethane

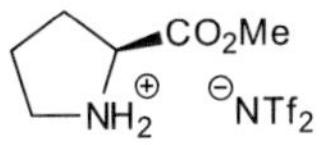

Figure 3.3 Chiral prolinium based ionic liquid.

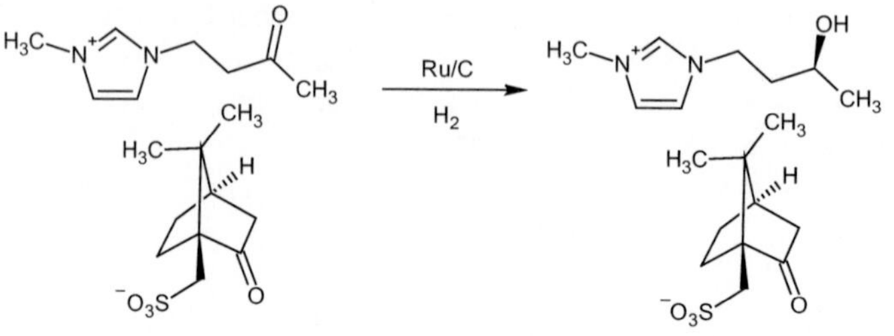

Scheme 3.7 Hydrogenation of an oxo tagged imidazolium (*S*)-camphorsulfonate ionic liquid.

to 41% (*S*) in the chiral IL. Moreover, the same system could also be employed for atropoisomeric ligands such as BIPHEP which does not possess any permanent chiral information.[65] Therein, the chiral ionic liquid was used as the exclusive source of chirality. For both the Rh-catalyzed hydrogenations of dimethylitaconate (see Scheme 3.5) and methyl acetamidoacrylate an improvement in ee was observed in the presence of NEt$_3$. These systems are two of the few examples of a chiral IL solvent having a significant positive effect in asymmetric catalysis and opens up another promising field for IL applications.

Effective chiral transfer in ILs through ion-pairing effects has recently been observed in the hydrogenation of a prochiral ketone (Scheme 3.7).[66,67] The transfer of chirality between the enantiopure (*R*) or (*S*)-camphorsulfonate anion to the imidazolium cation was observed whereby a Ru/C catalyst was able to selectively hydrogenate the carbonyl-tether to the corresponding alcohol with 94% ee. This enables the synthesis of chiral cation and anion containing ILs which could be exploited as chiral solvents or additives for asymmetric reactions and separations.

3.5 Oxidation Reactions using Ionic Liquids

Oxidation reactions are commonly applied in forming intermediates for the fine chemical industry; however, recycling and the use of toxic metals makes this process unappealing. In this regard, ILs have again emerged as promising candidates to combat this problem with catalysts based on Mn, Ru and V being the most widely applied with a range of oxidizing agents.

3.5.1 Asymmetric Alkene Oxidation

Epoxides and diols are of great importance as versatile and useful chiral building blocks in organic synthesis for pharmaceuticals and agrochemicals.[68] Asymmetric epoxidation of olefins is the most direct and effective approach to synthesizing enantio-enriched epoxides, and many research efforts have been devoted to the development of efficient catalysts for this process.[69] As a result of the intensive research in the field of catalytic asymmetric epoxidation, a

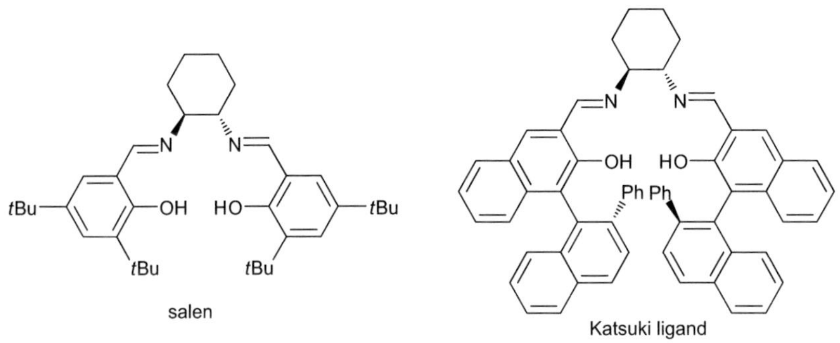

Figure 3.4　Salen and Katsuki ligands used in asymmetric epoxidation reactions.

series of transition metal catalysts based on titanium,[70] manganese,[71] vanadium[72] and molybdenum,[73] as well as organocatalysts such as chiral ketones,[74] have emerged as highly active systems. Due to the obvious advantages, many groups have concentrated their research on the heterogenization of these transition metal chiral catalysts. Some of the methods used for heterogenization include covalent attachment to solid supports,[75] steric occlusion in nanosized cages of zeolites,[76] entrapment in a polydimethylsiloxane membrane[77] and fluorous biphasic systems.[78] However, some of these immobilization techniques frequently require tedious ligand modifications and often lead to a marked decrease in both selectivity and activity.

Due to the fact that there has been already a well established catalyst for asymmetric epoxidation reactions, research involving ILs has progressed rapidly. There have been a significant number of reactions that proceed with high ee values and yields in both pure IL and/or IL biphasic media. Jacobsen's chiral Mn^{III}-salen epoxidation catalyst (Figure 3.4) was successfully immobilized in $[C_4C_1im][PF_6]/CH_2Cl_2$ and used in the asymmetric epoxidation of dimethylchromene derivatives using NaOCl.[79] In the co-solvent system, the reaction was shown to be much faster than in the homogeneous CH_2Cl_2 medium. The Mn^{III}-salen catalyst was recyclable, with a slight decrease in yield and ee from cycle 1 (yield 86%, 96% ee) to cycle 3 (yield 53%, 88% ee). UV-VIS spectroscopy[80] and cyclovoltammetry[81] indicated that the increase in rate was a reflection of the IL's ability to stabilize the active metal-oxo intermediate. Recently a series of imidazolium and tetra-alkyl-dimethylguanidium-based ionic liquids have also been used for the Mn^{III}-salen catalyzed epoxidation of substituted chromenes.[82] For reductions conducted in neat ILs, both the conversions and ee values were significantly lower than those found in dichloromethane indicating that a co-solvent is crucial for this particular system.

Immobilization of Katsuki-type catalyst in $[C_4C_1im][PF_6]$ gave comparable results with Mn^{III}-salen catalyst in the asymmetric epoxidation of dihydronaphthalene.[83] In the presence of an IL, higher reaction rates were observed and the overall stability of the catalyst was found to be higher on recycle compared with the Mn^{III}-salen catalyst (cycle 1: yield 81%, 96% ee; cycle 2:

yield 69%, 90% ee). Despite all these advantages, the critical factor in order to obtain high activity and selectivity was found to be the presence of a co-ligand, 4-phenylpyridine *N*-oxide and not the presence of the IL.

The asymmetric epoxidation of (*R*)-(+)-limonene using H_2O_2 has been performed in the presence of Mn^{III}-salen catalyst immobilized in $[C_4C_1im][BF_4]$ (Scheme 3.8).[84] Limonene was selectively converted into the corresponding epoxide with a diastereoisomeric excess of 74% and conversions up to 70%. Similar chemistries were observed in both IL and methanol/dichloromethane mixtures indicating that the configuration of both the substrate and catalyst was found to play an important role in stereochemical formation.

Sharpless catalytic asymmetric dihydroxylation has also been studied in $[C_4C_1im][PF_6]$ as a co-solvent rather than in the conventional system of *t*-butanol/H_2O.[85] Comparable or even higher yields and ee were found in ILs for a series of aryl and alkyl alkenes compared with the conventional molecular solvent system. The use of IL-water (biphasic) or IL/water/*t*-butanol (monophasic) solvent systems provided an efficient and reusable system with up to 9 cycles shown with only a 5% of yield reduction over the entire process. Additionally, the osmium content in the organic phase and in the aqueous phase was of the order of ≤ 7 ppb equivalent to $\leq 3\%$. Biphasic IL-scCO$_2$ systems have proved to be superior compared with conventional liquid–liquid systems for product extraction and catalyst retention in the osmium-catalyzed asymmetric dihydroxylation of *trans*-cinnamates.[86]

Asymmetric dihydroxylation has also been employed using $Rh_2(OAc)_4/$ $K_2OsO_2(OH)_4$ catalyst system in a chiral alkylguanidinium quinate IL (Scheme 3.9).[87] Styrene was found to dihydroxylate with 92% conversion and 74% ee and the aliphatic alkene, hex-1-ene, was converted to the corresponding *cis*-diol with 95% conversion with 85% ee. In general, for the conventional synthesis of diols using 4-methylmorpholine-*N*-oxide as co-oxidant, a slow addition of the olefin is required in order to eliminate the secondary catalytic cycle effect.

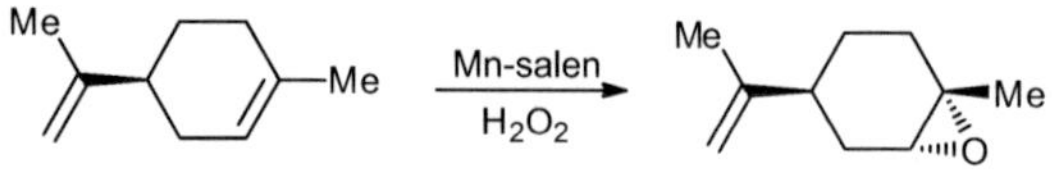

Scheme 3.8   Epoxidation of (*R*)-(+)-limonene.

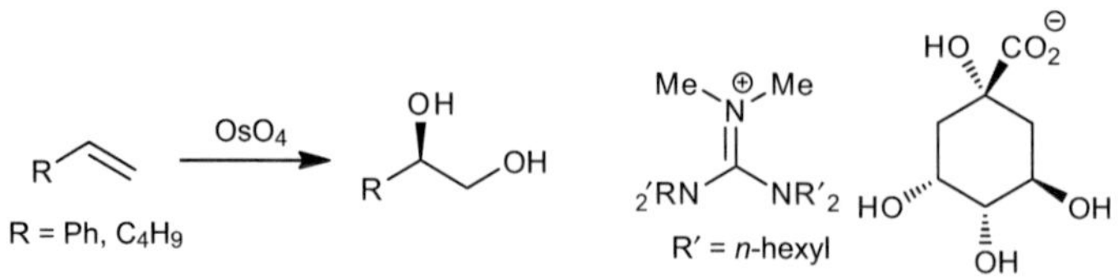

Scheme 3.9   Asymmetric dihydroxylation of alkenes in a chiral alkylguanidinium quinate IL.

Interestingly, using the chiral IL in the absence of the Sharpless chiral ligand, without slow addition of the olefin, the diols were obtained with conversions and ee values comparable to those obtained using the conventional system.

3.5.2 Supported IL Catalysts for Oxidation

Based on the results of Song *et al.*, Lou *et al.*[88] immobilized chiral Mn^{III}-salen complexes on an imidazolium-based modified mesoporous silicate MCM-48 (SILP type 2), with $[C_4C_1im][PF_6]$ used as the additional adsorbed IL (SILP type 3 system). The chiral heterogeneous catalysts were employed for the asymmetric epoxidation of unfunctionalized olefins. The immobilization of chiral Mn^{III}-salen catalysts produced lower conversions and ee values in comparison with the homogeneous catalysts for the epoxidation of styrene. However, in the case of α-methylstyrene, which is of a comparable size with styrene, the SILP type 2 catalyst exhibited notably higher enantioselectivity (99% ee) and comparable catalytic activity with the homogeneous catalyst (50% ee). The obtained conversions and ee values were markedly higher than those reported in the literatures related to inorganic supports.[89–91] The heterogeneous catalysts proved to be very active, maintaining high levels of ee values, when bulkier olefins such as 1-phenylcyclohexene and indene were epoxidized. Another important observation of this study was that the remarkably increase in ee was obtained with no obvious decrease in the conversions. Therein, this was proposed to be due to the combination of solution chemistry (IL) and the spatial effect originating from the heterogeneous carrier media MCM-48, which possesses a well-defined pore size and 3D topological structure. A highly enantioselective and recyclable SILP type 3 system was reported by Wei *et al.*[92] for the epoxidation of styrene using Jacobsen's catalyst $[C_4C_1im][BF_4]$ or $[C_4C_1im][PF_6]$ and silica gel (100-200 mesh) or siliceous earth. The Mn complex was grafted onto silica supports through the linkage of hydrogen bonds, and then dispersed into ILs. Reactions taking place in either the biphasic IL/H_2O or heterogeneous system $SILP/H_2O$ were faster than the reaction under conventional conditions, for example CH_2Cl_2/H_2O. The nature of IL was found to influence the ee, with the hydrophobic $[C_4C_1im][PF_6]$ giving slightly higher chiral induction than the hydrophilic $[C_4C_1im][BF_4]$. Although no explanation was provided, this could be due to the presence of trace amounts of residual halide. The recyclability of the catalyst as well as the ee was strongly dependent on the reaction medium. For example in a $[C_4C_1im][BF_4]/H_2O$ system the catalyst was recycled four times (with a small drop in ee from 64% in the first run to 52% in the fifth run). In order to maintain the high levels of ee the presence of a co-catalyst proved to be crucial.[93]

3.5.3 Oxidative Kinetic Resolution of Alcohols and Epoxides

Oxidative kinetic resolution of racemic alcohols is a viable approach to give optically active alcohols,[94,95] which are extremely useful starting materials and intermediates in synthetic organic chemistry and the pharmaceutical industry.[96]

Successful oxidative kinetic resolutions of racemic alcohols have been reported by a number of studies and these involve the aerobic oxidative kinetic resolution of secondary alcohols[97,98] catalyzed by (–)-sparteine/Pd^{II} or the use of chirally modified azaadamantane-type of organocatalysts[99] to produce enantiomerically enriched secondary alcohols.

Sun *et al.*[100] reported that kinetic resolution of secondary alcohols can be catalyzed by chiral Mn^{III}-salen complexes with excellent enantioselectivity (up to 98% ee) in water, in the presence of a phase-transfer catalyst with hypervalent iodine as the co-oxidant. Thereafter most reports have focused on homogeneous catalysis, with only a few concerning the heterogeneous oxidative kinetic resolution of secondary alcohols.[101–103] Some of these reports involve the immobilization of chiral Mn^{III} salen complexes with the aid of ILs.[103,104] Sahoo *et al.*[105] were the first to report the kinetic resolution of secondary alcohols with a chiral Mn^{III}-salen complex immobilized over mesoporous silica utilizing the SILP strategy (SILP type 2 system). The oxidative kinetic resolution of α-methylbenzyl alcohol (R = Ph, R′ = Me) as a model substrate used diacetoxy iodobenzene (PhI(OAc)$_2$) as the oxidant under biphasic conditions (Scheme 3.10). Hydrophilic additives such as [C$_4$C$_1$im]Br and [N$_{1111}$]Br have been shown to act both as a support and additive with promising conversions of 63–64%, ee values of 98–99%. Replacement of the bromide-based IL by the hydrophobic [C$_4$C$_1$im][PF$_6$] was used to generate a SILP type 2 catalyst. Using the hydrophobic IL to immobilize a Mn^{III}-salen complex into a mesoporous silica SBA-15 resulted in a highly active catalyst where both aromatic and aliphatic alcohols were oxidized with high yield and ee. Moreover, when different silica supports were studied (MCM-41, MCM-48, SBA-15 and amorphous silica), little or no difference was observed in terms of ee values.

Co^{III}-salen catalysts have also been used for the kinetic resolution of racemic epoxides, in the presence of [C$_4$C$_1$im][PF$_6$] and [C$_4$C$_1$im][NTf$_2$] ILs.[106] The hydrolytic kinetic resolution was successful for a range of substrates producing enantio-enriched epoxides (<98% ee) with the corresponding diols showing 79–92% ee (Scheme 3.11). The ILs stabilized the catalyst against reduction to

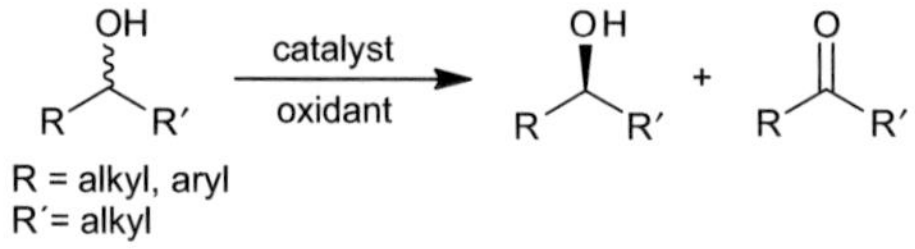

R = alkyl, aryl
R′= alkyl

Scheme 3.10   Kinetic resolution of dialkyl and aryl-alkyl alcohols.

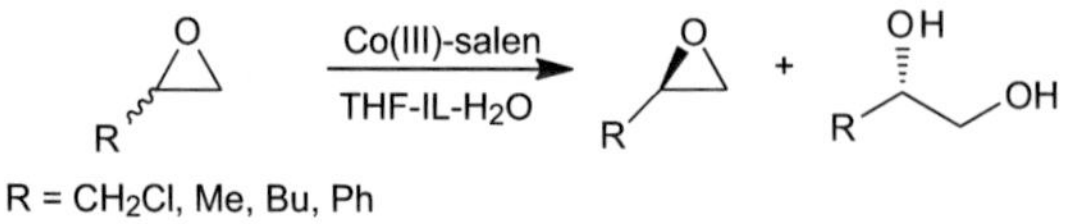

R = CH$_2$Cl, Me, Bu, Ph

Scheme 3.11   Hydrolytic kinetic resolution of epoxides.

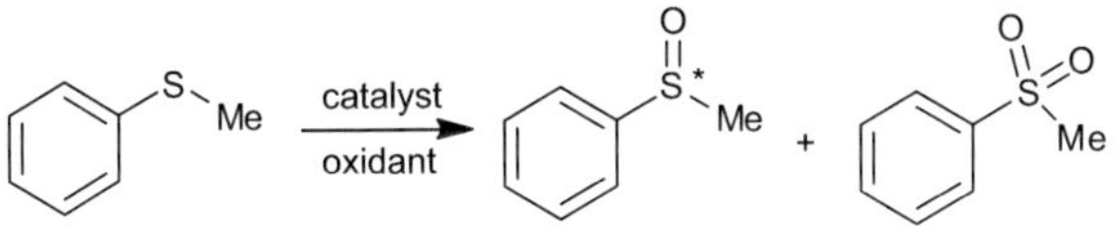

Scheme 3.12 Sulfoxidation of thioanisole.

Co^{II} complex thus enabling the reuse of the recovered catalyst in consecutive reactions without extra reoxidation.

3.5.4 Asymmetric Sulfoxidation

The selective asymmetric oxidation of organosulfides to the corresponding sulfoxide remains a challenging aspect of chemistry due to the large number of sulfoxides that exhibit pharmaceutical activity.[107] However, to date, there have been few studies examining this reaction utilizing asymmetric sulfoxidation catalysis in ILs. Titanium-salan type catalysts have been utilized for the sulfoxidation of thioanisole in a range of imidazolium, pyrollidinium and pyridinium ILs (Scheme 3.12).[108] Therein, *t*-butyl or cumyl hydroperoxide were found to give the highest conversion when used in [NTf$_2$]$^-$ hydrophobic ionic liquids, whilst H_2O_2 produced the highest conversions in hydrophilic [BF$_4$]$^-$ and [OTf]$^-$ ILs. In general, the titanium-salan complexes showed better activity in ILs compared with molecular solvents; however, a maximum 19% ee obtained in [C$_4$C$_1$im][NTf$_2$] compared with 51% ee obtained in dichloroethane.

Ionic liquids containing chiral tungstate(IV) anions have also been employed in the asymmetric sulfoxidation of thioanisole.[109] Tetralkylammonium and phosphonium ILs were used to improve the organic immiscibility of (*S*)-BINOL and (*S*)-mandelate tungsten complexes. Ee values up to 96% for sulfide oxidation to sulfoxides were possible at low sulfide conversions (>5%) when the CILs were employed as additives in molecular solvents such as dichloromethane. Conversions greater than 50% were observed in [C$_4$C$_1$pyr][NTf$_2$]; however, this had a detrimental effect on the ee, which was <10%.

3.6 Hydroformylation

The transformation of an alkene into an aldehyde by the addition of CO and H_2 (syngas) across an olefinic double bond represents one of the world's most important homogeneously catalyzed processes, and it is one of the largest volume processes in the chemical industry, with a world-wide oxoaldehyde production of 7.8×10^6 tonnes per year (1997).[110–112] Aqueous biphasic catalysis is employed on an industrial scale in the Ruhrchemie/Rhône-Poulenc process for the conversion of propene and syngas to *n*-butyraldehyde, with less than 4% *i*-butyraldehyde produced. The active catalyst is formed *in situ* from a rhodium salt and the sulfonated phosphine ligand TPPTS, thus achieving immobilization of the complex in the aqueous layer. This allows the product to be isolated simply by decanting the organic layer. TOFs up to 10 000 mol mol^{-1} h^{-1} have been achieved for the hydroformylation of propene using the water-soluble BINAS ligand, a sulfonated binaphthyl diphosphine.[113]

Due to the high interest in the mass production of linear aldehydes, heterogeneous catalysis has appeared as a more attractive alternative. Various approaches have been used for the immobilization of hydroformylation catalysts. These include:

1. anchoring the catalyst to a dendrimer,[114,115] polymer[116,117] or inorganic solid[118]
2. fluorous biphasic systems[119,120]
3. aqueous biphasic systems[121] or
4. use of ionic liquids in more studied biphasic systems[122,123] or SILP (batch mode[124,125] or continuous flow reactions[126,127]

As the heterogenization of Rh catalysts *via* classical methods, i.e. impregnation or covalent anchoring, has only been achieved with limited success, the Rh complex immobilization in an IL phase has been examined in detail.[128,129] Haumann and Riisager[130] have recently reviewed the development of room temperature ILs as alternative solvents for biphasic catalysis, especially for the hydroformylation reaction. Although hydroformylation in ILs has made considerable progress in the past decade using catalysts that are capable of achieving excellent linear-to-branched ratios together with good TOFs and negligible leaching, surprisingly, asymmetric hydroformylation in ILs has received little attention and is worthy of detailed investigation. Stereoselective biphasic hydroformylation in ILs was reported by Deng *et al.*[131] using asymmetric diphosphine ligands in [C$_4$C$_1$im][BF$_4$] for the enantioselective hydroformylation of vinyl acetate and styrene, as shown in Scheme 3.13.

The Rh-(R)-BINAPS-catalyzed hydroformylation of vinyl acetate resulted in the predominant formation of 2-acetoxypropanal with ee values up to 50% higher in the ILs compared with molecular solvents. By comparison with the results obtained in aqueous and homogeneous systems, it was concluded that similar coordination chemistry was taking place in the biphasic IL-toluene system.[132] The IL-biphasic system with Rh-(R)-BINAPS chiral complex was recycled six consecutive times without significant loss in activity or selectivity. The hydroformylation of styrene under the same reaction conditions used the Rh-(R)-BINAP and Rh-(R)-BINAPS catalysts. In the hydrophilic [C$_4$C$_1$im][BF$_4$] with the Rh-(R)-BINAP catalyst, the conversion was 79% with high regioselectivity for 2-phenyl-1-propanal while the ee was moderate at around 22%. The same

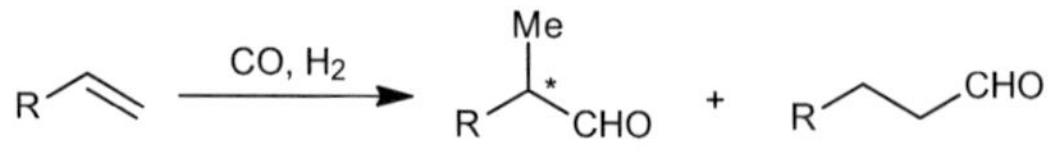

Scheme 3.13 Asymmetric hydroformylation of vinyl acetate (R = CH$_3$C(O)O) and styrene (R = Ph).

catalyst only gave 64% conversion and an ee of only 6% in the hydrophobic [C$_4$C$_1$im][PF$_6$]. Using the sulfonated (*R*)-BINAPS ligand, the ee was low in both ILs (~8%). Although the results in the biphasic system were encouraging, no further studies have been done to heterogenize the catalysts.

3.7 Pericyclic Reactions

Pericyclic reactions represent an important class of carbon–carbon bond-forming atom-efficient reactions. Conventionally, these concerted systems generally give predictable stereo- and regiochemical control of the product whereby the choice of molecular solvent has little influence in the overall chemistry.[133]

3.7.1 Diels–Alder Reactions

Diels–Alder chemistry has been studied extensively in ionic liquids and has found that using alternative solvents such as water or lithium perchlorate-diethyl ether mixtures compared with pure organic solvents leads to significant changes in the reaction rate and *endo/exo* selectivity.[134–137] For example, *endo* selectivities as high as 91% has been reported in [C$_4$C$_1$im]Cl/AlCl$_3$ systems[138] and as low as 76% in [C$_4$C$_1$im][BF$_4$].[139]

Recently, both asymmetric Diels–Alder reactions using organocatalysts[140] and organometallic catalyzed reactions in ILs have been studied. Meracz and Oh[141] observed an ee of 96% for the cycloadduct between *N*-crotonoylox-azolidinone (R = Me) and cyclopentadiene at room temperature using a rigid copper(II) BOX (R′ = CMe$_3$)-based chiral Lewis acid (Scheme 3.14) with a yield of 65% in [C$_4$C$_4$im][BF$_4$]. Large rate enhancements and increases in ee were observed in IL compared with CH$_2$Cl$_2$, which showed only 76% ee with a

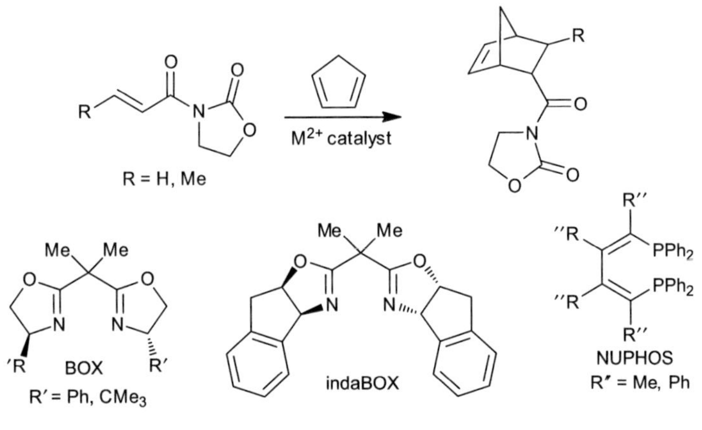

Scheme 3.14 Diels–Alder reaction between *N*-substituted oxazolidinones and cyclopentadiene using various BOX and NUPHOS ligands.

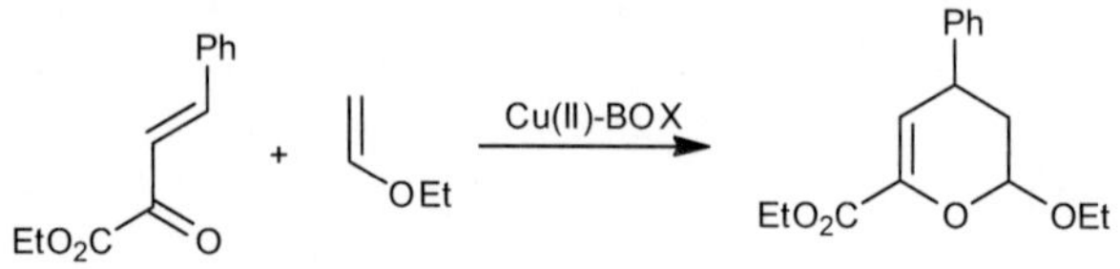

Scheme 3.15 Hetero Diels–Alder reaction.

yield of only 4%. Similarly, an Cu^{II}-indaBOX complex was also found to be an excellent catalyst for the reaction of *N*-acryloyloxazolidinone (R = H) in a range of $[C_4C_1im]^+$ ILs at room temperature.[142] Moderate yields of 72–76% were recorded in hydrophilic [OTf]⁻ and $[BF_4]^-$ ILs whereas complete conversion was observed in the hydrophobic $[PF_6]^-$ and $[SbF_6]^-$ based ILs. However, the most significant change was the lack of asymmetric induction in the hydrophilic ILs compared with 94% ee obtained in $[C_4C_1im][SbF_6]$. This lack of induction maybe in part due to the large amounts of halide (~1000 ppm) present in the hydrophilic ILs. Due to the increased reactivity in hydrophobic ILs, the amount of the metal catalyst required could be reduced to 0.6 mol% with respect to the substrate without any significant compromise in the selectivity achieved. Furthermore, recycling of the ligand–metal complex was achieved over 18 reaction cycles.

Hydrophobic ionic liquids such as $[C_4C_1im][PF_6]$ and $[C_4C_1im][SbF_6]$ were used successfully as powerful media for Cu^{II}-BOX (R′ = CMe_3)-catalyzed asymmetric hetero-Diels–Alder reaction (Scheme 3.15).[143] As found with other Diels–Alder reactions, accelerated reaction rates were observed in ILs and high ee values (94%) could be obtained at 3 °C in ILs whereas as comparable ee values were only obtained at –78 °C in CH_2Cl_2. The recovered catalyst could be recycled up to eight times exhibiting almost the same reactivity and selectivity.

In^{III}-BINOL complex with allylstannane as an additive was developed producing enantiomerically enriched Diels–Alder adducts in $[C_6C_1im][PF_6]$.[144] The cycloaddition of a variety of dienes with 2-methacrolein and 2-bromoacrolein also resulted in good yields and high enantioselectivities (98% ee) without the need to recourse to low temperatures. The catalyst could be recycled up to seven times maintaining the initial high yields and ee values. Moreover, the stringent anhydrous conditions employed during work-up involving molecular solvents were not required. Doherty *et al.* also reported that ILs can increase the ee and yield for the reaction between oxazolidinones and cyclopentadiene using Pt^{II} complexes of BINAP as well as conformationally flexible NUPHOS-type diphosphines (Scheme 3.14).[145] Significant enhancements in the enantioselectivity (up to an increase of 20% ee) as well as reaction rate were achieved in ILs compared with the organic media. In this case, the IL was thought to limit the extent of Pt-NUPHOS atropinversion observed in dichloromethane (Figure 3.5). A further increase in ee was observed when employing ILs under biphasic conditions using IL/diethyl ether mixtures. The lower concentration of cyclopentadiene and other reagents/products in the IL phase increased the ee; however, some decrease in the rate was observed, as expected. Again, the IL

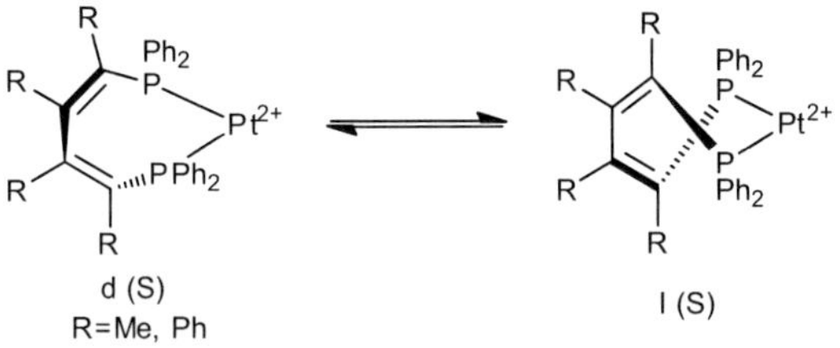

Figure 3.5 Atropinversion of Pt^{2+} NUPHOS complex.

allowed the catalyst to be recycled in air without hydrolysis or oxidation of the phosphine ligand.

Pd-BINAP catalysts immobilized in ILs have also shown excellent asymmetric catalytic activity in the Diels–Alder reactions between acryloyl oxazolidinone and a range of carbo- and heterocyclic dienes.[146] Therein, moderate conversions were obtained but excellent enantioselectivities were reported (between 94 and 98% ee). This was extended to cationic palladium-phosphinooxazolidine catalysts for the asymmetric Diels–Alder reaction between acryloyl oxazolidinone (R = H) and cyclopentadiene.[147] Moderate ee values were obtained in $[C_4C_1im][BF_4]$ (71–88% ee) at room temperature; however, this was improved using a mixture of IL and CH_2Cl_2 at –40 °C with ee values between 91 and 94% obtained.

Goodrich *et al.*[148,149] reported the optimization and the comparison of homogeneous and heterogenized (SILP type 2) BOX complexes of Mg^{II}, Cu^{II} and Zn^{II} for the asymmetric Diels–Alder reaction between N-acryloyloxazolidinone (R = H) and cyclopentadiene. Under homogeneous conditions in $[C_2C_1im]$ $[NTf_2]$ all metal complexes showed accelerated rates of reaction with ee values higher than reactions conducted in molecular solvents. In some catalyst systems where the BOX chiral ligand (R' = Ph) was complexed with the metal, a reversal in product configuration was also noted upon switching from CH_2Cl_2 to IL and was attributed to a change in catalyst geometry. ^1H NMR experiments were consistent with the IL stabilizing an octahedral geometry around the metal centre whereas a tetrahedral geometry was found in CH_2Cl_2. The use of the corresponding SILP type 2 catalysts was also employed in the Diels–Alder reaction using various silica (SiO_2, MCM-41, SBA-15) and carbon supports (activated carbon, graphite and carbon nanotubes). While Cu and Mg catalysts showed small enhancements in ee between SILP and homogeneous IL reactions, the use of Zn-based SILP catalyst showed significant ee enhancement. For example, ee values up to 25–30% higher were observed for the Zn catalyst supported on high surface area MCM-41 and low surface area graphite highlighting the complex support/catalyst interactions even in the presence of an IL.

3.7.2 Carbonyl-ene Reactions

Similar to the Diels–Alder reaction, the carbonyl-ene is a pericyclic reaction initiated by heat or Lewis acid catalysts. Lewis acid complexes of

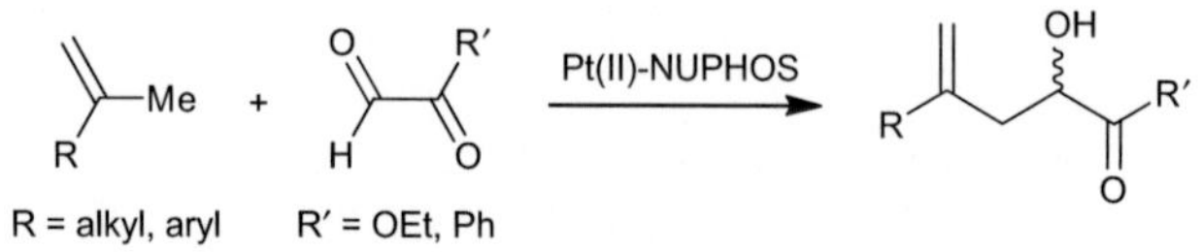

Scheme 3.16   Carbonyl-ene reaction of 1,1'-disubstituted alkenes and glyoxyals.

conformationally flexible NUPHOS catalysts have also shown to be efficient as catalysts for reactions involving unsymmetrical 1,1-disubstituted alkenes and glyoxals (Scheme 3.16).[150] Up to 95% ee and >90% conversions were obtained in [C_2C_1im][NTf$_2$]. In all cases the product ee was higher or comparable with those obtained in CH$_2$Cl$_2$. As highlighted for similar Diels–Alder reactions, the prevention of catalyst racemization in the IL compared with CH$_2$Cl$_2$ was thought to be responsible for the improvement in ee.

Similar studies using Pd-BINAP complexes in [C_4C_1im][PF$_6$][151] and [$C_4C_1C_1$im][NTf$_2$][152] resulted in comparable or higher ee values to those obtained in molecular solvents. A highly enantioselective carbonyl-ene reaction of trifluoropyruvate catalyzed by a recyclable indium(III)-pyBOX complex in the IL afforded trifluoromethyl-containing tertiary homoallylic alcohols with excellent yields (up to 98%) and enantioselectivities (up to 98% ee). Notably, this catalytic system can be recycled up to seven cycles.[153]

3.8   Nucleophilic Additions

3.8.1   Aldol Reaction

Another important reaction used in the construction of new C–C bonds in organic synthesis is the asymmetric aldol reaction. ILs have been successfully employed as asymmetric organocatalysts;[154–156] however, the Mukaiyama-aldol reaction has received little attention. In this reaction, enantiomerically enriched secondary and tertiary β-hydroxy esters are formed by coupling silyl enolates or silylketene acetals with aldehydes and activated ketones, respectively. Common immobilization strategies include anchoring the catalyst to a polymer[157] or grafting it onto silica.[158] In general, these approaches have met with limited success primarily due to significant leaching of the ligand during the extraction process.

Supported IL systems in the presence of a copper Lewis acid have been successful for the asymmetric Mukaiyama-aldol reaction between 1-phenyl-1-trimethylsiloxyethene and methyl pyruvate (Scheme 3.17).[159] Accelerated reaction rates were observed in a range of imidazolium and ammonium [NTf$_2$]$^-$ based ILs compared with dichloromethane. Despite the higher reaction rates, lower chemoselectivity was observed due to the formation of a by-product resulting from the Mukaiyama-aldol reaction of 1-phenyl-1-trimethylsiloxy-ethene and acetophenone (Scheme 3.18). Formation of this by-product was

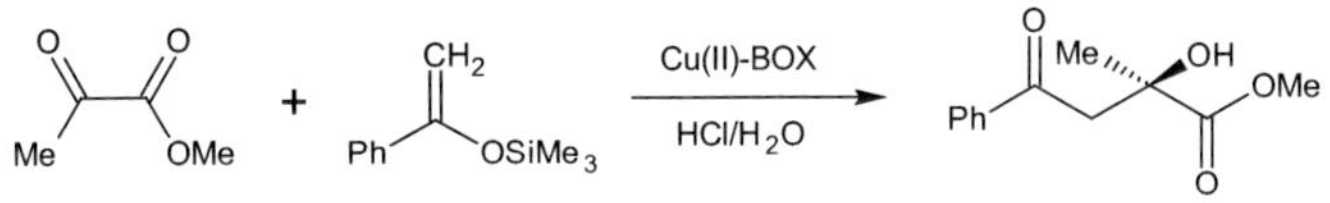

Scheme 3.17 The asymmetric Mukaiyama-aldol condenstation between methyl pyruvate and 1-phenyl-1-trimethylsiloxyethene.

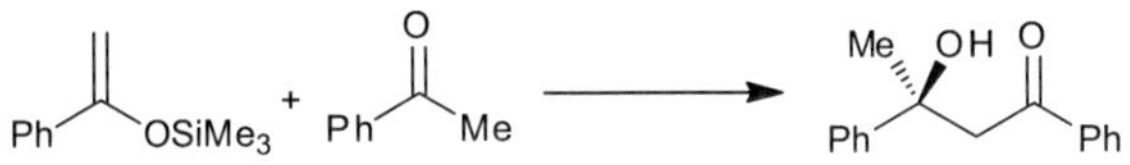

Scheme 3.18 Self-Mukaiyama-aldol by-product.

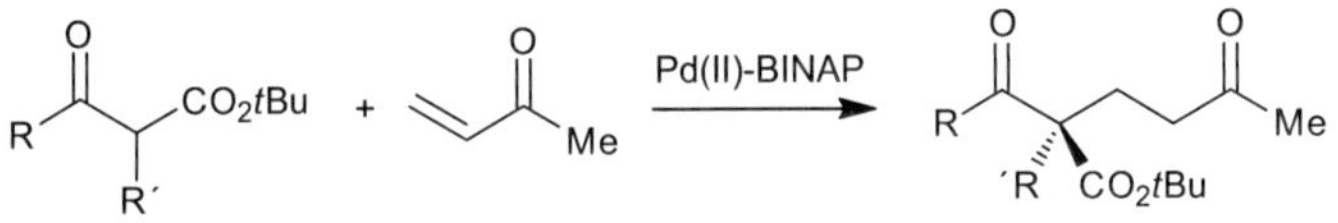

Scheme 3.19 Michael addition of β-keto esters with MVK.

suppressed, without any reduction in the enantioselectivity, by supporting the catalyst on SILP type 1 or type 2 systems.

The heterogenized systems were less active than the homogeneous in ILs, but more active than the analogous system in dichloromethane. Importantly, the use of the IL-grafted silica SILP type 2 was not necessary for high conversions and ee values, provided that a supporting IL film on the silica was employed. In these cases, ligand leaching leading to loss in activity was observed; however, by attaching the ligand to the IL, the leaching was significantly reduced.

Palladium BINAP complexes have also been immobilized in [C$_4$C$_1$im][OTf] and their applications to catalytic reaction of β-keto esters with MVK were successfully demonstrated (Scheme 3.19).[160] This immobilization in IL enabled the reuse of the catalysts five times in the Michael reaction with 84% ee which is comparable to those obtained in organic solvents.

3.8.2 Organometallic Reagents

The ionic liquid [C$_6$C$_1$im][PF$_6$] has been demonstrated as an efficient reaction medium for the enantioselective allylation of aldehydes *via* a chiral indium(III) BINOL complex (Scheme 3.20).[161] The allylation of a variety of aromatic, α,β-unsaturated and aliphatic aldehydes resulted in moderate to good yields and enantioselectivities up to 92% ee. Despite the high activity the IL catalyst system could not be efficiently recycled due to decomposition during work-up.

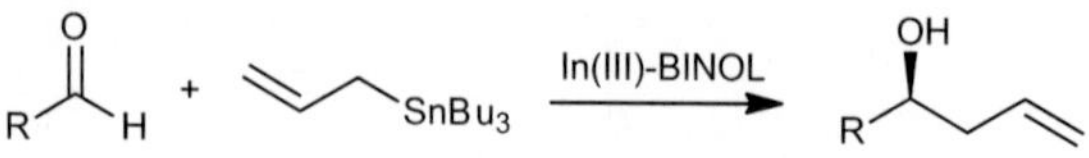

Scheme 3.20 Enantioselective allylation of aldehydes.

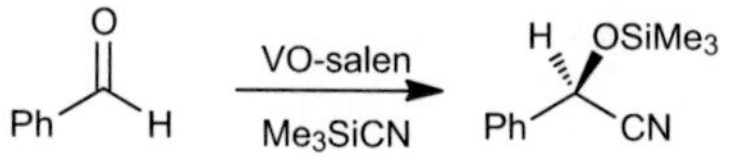

Scheme 3.21 Enantioselective cyanosilylation of benzaldehyde.

support = SiO_2, AC, SWCNT, IL

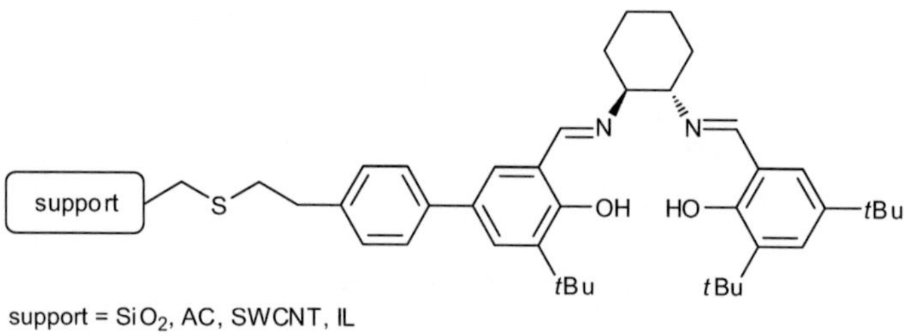

Figure 3.6 Supported salen complex for the enantioselective cyanisilylation of benzaldehyde.

3.8.3 Cyanosilylation of Benzaldehyde

As well as finding extensive deployment in oxidation reactions, immobilized chiral, metal-salen have also been used for the asymmetric cyanosilylation of benzaldehyde (Scheme 3.21).[162]

The catalytic activity and the asymmetric induction ability of vanadyl-salen complexes was examined under liquid–liquid and solid–liquid systems. It was found that the liquid–liquid systems employing the IL-tagged salen complex (Figure 3.6) dissolved in $[C_4C_1im][PF_6]$ was five to six times more active than the non-tagged salen complex dissolved in IL or those of the supported silica, single-wall carbon nanotube and activated carbon systems.

However, the enantioselectivity of IL-tagged catalyst (57% ee) was lower than silica-based systems (89% ee). A possible explanation for this reduced asymmetric induction is the interaction of the vanadyl group of the complex with the associated chloride anion present in ionic liquid; however, this was not verified by exchanging chloride to a non-co-ordinating anion. Both the carbon supports also showed a significant decrease in ee (48–66%) compared with the silica system and this was thought to be due to the ill-defined structure and/or the presence of heteroatoms on the support. Surprisingly, the immobilization of the salen complex using a SILP concept was not attempted and would make an interesting addition to the results.

3.9 Miscellaneous Reactions

3.9.1 Cyclopropanations

Chiral copper(II) BOX complexes have also been easily immobilized onto anionic solids or in ILs and used as catalysts in cyclopropanation reactions (Scheme 3.22).[163]

Significant differences between the homogeneous reaction in dichloromethane, the use of heterogeneous laponite support and the reaction in ILs were observed. While little change in the *trans/cis* ratio was reported, the (1*R*) ee, efficiency and recovery of the catalysts were all found to be strongly dependent on the reaction system employed. For reactions where the chiral Cu^{II}-BOX catalyst was supported on laponite, lower ee values of 69% were reported compared with those in the homogeneous reaction (94% ee). The reason for this behaviour is the existence of an equilibrium between free and complexed copper, leading to the formation of non-chiral catalytic sites (Figure 3.7). This was further evidenced by an additional reduction in ee upon recycle.

In contrast, these support interactions were eliminated for experiments conducted in neat $[C_2C_1im][OTf]$, providing a more strongly complexed catalyst which could be recycled up to five times with only a slight reduction in ee (85% to 63%) due to ligand leaching into the organic extraction phase. Addition of fresh portion of ligand on the sixth recycle resulted in initial catalytic activity and performance.

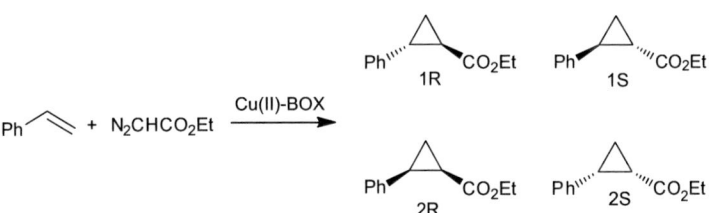

Scheme 3.22 Cylopropanation reaction between styrene and ethyl diazoacetate.

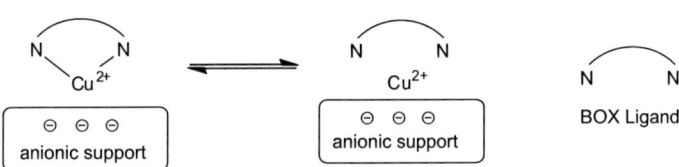

Figure 3.7 Equilibrium between the free and copper-complexed ligand on an anionic support.

3.9.2 Heck and Related Reactions

The Heck reaction is one of the most widely studied reactions in ILs under both homogeneous and heterogeneous systems. Therein, ILs have proven to be excellent media, in terms of yield and with excellent control over selectivity and recyclability. Despite the many reviews highlighting the merits of ILs,[164] to date only one asymmetric Heck reaction has been attempted (Scheme 3.23).[165] The Heck oxyarylation of a benzopyrene using a Pd BINAP catalyst in a chiral imidazolium [PF$_6$]⁻ based IL formed a pterocarpans, a natural plant product. Despite high yields only poor ee values (<10%) were obtained using a range of chiral phosphine ligands. This was thought to be due to the reaction mechanism which can go *via* a Pd^0 chiral intermediate whereby any chiral information imparted to the substrate from the diphosphine ligand is lost.[166] Moreover, there was no evidence for the chiral IL having any influence on the reaction.

Toma *et al.* studied the enantioselective Pd-catalyzed allylic substitution of (*rac*)-(*E*)-1,3-diphenyl-3-acetoxyprop-1-ene with dimethyl malonate [C$_4$C$_1$im][PF$_6$].[167] The reactions were catalyzed by Pd^0 complexes of homochiral ferrocenylphosphine ligands (Scheme 3.24). Significant improvements in ee were observed upon switching from THF (40%) to the IL (74%). A further improvement to 84% ee could be achieved in IL by using a ferrocenylphospherox type ligand.[168]

The enantioselective alkyne–imine addition can be efficiently performed in the ionic liquid [C$_4$C$_1$im][NTf$_2$] (Scheme 3.25).[169] Several substrates were tested

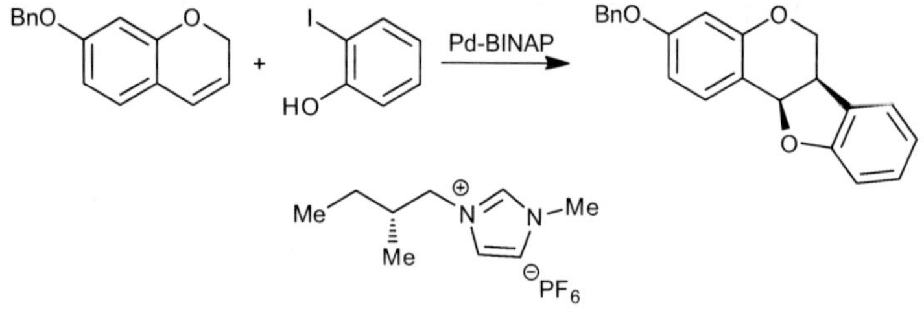

Scheme 3.23   Enantioselective Heck oxyarylation.

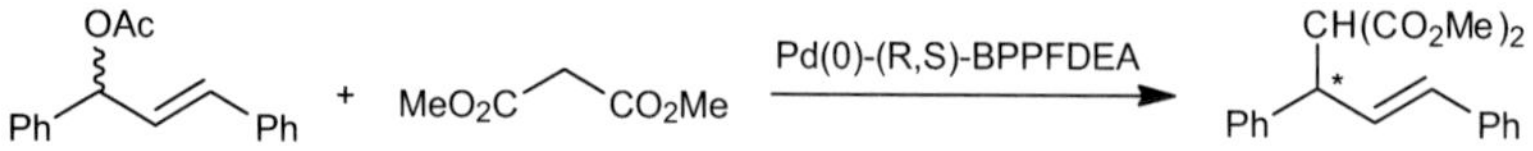

Scheme 3.24   Allylic substitution reaction.

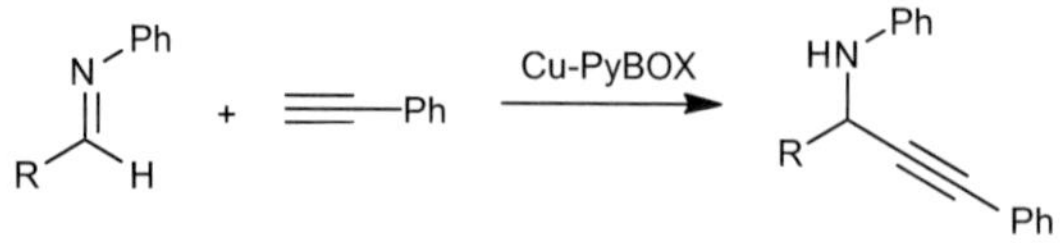

Scheme 3.25   Asymmetric addition of alkynes to imines.

and the reuse of the catalytic Cu^I-pyBOX was carried out for six cycles allowing high overall yield (87%) and ee values (first cycle, 94%; sixth cycle, 89%). Although it is thought that the Cu^I catalyst is formed *in situ*, this has not been confirmed and Cu^0 may be the active catalyst.

3.9.3 Fluorination Reaction

The enantioselective fluorination of ketophosphonates was achieved in ILs using a palladium BINAP complex immobilized in $[C_4C_1im][BF_4]$ (Scheme 3.26).[170] The transformations proceeded with good selectivity and ee values above 90% were obtained in most cases.

3.9.4 Strecker Reaction

A self-assembled ionic liquid phase (SAILP) has been used to immobilize Yb^{III}-pybox catalyst which has been successful in the asymmetric Strecker hydro-cyanation of aldimines (Scheme 3.27).[171] Similar conversions were comparable to the homogeneous systems; however, a decrease in ee from 80% to 53% was observed. This catalytic system was found to be reusable, however, with the catalyst showing no significant loss of activity or enantioselectivity over six cycles.

3.9.5 Asymmetric Polymerization

A Pd-catalyzed stereoselective co-polymerization of propene and carbon monoxide used chiral phosphine ligands, such as (2S,3S)-DIOP and (R)-P-Phos

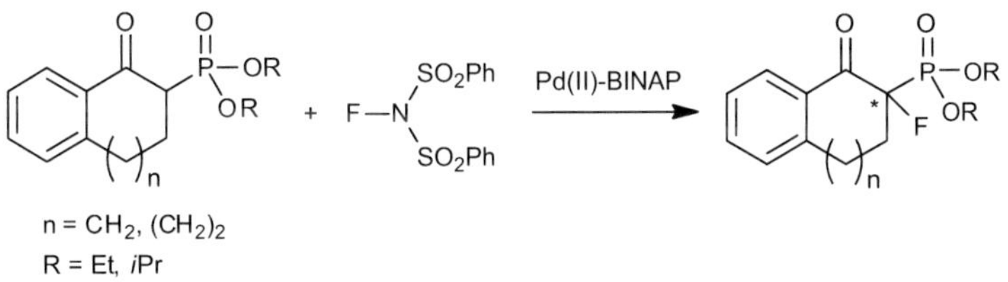

$$n = CH_2, (CH_2)_2$$
$$R = Et, iPr$$

Scheme 3.26 Asymmetric fluorination of ketophosphates.

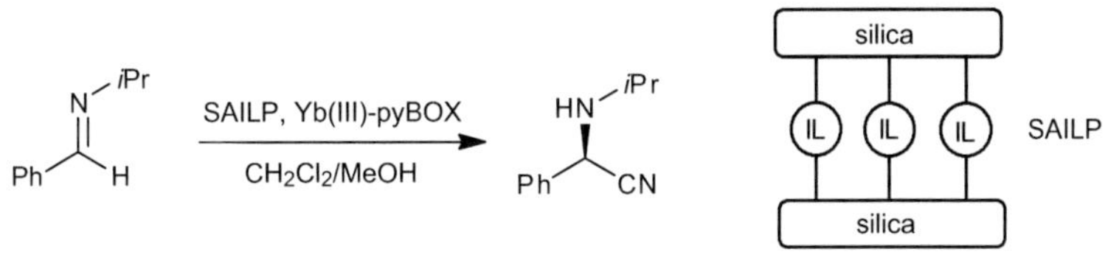

Scheme 3.27 Asymmetric Strecker reaction.

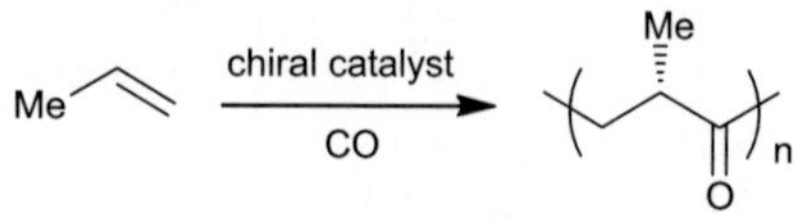

Scheme 3.28   Asymmetric co-polymerization.

in $[C_nC_1im][PF_6]$ ($n = 4–8$) ILs.[172] Using (2S,3S)-DIOP as chiral ligand and $[C_4mim][PF_6]$ as medium, the Pd-catalyzed co-polymerization of propene and CO gave almost completely regioregular polyketones, and the product polymer showed moderate stereoregularity (61% of ℓ-diads). The highest molar optical rotation $= +15.9°$ and polydispersity $= 1.2$ were attained when (R)-P-Phos was used as the ligand and $[C_6mim][PF_6]$ as the solvent (Scheme 3.28).

3.10   Summary and Outlook

The use of ILs in the area of enantioselective catalysis has shown significant improvements may be obtained in enantioselectivity and recyclability over a wide range of reactions. In general, the deployment of an IL onto a support regardless of its nature has, for the most part, resulted in a positive change in reaction activity. There only a few examples where detrimental effects are observed from going from bulk solvents, through to biphasic liquid–liquid and further onto solid–liquid and solid–gas systems. With a better understanding of ILs and catalyst systems, the reactivity, stability and reaction pathways can be influenced by the choice of cations, anions, supports and other additives.

References

1. J. A. Boon, J. A. Levinsky, J. I. Pflug and J. S. Wilkes, *J. Org. Chem.*, 1986, **51**, 480.
2. M. J. Earle, S. P. Katdare and K. R. Seddon, *Org. Lett.*, 2004, **6**, 707.
3. S. Doherty, P. Goodrich, C. Hardacre, J. G. Knight, M. T. Nguyen, V. I. Parvulescu and C. Paun, *Adv. Synth. Catal.*, 2007, **349**, 951.
4. J. S. Wilkes, P. Wasserscheid and T. Welton, in *Ionic Liquids in Synthesis*, Wiley-VCH, Weinheim, 2003.
5. *Ionic Liquids IIIA and B: Fundamentals, Progress, Challenges, and Opportunities – Properties and Structure*, ed. R. D. Rogers and K. R. Seddon, American Chemical Society, Washington, DC, 2005.
6. *Ionic Liquids in Synthesis*, ed. P. Wasserscheid and T. Welton, Wiley-VCH, Weinheim, 2003.
7. C. Hardacre and V. Parvelscu, *Chem. Rev.*, 2007, **107**, 2615.
8. H. Olivier-Bourbigou, L. Magna and D. Morvan, *Appl. Catal. A: General*, 2010, **373**, 1.
9. K. R. Seddon, A. Stark and M.-J. Torres, *Pure Appl. Chem.*, 2000, **72**, 2275.

10. (a) J. D. Holbrey and K. R. Seddon, *Clean Prod. Proc.*, 1999, **1**, 223; (b) P. Wasserscheid and W. Keim, *Angew. Chem., Int. Ed. Engl.*, 2000, **3**, 3772; (c) T. Welton, *Coord. Chem. Rev.*, 2004, **248**, 2459; (d) M. Freemantle, *Chem. Eng. News*, 1998, **76**, 32; (e) P. A. Z. Suarez *et al.*, *Inorg. Chim. Acta*, 1997, **255**, 207.

11. J. D. Holbrey, W. M. Reichert, R. P. Swatloski, G. A. Broker, W. R. Pitner, K. R. Seddon and R. D. Rogers, *Green Chem.*, 2002, **4**, 407.

12. P. Wasserscheid, R. van Hal and A. Bosmann, *Green Chem.*, 2002, **4**, 400.

13. For example: (a) E. J. Amigues, C. Hardacre, G. Keane, M. E. Migaud, M. O'Neill, *Chem. Commun.*, 2006, 72; (b) E. J. Amigues, C. Hardacre, G. Keane and M. E. Migaud, *Green Chem.*, 2008, **10**, 660.

14. T. Welton, *Chem. Rev.*, 1999, **99**, 2071.

15. R. A. Sheldon, *Chem. Commun.*, 2001, **23**, 2399.

16. R. P. J. Bronger, S. M. Silva, P. C. J. Kamer, P. W. N. M van Leeuwen, *Chem. Commun.*, 2002, 3044.

17. C. C. Brasse, U. Englert, A. Salzer, H. Waffenschmidt and P. Wasserscheid, *Organometallics*, 2000, **19**, 3818.

18. Q. Yao and Y. Zhang, *Angew. Chem., Int. Ed. Engl.*, 2003, **42**, 3395.

19. M. H. Valkenberg, C. de Castro and W. F. Holderich, *Green Chem.*, 2002, **4**, 88.

20. C. P. Mehnert, R. A. Cook, N. C. Dispenziere and M. Afeworki, *J. Am. Chem. Soc.*, 2002, **124**, 12932.

21. C. P. Mehnert, E. J. Mozeleski and R. A. Cook, *Chem. Commun.*, 2002, 3010.

22. S. Sowmiah, V. Srinivasadesikan, M.-C. Tseng and Y.-H. Chu, *Molecules*, 2009, **14**, 3780.

23. X. Wang and K. Ding, *J. Am. Chem. Soc.*, 2004, **126**, 10524.

24. Y. Liang, Q. Jing, X. Li, L. Shi and K. Ding, *J. Am. Chem. Soc.*, 2005, **127**, 7694.

25. C. Maillet, P. Janvier, M. Pipelier, T. Praveen, Y. Andres and B. Bujoli, *Chem. Mater.*, 2001, **13**, 2879.

26. K. Aoki, T. Shimada and T. Hayashi, *Tetrahedron: Asymmetry*, 2004, **15**, 1771.

27. C. E. Song and S.-G. Lee, *Chem. Rev.*, 2002, **102**, 3495.

28. X. Li, W. Chen, W. Hems, F. King and J. Xiao, *Org. Lett.*, 2003, **5**, 4559.

29. V. Neff, T. E. Müller and J. A. Lercher, *J. Chem. Soc., Chem. Commun.*, 2002, **8**, 906.

30. H. L. Ngo, A. Hu and W. Lin, *Tetrahedron Lett.*, 2005, **46**, 595.

31. H. L. Ngo, K. LeCompte, L. Hargens and A. B. McEwen, *Thermochim. Acta*, 2000, **357**, 97.

32. L. A. Blanchard,. D. Hancu, E. J. Beckman and J. F. Brennecke, *Nature*, 1999, **399**, 28.

33. L. A. Blanchard, Z. Gu and J. F. Brennecke, *J. Phys. Chem. B*, 2001, **105**, 2437.

34. C. S. Santos and S. Baldelli, *Chem. Soc. Rev.*, 2010, **39**, 2136.

35. F. Endres and S. Z. El Abedin, *Phys. Chem. Chem. Phys.*, 2006, **8**, 2101.

36. C. S. Santos and S. Baldelli, *J. Phys. Chem. B*, 2009, **113**, 923.
37. C. Aliaga, C. S. Santos and S. Baldelli, *Phys. Chem. Chem. Phys.*, 2007, **9**, 3683.
38. Y. Jeon, J. Sung, W. Bu, D. Vaknin, Y. Ouchi and D. Kim, *J. Phys. Chem. C*, 2008, **112**, 19649.
39. Y. F. Yano and H. Yamada, *Anal. Sci.*, 2008, **24**, 1269.
40. W. Jiang, Y. T. Wang, T. Y. Yan and G. A. Voth, *J. Phys. Chem. C*, 2008, **112**, 1132.
41. H. P. Steinrück, *Surf. Sci.*, 2010, **604**, 481.
42. F. Maier, J. M. Gottfried, J. Rossa, D. Gerhard, P. S. Schulz, W. Schwieger, P. Wasserscheid and H. P. Steinruck, *Angew. Chem., Int. Ed. Engl.*, 2006, **45**, 7778.
43. V. Lockett, R. Sedev, C. Bassell and J. Ralston, *Phys. Chem. Chem. Phys.*, 2008, **10**, 1330.
44. P. J. Dyson, T. Geldbach, F. Moro, C. Taeschler and D. Zhao. *Hydrogenation Reactions in Ionic Liquids: Finding Solutions for Tomorrow's World*, 2005, pp. 322–333.
45. P. J. Dyson, G. Laurenczy, C. A. Ohlin, J. Vallance and T. Welton, *Chem. Commun.*, 2003, 2418.
46. B. Pugin, M. Studer, E. Kuesters, G. Sedelmeier and X. Feng, *Adv. Synth. Catal.*, 2007, **349**, 1803.
47. J. Wang, J. Feng, R. Qin, H. Fua, M. Yuana, H. Chen and X. Li, *Tetrahedron: Asymmetry*, 2007, **18**, 1643.
48. J. Wang, R. Qin, H. Fu, J. Chen, J. Feng, H. Chen and X. Li, *Tetrahedron: Asymmetry*, 2007, **18**, 847.
49. P. G. Jessop, R. R. Stanley, R. A. Brown, C. A. Eckert, C. L. Liotta, T. T. Ngo and P. Pollet, *Green Chem.*, 2003, **5**, 123.
50. R. Giernoth and M. S. Krumm, *Adv. Synth. Catal.*, 2004, **346**, 989.
51. H. Zhou, Z. Li, Z. Wang, T. Wang, L. Xu, Y. Hel, Q.-H. Fan, J. Pan, L. Gu and A. S. C. Chan, *Angew. Chem.*, 2008, **120**, 8592.
52. M. Solinas, A. Pfaltz, P. Giorgio Cozzi and W. Leitner, *J. Am. Chem. Soc.*, 2004, **126**, 16142.
53. A. Wolfson, J. F. J. Vankelecom and P. A. Jacobs, *Tetrahedron Lett.*, 2003, **44**, 1195.
54. K. L. Fow, S. Jaenicke, T. E. Müller and C. Sievers, *J. Mol. Catal. A: Chemical*, 2008, **279**, 239.
55. C. Sievers, O. Jimenez, T. E. Müller, S. Steuernagel and J. A. Lercher, *J. Am. Chem. Soc.*, 2006, **128**, 13990.
56. L.-L. Lou, X. Peng, K. Yu and S. Liu, *Catal. Commun.*, 2008, **9**, 1891.
57. U. Hintermair, T. Höfener, T. Pullmann, G. Franciò and W. Leitner, *ChemCatChem.*, 2010, **2**, 150.
58. S. Shi, X. Wang, C. A. Sandoval, Z. Wang, H. Li, J. Wu, L. Yu and K. Ding, *Chem. Eur. J.*, 2009, **15**, 9855.
59. I. Kawasaki, K. Tsunoda, T. Tsuji, T. Yamaguchi, H. Shibuta, N. Uchida, M. Yamashita and S. Ohta, *Chem. Commun.*, 2005, 2134.
60. J.-M. Joerger, J.-M. Paris and M. Vaultier, *Arkivoc*, 2006, 152.

61. D. Seebach and H.-A. Oei, *Angew. Chem., Int. Ed. Engl.*, 1975, **14**, 634.
62. F. Furia, G. Modena and R. Curci, *Tetrahedron Lett.*, 1976, **50**, 4637.
63. S. H. Hüttenhain, M. U. Schmidt, F. R. Schoepke and M. Rueping, *Tetrahedron*, 2006, **62**, 12420.
64. D. Chen, M. Schmitkamp, G. Francio, J. Klankermayer and W. Leitner, *Angew. Chem., Int. Ed. Engl.*, 2008, **47**, 7339.
65. M. Schmitkamp, D. Chen, W. Leitner, J. Klankermayer and G. Francio, *Chem. Commun.*, 2007, 4012.
66. P. S. Schulz, N. Müller, A. Bösmann and P. Wasserscheid, *Angew. Chem., Int. Ed. Engl.*, 2007, **46**, 1293.
67. S. Schneiders, A. Bösmann, P. S. Schulz and P. Wasserscheid, *Adv. Synth. Catal.*, 2009, **351**, 432.
68. A. K. Yudin (ed.), *Aziridines and Epoxides in Organic Synthesis*, Wiley-VCH, Weinheim, 2006.
69. A. Berkessel and H. Gröger (eds.), *Asymmetric Organocatalysis*, Wiley-VCH, Weinheim, 2005.
70. (a) T. Katsuki and K. B. Sharpless, *J. Am. Chem. Soc.*, 1980, **102**, 5974; (b) T. Katsuki and V. S. Martin, *Org. React.*, 1996, **48**, 1.
71. (a) W. Zhang, J. L. Loebach, S. R. Wilson and E. N. Jacobsen, *J. Am. Chem. Soc.*, 1990, **112**, 2801; (b) R. Irie, K. Noda, Y. Ito, N. Matsumoto, T. Katsuki, *Tetrahedron Lett.*, 1990, **31**, 7345.
72. (a) W. Zhang, A. Basak, Y. Kosugi, Y. Hoshino and H. Yamamoto, *Angew. Chem.*, 2005, **117**, 4463; *Angew. Chem., Int. Ed. Engl.*, 2005, **44**, 4389; (b)W. Zhang and H. Yamamoto, *J. Am. Chem. Soc.*, 2007, **129**, 286.
73. A. U. Barlan, A. Basak and H. Yamamoto, *Angew. Chem.*, 2006, **118**, 5981; *Angew. Chem., Int. Ed. Engl.*, 2006, **45**, 5849.
74. (a) Y. Shi, *Acc. Chem. Res.*, 2004, **37**, 488; (b) D. Yang, *Acc. Chem. Res.*, 2004, **37**, 497.
75. M. J. Sabater, A. Corma, A. Domenech, V. Fornes and H. Garcia, *Chem. Commun.*, 1997, 1285.
76. (a) Y. Zhang, D. H. Yin, Z. H. Fu, C. Z. Li and D. L. Yin,. *J. Catal.*, 2003, **24**, 942; (b) D. Chatterjee and A. Mitra, *J. Mol. Catal., A*, 1999, **144**, 363; (c) P. Piaggio, P. McMorn, D. Murphy, D. Bethell, P. C. B. Page, F. E. Hancock, C. Sly, O. J. Kerton and G. J. Hutchings, *Perkin Trans. 2*, 2000, 2008.
77. (a) I. F. J. Vankelecom, K. A. L. Vercruysse, P. E. Neys, D. W. A. Tas, K. B. M. Janssen, P. P. Knops-Gerrits and P. A. Jacobs, *Top. Catal.*, 1998, **5**, 125; (b) I. F. J. Vankelecom, N. M. F. Moens, K. A. L. Vercruysse, R. F. Parton and P. A. Jacobs, *Stud. Surf. Sci. Catal.*, 1997, **108**, 437.
78. G. Pozzi, F. Cinato, F. Montanari and S. Quici, *J. Chem. Soc., Chem. Commun.*, 1998, 877.
79. C. E. Song and E. J. Roh, *Chem. Commun.*, 2000, 837.
80. Z. Li, C. G. Xia and M. Ji, *Appl. Catal., A*, 2003, **252**, 17.
81. L. Gaillon and F. Bedioui, *Chem. Commun.*, 2001, 1458.
82. J. Teixeira, A. R. Silva, L. C. Branco, C. A. M. Afonso and C. Freire, *Inorg. Chim. Acta*, 2010, **363**, 3321.

83. K. Smith, S. Liu and G. A. El-Hiti, *Catal. Lett.*, 2004, **98**, 95.
84. L. D. Pinto, J. Dupont, R. F. de Souza and K. Bernardo-Gusmão, *Catal. Commun.*, 2008, **9**, 135.
85. L. C. Branco and C. A. M. Afonso, *J. Org. Chem.*, 2004, **69**, 4381.
86. A. Serbanovic, L. C. Branco, M. N. da Ponte and C. A. M. Afonso, *J. Organomet. Chem.*, 2005, **690**, 3600.
87. L. C. Branco, P. M. P. Gois, N. M. T. Lourenço, V. B. Kurteva and C. A. M. Afonso, *Chem. Commun.*, 2006, 2371.
88. L.-L. Lou, K. Yu, F. Ding, W. Zhou, X. Peng and S. Liu, *Tetrahedron Lett.*, 2006, **47**, 6513.
89. G.-J. Kim and J.-H. Shin, *Tetrahedron Lett.*, 1999, **40**, 6827.
90. S. Xiang, Y. Shang, Q. Xin and C. Li, *Chem. Commun.*, 2002, 2696.
91. H. Zhang, S. Xiang and C. Li, *Chem. Commun.*, 2005, 1209.
92. S. Wei. Y. Tang, G. Xu, X. Tang, Y. Ling, R. Li and Y. Sun, *React. Kinet. Catal. Lett.*, 2009, **97**, 329.
93. P. Pietikänen, *Tetrahedron*, 1998, **54**, 4319.
94. B. Morgan, A. C. Oehlschlager and T. M. Stokes, *Tetrahedron*, 1991, **47**, 1611.
95. T. M. Stokes and A. C. Oehlschlager, *Tetrahedron Lett.*, 1987, **28**, 2091.
96. Y. Y. Li, X. Zhang, Z. R. Dong, W. Y. Shen, G. Chen and J. X. Gao, *Org. Lett.*, 2006, **8**, 5565.
97. M. S. Sigman and D. R. Jensen, *Acc. Chem. Res.*, 2006, **39**, 22.
98. B. M. Stoltz, *Chem. Lett.*, 2004, **33**, 362.
99. M. Tomizawa, M. Shibuya and Y. Iwabuchi, *Org. Lett.*, 2009, **11**, 1829.
100. W. Sun, H. Wang, C. Xia, J. Li and P. Zhao, *Angew. Chem., Int. Ed. Engl.*, 2003, **42**, 1042.
101. K. Pathak, I. Ahmad, S. H. R. Abdi, R. I. Kureshy, N. H. Khan and R. V. Jasra, *J. Mol. Catal. A*, 2007, **274**, 120.
102. W. Sun, X. Wu and C. Xia, *Helv. Chim. Acta*, 2007, **90**, 623.
103. M. Lakshmi Kantama, T. Ramania, L. Chakrapani and B. M. Choudary, *J. Mol. Catal. A*, 2007, **274**, 11.
104. K. Pathak, I. Ahmad, S. H. R. Abdi, R. I. Kureshy, N. H. Khan and R. V. Jasra, *J. Mol. Catal. A*, 2007, **274**, 120.
105. S. Sahoo, P. Kumar, F. Lefebvre and S. B. Halligudi, *Tetrahedron Lett.*, 2008, **49**, 4865.
106. C. R. Oh, D. J. Choo, W. H. Shim, D. H. Lee, E. J. Roh, S.-G. Lee and C. E. Song, *Chem. Commun.*, 2003, 1100.
107. H. Ringsdorf, *J. Polym. Sci.: Polym. Symp.*, 1975, **51**, 135.
108. P. Adão, F. Avecilla, M. Bonchio, M. Carraro, J. C. Pessoa and I. Correia, *Eur. J. Inorg. Chem.*, 2010, **35**, 5568.
109. F. Bigi, H. Q. N. Gunaratne, C. Quarantelli and K. R. Seddon, *C. R. Chimie*, 2010, doi:10.1016/j.crci.2010.09.003.
110. A. M. Trzeciak and J. J. Ziolkowski, *Coord. Chem. Rev.*, 1999, **190**, 883.
111. C. D. Frohning and C. W. Kohlpaintner, in *Applied Homogeneous Catalysis with Organometallic Compounds*, ed. B. Cornils and W. A. Herrmann, VCH, Weinheim, 1996, Vol. 1, pp. 27–104.

112. P. W. N. M. van Leeuwen and C. Claver (eds.), *Rhodium Catalyzed Hydroformylation*, Kluwer Academic Publishers, Dordrecht, 2000.
113. H. Bahrmann, H. Bach, C. D. Frohning, H. J. Kleiner, P. Lappe, D. Peters, D. Regnat and W. A. Herrmann, *J. Mol. Catal. A*, 1997, **116**, 49.
114. M. T. Reetz, G. Lohmer and R. Schwickardi, *Angew. Chem., Int. Ed. Engl.*, 1997, **36**, 1526.
115. S. M. Lu and H. Alper, *J. Am. Chem. Soc.*, 2003, **125**, 13126.
116. F. Shibahara, K. Nozaki and T. Hiyama, *J. Am. Chem. Soc.*, 2003, **125**, 8555.
117. N. Yoneda, T. Minami, Y. Shiroto, K. Hamato and Y. Hosono, *J. Jpn. Pet. Inst.*, 2003, **46**, 229.
118. P. van Leeuwen, A. J. Sandee, J. N. H. Reek and P. C. J. Kamer, *J. Mol. Catal. A*, 2002, **182**, 107.
119. W. P. Chen,. L. J. Xu and J. L. Xiao, *Chem. Commun.*, 2000, 839.
120. T. Mathivet, E. Monflier, Y. Castanet, A. Mortreux and J. L. Couturier, *C. R. Chimie*, 2002, **5**, 417.
121. B. Cornils, W. A. Herrmann (eds.), *Aqueous-Phase Organometallic Catalysis*, Wiley-VCH, Weinheim, 1998.
122. Y. Chauvin, L. Mussmann and H. Olivier, *Angew. Chem., Int. Ed. Engl.*, 1995, **34**, 2698.
123. L. Leclercq, I. Suisse and F. Agbossou-Niedercorn, *Chem. Commun.*, 2008, 311.
124. C. P. Mehnert, R. A. Cook, N. C. Dispenziere and M. Afeworki, *J. Am. Chem. Soc.*, 2002, **124**, 12932.
125. Y. Yang, H. Lin, C. Deng, J. She and Y. Yuan, *Chem. Lett.*, 2005, **34**, 220.
126. A. Riisager, K. M. Eriksen, P. Wasserscheid and R. Fehrmann, *Catal. Lett.*, 2003, **90**, 149.
127. A. Riisager, P. Wasserscheid, R. van Hal and R. Fehrmann, *J. Catal.*, 2003, **219**, 452.
128. J. P. Arhancet, M. E. Davies, J. S. Merola and B. E. Hanson, *Nature*, 1989, **339**, 454.
129. P. Tundo and A. Perosa, *Chem. Soc. Rev.*, 2007, **36**, 532.
130. M. Haumann and A. Riisager, *Chem. Rev.*, 2008, **108**, 1474.
131. C. Deng, G. Ou, J. She and Y. Yuan, *J. Mol. Catal. A*, 2007, **270**, 76.
132. A. Köckritz, S. Bischoff, M. Kant and R. Siefken, *J. Mol. Catal. A*, 2001, **174**, 119.
133. W. Carruthers and I. Coldham, *Modern Methods of Organic Synthesis*, 4th edn., Cambridge University Press, New York, 2004, Chapter 3.
134. R. Herter and B. Föhlisch, *Synthesis*, 1982, 976.
135. H. Olivier, *J. Mol. Catal. A*, 1999, **146**, 285.
136. A. Vidiš, E. Küsters, G. Sedelmeier and P. J. Dyson, *J. Phys. Org. Chem.*, 2008, **21**, 264.
137. P. A. Abbott, G. Capper, D. L. Davies, R. K. Rasheed and V. Tambyrajah, *Green Chem.*, 2002, **4**, 24.
138. C. Lee, *Tetrahedron Lett.*, 1999, **40**, 2461.

139. A. Vidiš, C. A. Ohlin, G. Laurenczy, E. Küsters, G. Sedelmeier and P. J. Dyson, *Adv. Synth. Catal.*, 2005, **347**, 266.

140. H. Hagiwara, T. Kuroda, T. Hoshi and T. Suzuki, *Adv. Synth. Catal.*, 2010, **352**, 909.

141. I. Meracz and T. Oh, *Tetrahedron Lett.*, 2003, **44**, 6465.

142. C. Yeom, H. W. Kim, Y. J. Shin and B. Moon Kim, *Tetrahedron Lett.*, 2007, **48**, 9035.

143. Y. J. Shin, C.-E. Yeom, M. J. Kim and B. M. Kim, *Synlett*, 2008, **1**, 89.

144. F. Fu, Y.-C. Teo and T.-P. Loh, *Org. Lett.*, 2006, **8**, 5999.

145. S. Doherty, P. Goodrich, C. Hardacre, H.-K. Luo, D. W. Rooney, K. R. Seddon and P. Styring, *Green Chem.*, 2004, **6**, 63.

146. Y. Nishiuchi, H. Nakano, Y. Araki, R. Sato, R. Fujita, K. Uwai and M. Takeshita, *Heterocycles*, 2009, **77**, 1323.

147. H. Nakano, Y. Nishiuchi, K. Takahashi, R. Fujita, K. Uwai and M. Takeshita, *Heterocycles*, 2008, **76**, 381.

148. P. Goodrich, C. Hardacre, C. Paun, V. I. Parvulescu and I. Podolean, *Adv. Synth. Catal.*, 2008, **350**, 2473.

149. P. Goodrich, C. Hardacre, C. Paun, A. Ribeiro, S. Kennedy, M. J. V. Lourenço, H. Manyar, C. A. Nieto de Castro, M. Besnea and V. I. Pârvulescu, *Adv. Synth. Catal.*, 2011, **353**, 995.

150. S. Doherty, P. Goodrich, C. Hardacre, H. K. Luo, M. Nieuwenhuyzen and R. K. Rath, *Organometallics*, 2005, **24**, 5945.

151. X. J. He, Z. L. Shen, W. M. Mo, B. X. Hu and N. Sun, *Int. J. Mol. Sci.*, 2007, **8**, 553.

152. H.-K. Luo, L. B. Khim, H. Schumann, C. Lim, T. X. Jie and H.-Y. Yang, *Adv. Synth. Catal.*, 2007, **349**, 1781.

153. J. F. Zhao, B. H. Tan, M. K. Zhu, T. B. W. Tjan and T. P. Loh, *Adv. Synth. Catal.*, 2010, **352**, 2085.

154. W. Miao and T. H. Chan, *Adv. Synth. Catal.*, 2006, **348**, 1711.

155. M. Lombardo, F. Pasi, S. Easwar and C. Trombini, *Adv. Synth. Catal.*, 2007, **349**, 2061.

156. D. E. Siyutkin, A. S. Kucherenko, M. I. Struchkova and S. G. Zlotin, *Tetrahedron Lett.*, 2008, **49**, 1212.

157. (a) S. Lundgren, S. Lutsenko, C. Jonsson and C. Moberg, *Org. Lett.*, 2003, **5**, 3663; (b) M. I. Burguete, J. M. Fraile, J. I. Garcia, E. Garcia-Verdugo, S. V. Luis and J. A. Mayoral, *Org. Lett.*, 2000, **2**, 3905; (c) M. I. Burguete, J. M. Fraile, E. Garcia-Verdugo, S. V. Luis, V. Martinez-Merino and J. A. Mayoral, *Ind. Eng. Chem. Res.*, 2005, **44**, 8580.

158. (a) M. I. Burguete, J. M. Fraile, J. I. Garcia, E. GarciaVerdugo, C. I. Herrerias, S. V. Luis and J. A. Mayoral, *J. Org. Chem.*, 2001, **66**, 8893; (b) D. Rechavi and M. Lemaire, *Org. Lett.*, 2001, **3**, 2493; (c) H. Wang, J. Liu, P. Liu, Q. Yang, J. Xiao and C. Li, *Chin. J. Catal.*, 2006, **27**, 946; (d) N. Debono, L. Djakovitch and C. Pinel, *J. Organometallic Chem.*, 2006, **691**, 741.

159. S. Doherty, P. Goodrich, C. Hardacre, V. I. Parvulescu and C. Paun, *Adv. Synth. Catal.*, 2008, **350**, 302.

160. Y. Hamashima, H. Takano, D. Hotta and M. Sodeoka, *Org. Lett.*, 2003, **5**, 3225.
161. Y.-C. Teo, E.-L. Goh and T.-P. Loh, *Tetrahedron Lett.*, 2005, **46**, 4573.
162. C. Baleizão, B. Gigante, H. Garcia and A. Corma, *Tetrahedron*, 2004, **60**, 10461.
163. J. M. Fraile, J. I. García, C. I. Herrerías, J. A. Mayoral, S. Gmough, M. Vaultier, *Green Chem.*, 2004, **6**, 93.
164. (a) F. Bellina and C. Chiappe, *Molecules*, 2010, **15**, 2211; (b) N. T. S. Phan, M. Van der Sluys and C. W. Jones, *Adv. Synth. Catal.*, 2006, **348**, 609; (c) I. P. Beletskaya and A. V. Cheprakov, *Chem. Rev.*, 2000, **100**, 3009.
165. L. Kiss, T. Kurtàn, S. Antus and H. Brunner, *Arkivoc*, 2003, **5**, 69.
166. G. Kerti, T. Kurtán and S. Antus, *Arkivoc*, 2009, **6**, 103.
167. Š. Toma, B. Gotov, I. Kmentová and E. Solčániová, *Green Chem.*, 2000, **2**, 149.
168. I. Kmentová, B. Gotov, E. Solčániová and Š. Toma, *Green Chem.*, 2002, **4**, 103.
169. J. N. Rosa, A. G. Santos and C. A. M. Afonso, *J. Mol. Catal. A: Chem.*, 2004, **214**, 161.
170. S. M. Kim, Y. K. Kang, K. S. Lee, J. Y Mang and D. Y. Kim, *Bull. Korean Chem. Soc.*, 2006, **27**, 423.
171. B. Karimi, A. Maleki, D. Elhamifar, J. H. Clark and A. J. Hunt, *Chem. Commun.*, 2010, **46**, 4947.
172. H.-J. Wang, L.-L. Wang, W.-S. Lam, W.-Y. Yu and A. S. C. Chan, *Tetrahedron: Asymmetry*, 2006, **17**, 7.

CHAPTER 4

Metal Catalysts on Soluble Polymers

MARCO BANDINI

Alma Mater Studiorum – Università di Bologna, Dipartimento di Chimica
'G. Ciamician', via Selmi 2, 40126 Bologna, Italy

4.1 Introduction

Chemistry has a central role in science, and synthesis has a central role in chemistry.

R. Noyori, *Nat. Chem.*, 2009

Asymmetric catalysis over the past 40 years has reached an undoubted level of maturity with an astonishing development in terms of generality, reliability, selectivity and applicability.[1] Currently, more than 31 000 articles and 2000 patents deal with this area, making asymmetric catalysis one of the most productive fields of the modern organic chemistry.[2] Over the past few years the growing demand for chemical and economic sustainability has forced chemists to enterprise new and more facilitated synthetic tools.

In this direction, chemists have targeted many aspects such as: the contraction of the costs related to the process, the simplification of the reaction work-ups and minimization of use, production and subsequently disposal of hazardous chemicals.

Reduce, recycle and reuse (the three 'R's)[3] are becoming mandatory requisites to be satisfied during the design/planning of synthetic sequences. Here, the

RSC Green Chemistry No. 15
Enantioselective Homogeneous Supported Catalysis
Edited by Radovan Šebesta
© Royal Society of Chemistry 2012
Published by the Royal Society of Chemistry, www.rsc.org

chemical modification of a certain catalyst that would allow for its easy recoverability/reusability with unchanged activity, directly addresses all the crucial requirements aforementioned. In particular, the covalent grafting of chiral catalytically active species onto soluble polymers is a reliable tool to reinforce the concept of chemical sustainability in asymmetric transformations. As a matter of fact, the facile recovering of the active species by means of simple work-up procedures (heterogeneous processes) will match the high productivity/selectivity owned by homogenous reaction conditions.

A seminal report on homogeneous supported catalysis dates back to 1972 when Kagan and co-workers elegantly introduced the possibility to anchor chiral organic ligand DIOP to a Merrifield resin. The derived soluble catalytic system was then efficiently used in the enantioselective hydrosilylation of prochiral carbonyl compounds.[4] Since then, a plethora of soluble polymeric modifiers were developed, their covalent tethering to the chiraphoric units (i.e. chiral organic ligands) optimized, with the consequent opportunity to extend such a methodology to a massive number of metal-catalyzed asymmetric protocols.

Focusing on the polymeric matrix, desired solubility in conventional organic solvents is generally realized by means of non-cross-linked polymeric architectures (i.e. poly(ethylene glycol), non-cross-linked polystyrene) and saturated hydrocarbon oligomers are generally preferred. However, more articulated structures such as polysiloxanes have been successfully utilized as well. Finally, commercial availability, high chemical and mechanical stability, presence of functional groups for attachment of organic groups and suitability for analysis, by means of conventional solution-state methodologies, are some of the main requirements for ideal polymeric soluble supports.

A pre-modification of the original chiral organic ligand is performed, aiming at the robust covalent anchoring between the soluble polymer and the metal catalyst. It is worth noting that the engineering of the catalytic structure that must be tightly related to the type of stereoselective transformation targeted by the modified catalytic system. In some cases, in order to prevent undesired perturbations of the catalytic performances, the binding site(s) in the chiral ligand is(are) located in a peripheral position with respect the coordinating atoms to the metal. Contrarily, in certain cases the steric congestion created by the polymeric matrix adjacent to the reaction site improved the overall activity of the catalytic system by minimizing undesired quenching of the catalyst occurring when 'monomeric' species are utilized.[5]

The recovering of the catalyst is a mandatory requirement and since monophasic conditions are operating, this event must occur at the end of the chemical process. In this direction different methodologies have been proposed: use of selective membrane; solid/liquid separation of the polymer–catalyst after precipitation; separation of different density liquid phases.[6] Among them, mechanical removal by filtration (perhaps combined with centrifugation) is the most popular strategy which exploits the selective solubility of chemical modified organic polymers in some specific solvents and their insolubility in others.

Nowadays, the field is so diversified that to collect all the reports in a single chapter would be unrealizable. Therefore, each section will be discussed

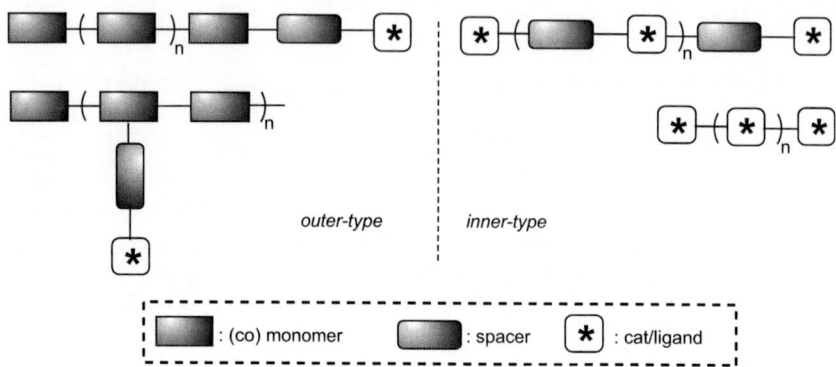

outer-type   inner-type

: (co) monomer   : spacer   * : cat/ligand

Scheme 4.1 Pictorial representation of the two classes of polymer-supported chiral metal catalysts.

through a selection of the most representative/recent examples, pointing out crucial issues such as chemo- and stereochemical performances, scope of applicability and limitations.

The chapter has been organized according to the molecular organization of the covalently bound chiral metal catalysts. In this context, two macro areas were identified namely: 'inner-sphere type chiral metal polymers' and 'outer-sphere type chiral metal polymers', featuring different disposition of the chiral ligand into the polymeric architectures. In particular, while structures belonging to the first type deal with the embedding of the ligand as pendants through the use of adequate spacers or end-capping units, in the latter family of polymers, the catalytically active site is directly involved in the construction of the main polymeric chain (Scheme 4.1).

Each example will be supplied with a short table summarizing the productivity of the catalytic system such as: number of iterations, way of recovering, efficiency in reuse. Last, but not least, control experiments to assess leaching of metal species in the reaction mixture will be discussed when available.[7]

4.2 Outer-type Chiral Metal Polymers

4.2.1 PEG-supported Catalysts

Poly(ethylene glycol) (PEG) is probably, with polystyrene, among the most utilized organic derivatizing agents to obtain soluble chiral catalysts by interposing chemical spacers between the terminal site and the catalytically active unit.[8] It is an expensive linear polymer comprising ethylene oxide as the repeating unit, with one or two functionalized (–OH) termini with large availability through commercial sources. Monomethyl ether PEG polymer (MeO-PEG) is also commercially available (–OMe, –OH terminal positions) allowing a synthetically useful 1:1 support/catalyst ratio to be realizable

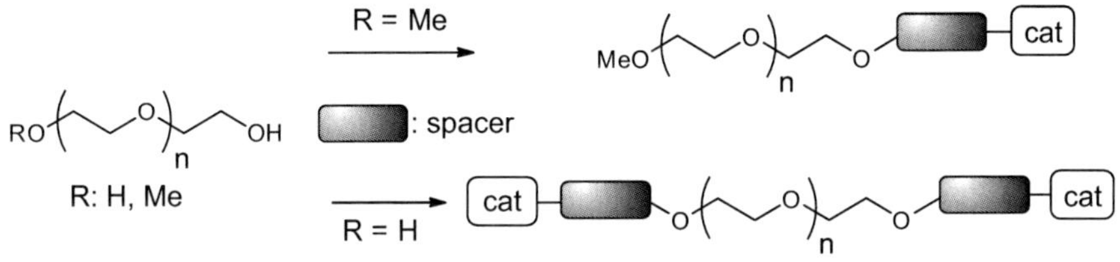

Figure 4.1   Types of PEG-supported chiral catalysts.

(Figure 4.1). Most importantly, chemical modifications of the PEG polymers can be easily monitored by NMR analysis, by exploiting also its high solubility in organic halogenated solvents (CH_2Cl_2, $CHCl_3$). On the contrary, insolubility in Et_2O, TBME (*t*-butyl methyl ether) and IPA (*i*-propyl alcohol) suggests an effective recovering route of the modified catalyst by means of selective precipitation and subsequent filtration.[9]

As a general consideration, the prolonged reaction times observed with PEG-supported ligands, with comparison to the non-immobilized versions, can be ascribed to the overall reducing in catalyst concentration deriving from the higher molecular weight of the tethered catalyst. This aspect is particularly relevant with PEGs that are generally characterized by a low loading capacity (i.e. $MeOPEG_{5000}$: 1 mol of catalyst per 5 kg of polymer is approx. 0.2 mmol/g).[10]

4.2.1.1   PEG-supported BOX Ligands

C_2-Symmetric bis-oxazolines (BOXs) are a widely utilized family of privileged ligands for enantioselective transformations.[11] Ready availability from natural amino acids and fine-tuneability comprising bi- and tridentate coordination modes (tridentate bis-oxazolines are generally referred as PyBOX, due to the presence of a pyridyl ring in the chiral scaffold) made BOX an ideal chiraphoric unit for a plethora of transition metal species. PEG-supported bisoxazoline ligands are commonly prepared *via* chemical manipulation of the bridging framework between the oxazoline rings or the analogous pyridine system.[12] In this area, Benaglia and Cozzi pioneered the use of a new type of PEG_{5000}-supported BOX **1** with excellent catalytic performances in Cu-catalyzed cyclopropanation and ene-type reactions.[13] More recently, the same team employed ligands **1a–d** in the Mukaiyma-aldol condensation in the presence of $Cu(OTf)_2$, obtaining the corresponding 3-hydroxy propanoates **3** in moderate enantiomeric excess (ee up to 60%).[14] The water solubility conferred to the BOX ligand, by the presence of the linear polymer, allowed either the use of water as the reaction media and the facile recovering/reuse of the whole species *via* extraction of the reaction products with diethyl ether (Figure 4.2 and Table 4.1).

Reiser and co-workers developed a new class of bis-oxazoline ligands (i.e. aza-BOXs, **4**) bearing a nitrogen atom bridging the two oxazoline rings. Despite the lower Lewis acidity displayed by the corresponding metal

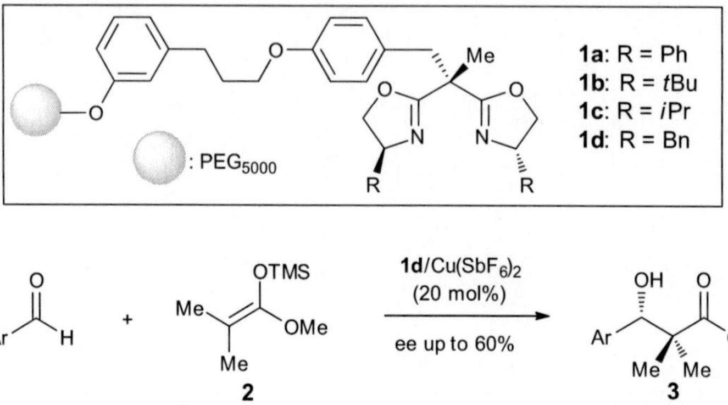

Figure 4.2 Enantioselective Mukaiyama condensation promoted by PEG-BOX-Cu(OTf)$_2$ complexes.

Table 4.1

Catalyst	Number of cycles	Method recovering	Reusability	Ref.
1d	3	Extraction with Et$_2$O and addition of fresh reagents to the aqueous solution	High	14

complexes, with respect to BOX analogues, the catalysts were efficiently utilized in Michael additions[15a] and cyclopropanation reactions.[15b] Moreover, the nitrogen atom offers a reliable binding site for anchoring PEG polymers as pendants.[16] As a proof of concept the PEG-aza-BOX **5** was synthesized with moderate catalyst loading and utilized in the copper-catalyzed kinetic benzoylation of racemic 1,2-diols. Stereochemical outcomes (ee up to 98%) i analogous with respect to non-immobilized complexes, were obtained with the possibility to recover and reuse the whole organometallic species five times (Figure 4.3 and Table 4.2).[17a] Subsequently, the same catalytic species found efficient application also in the enantioselective cyclopropanation reactions.[17b]

4.2.1.2   PEG-supported Salen Ligands

Salen ligand[18] was firstly supported on soluble MeO-PEG polymers by Janda (2000)[19] and subsequently by Venkataraman (2003).[20] In both cases, the new class of C_1-symmetry ligands was realized by tethering the organic polymer to one of the salicyl aldehyde units with or without spacers. In the former case, the enantioselective addition of Et$_2$Zn to aromatic aldehydes catalyzed by chiral salen-Zn complexes was screened. Ligand **8**, carrying a glutarate spacer, proved to be the ligand of choice, leading to the final secondary alcohol **9a** with

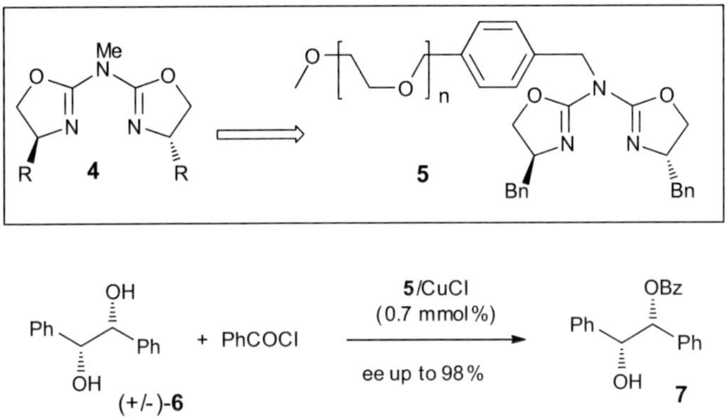

Figure 4.3 PEG-supported aza-BOX ligands for copper-catalyzed kinetic resolution of racemic 1,2-diols.

Table 4.2

Catalyst	Number of cycles	Method recovering	Reusability	Ref.
5	5	Precipitation from Et$_2$O and filtration	Very high	17a

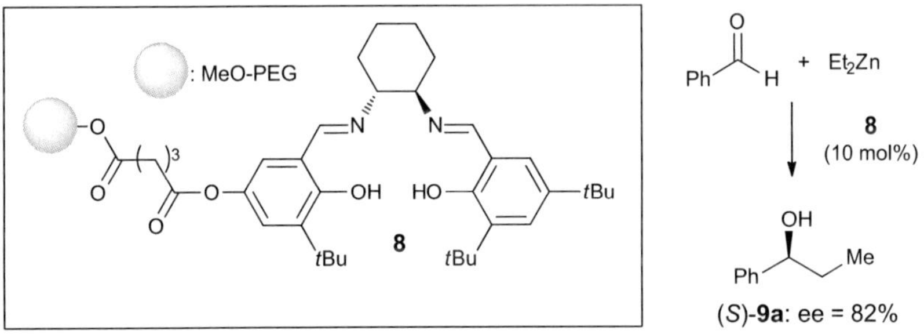

Figure 4.4 Soluble PEG-supported salen ligand for the enantioselective addition of Et$_2$Zn to aldehydes.

enantiomeric excess up to 82%. From experimental controls emerged clearly the role of the linker in determining the final stereocontrol of the process (Figure 4.4 and Table 4.3).

Subsequently, the same authors expanded the applicability of PEG-supported salen metal catalysts, highlighting the effectiveness of 8-TiCl$_2$ complex in the silylcyanation of benzaldehyde. High enantiomeric excesses (up to 86%) were recorded, in combination with the recoverability of the whole modified catalyst without any appreciable loss of activity.[21]

Table 4.3

Catalyst	Number of cycles	Method recovering	Reusability	Ref.
8	3	Precipitation from Et$_2$O and filtration	Very high	20

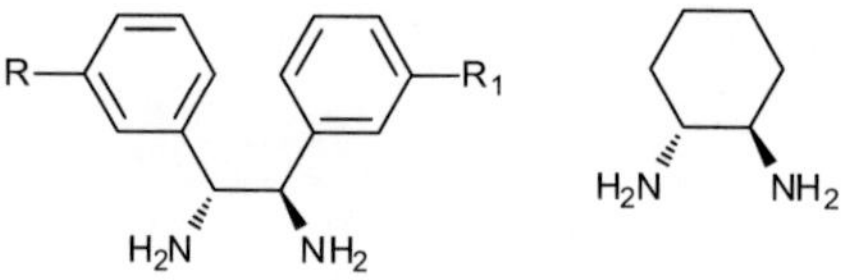

10a, R = R$_1$ = H (DPEN) **11**: DACH
10b, R = H, R$_1$ = OH
10c, R = R$_1$ = OH

Figure 4.5 1,2-Diamino-based chiral scaffolds in asymmetric catalysis.

4.2.1.3 PEG-supported Chiral Diamines

1,2-Diphenylethylenediamine (DPEN, **10a**), 1,2-diaminocyclohexane (DACH, **11**) and analogues are prominent chiral scaffolds for the realization of chiral organic ligands in asymmetric synthesis (Figure 4.5).[22]

The anchoring of the DPEN ligand onto PEG (i.e. PEG$_{2000}$) was reported by Xiao and co-workers, which condensed C_2-symmetric 3,3'-dihydroxyl derivative (**10c**) with activated MeO-PEG-OMs through conversional S$_N$2 reactions. After adequate chemical manipulations the corresponding chiral ligands **10d** and **10e** were utilized in assisting the Ru^{II}-catalyzed reduction and asymmetric transfer hydrogenation of prochiral ketones, respectively (Figure 4.6).[23]

Despite the excellent results recorded in the first run (ee higher than 90%), chemical and optical yields dropped quickly in the subsequent recycling experiments for both catalytic systems (Table 4.4). It should be noted that the authors subjected the washing phases to analysis for the ruthenium content determination. In the case of ligand **10d**, significant leaching of metal was found (5–10% of the charged Ru).

Shortly after, Bandini and Benaglia addressed the modification of the chiral diamino-oligothiophene ligands (DAT2, **11b**) with MeO-PEG$_{5000}$, by means of replacing one of the outer thienyl units with a *p*-phenol ring. The corresponding soluble ligand **11c** was chosen as ligand of choice in the Pd^0-catalyzed asymmetric allylic alkylation (AAA)[24a] and enantioselective nitroaldol condensation in the presence of Cu(OAc)$_2$ (Figure 4.7).[24b] Interestingly, AAA reaction worked smoothly in CH$_2$Cl$_2$ and THF, while the Henry protocol required a substantial excess of MeNO$_2$ (40:1), with respect to the aldehydic compounds, in order to obtain perfect homogeneous conditions and synthetically acceptable reaction times (Table 4.5). The comparable results obtained with **11c** and unmodified DAT2 proved the negligible role of the PEG polymer over the chemical outcomes of the reactions investigated.

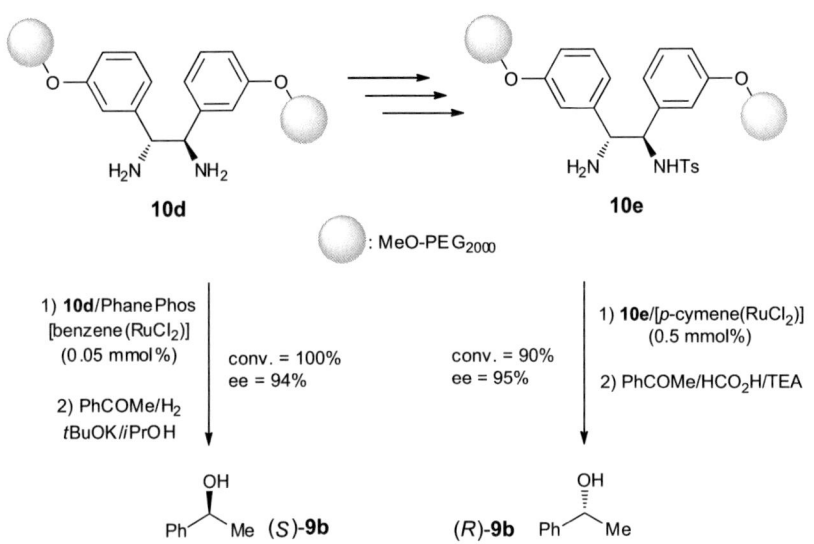

Figure 4.6 Polymer-bound DPEN derivatives as ligands in enantioselective reduction of ketones.

Table 4.4

Catalyst	Number of cycles	Method recovering	Reusability	Ref.
10d	3	Precipitation from Et₂O and product removal by syringe	Very high	23a
10e	3	Precipitation from Et₂O and product removal by syringe	Significant drop at the third cycle	23b

4.2.1.4 PEG-supported Phosphine Ligands

Nowadays, chiral PEG-supported phosphines are receiving growing attention after the large volume of literature appeared on supported achiral P-based ligands from the 1960s onwards. In this context, Ding and co-workers described a new class of chiral C_2-symmetric biphosphines **15** carrying a cyclobutyl scaffold with great potential in the Pd-catalyzed asymmetric allylic alkylation.[25] A subsequent report by the same team addressed the recoverability of such ligands by developing a MeO-PEG-bound ligand **16** in which one of the carboxylic groups of the backbone acted as linking site for the MeO-PEG$_{2000}$-OH polymer.[26] Interestingly, the cationic palladium complex, obtained *in situ* from (*S,S,S,S*)-**16** and [Pd(π-C₃H₅)Cl]₂, promoted the condensation of dimethyl malonate and benzylamine to racemic acetate **12b** in quantitative yield and 94.6% and 97.2% ee, respectively (Figure 4.8).

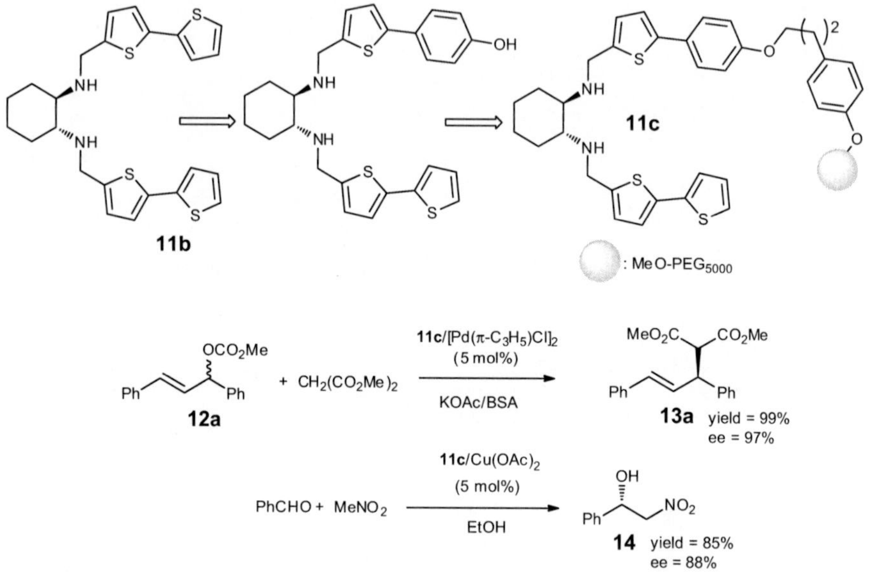

Figure 4.7 Use of PEG-modified DAT ligands for Pd and Cu-catalyzed stereoselective transformations.

Table 4.5

Catalyst	Number of cycles	Method recovering	Reusability	Ref.
11c (AAA)	3	Precipitation from Et_2O and product removal by syringe	High (Pd addition every run required)	24a
11c (Henry)	5	Precipitation from Et_2O and product removal by syringe	Very high	24b

No significant differences were observed with respect to the non-immobilized
counterpart **15** and attempts of recycling the catalyst highlighted a progressive
drop in catalytic activity after the fifth run (Table 4.6).

Wang and co-workers contributed substantially in expanding the applicability of PEG-supported biphosphine ligands in asymmetric catalysis. In
particular, the team developed a new C_2-symmetric soluble polymer (MeO-
PEG)-supported biphenylbisphosphine ligand (MeO-BIPHEP) **17** that was
obtained from MeO-BIPHEP through simple chemical manipulations
involving demethylation and subsequent condensation with MeO-PEG$_{2000}$-
OMs. The corresponding Noyori-type complex (MeO-PEG)$_2$-(MeO)BIPHEP-
RuCl$_2$-DPEN was efficiently utilized in the enantioselective hydrogenation of aromatic ketones ($H_2 = 20$ bar, Figure 4.9)[27a] and β-keto esters
($H_2 = 1$ bar).[27b]

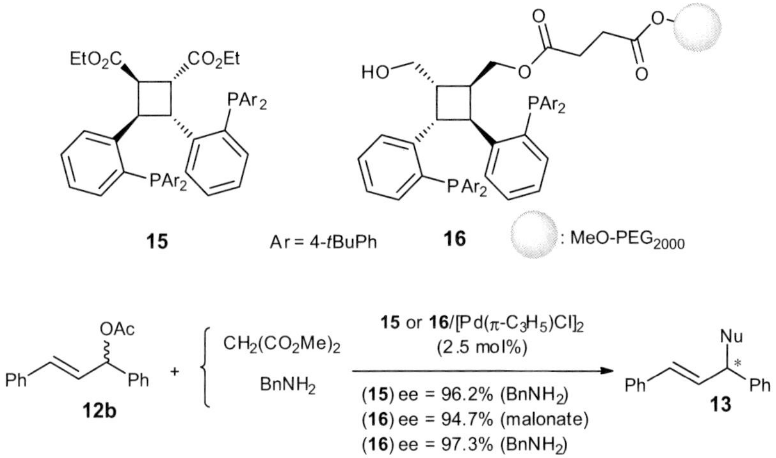

Figure 4.8 New PEG-supported cyclobutyl diphosphine ligands for enantioselective nucleophilic allylic substitution.

Table 4.6

Catalyst	Number of cycles	Method recovering	Reusability	Ref.
16 (BnNH$_2$)	9	Precipitation from Et$_2$O and filtration (>90% catalyst recovered each run)	High, with gradual drop from the fifth cycle	26

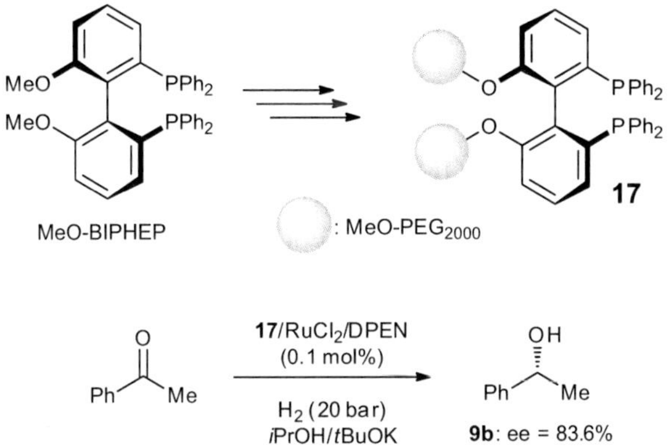

Figure 4.9 C_2-symmetric methoxy-BIPHEP supported onto PEG$_{2000}$.

Recycling experiments were carried out as summarized in Table 4.7 and IPC analysis of the filtrate revealed a content of Ru lower than 1 ppm (Table 4.7).

Table 4.7

Catalyst	Number of cycles	Method recovering	Reusability	Ref.
17	6	Precipitation from Et$_2$O and filtration	Sharp drop at the sixth cycle	27a

4.2.2 Non-cross-linked Polystyrene-supported Ligands

Polystyrene is a linear polymer that shares many features with the previously described PEG. Non-cross-linked polystyrenes (NCPS) are generally soluble in many common organic solvents (i.e. THF, CH$_2$Cl$_2$, CHCl$_3$) and insoluble in *n*-hexane and MeOH. As a consequence, post-reaction precipitation/filtration procedures are usually applied for the recovering of the NCPS-supported species.[8] Contrarily to PEG polymers, NCPS allows different loading capacities to be realized by controlling the ratio of co-monomers in the polymerization step or by a tailor-made post-chemical modification of the final polymer. Last but not least, also in this case, the loading of ligand/catalyst can be determined by NMR spectroscopy in solution.[28]

4.2.2.1 NCPS-supported Phosphorous-based Ligands

Chiral phosphoroamidites are a class of ligands largely employed in asymmetric transformations featuring high tuneability and flexible coordination modes with a particular preference for metal-catalyzed hydrogenation reactions and C–C bond-forming processes.[29] Doherty and co-workers identified in the readily available phosphoroamidite **18** a suitable monomer to co-polymerize with styrene.[30] Therefore, the soluble non-cross-linked styrene-based co-polymer **19**, obtained in the presence of 2,2'-azobis(2-methyl)propionitrile (AIBN) as radical initiator ($M_w = 20\,100$ Da.), was transformed into the corresponding rhodium complex **19**-[Rh] through condensation with [Rh(COD)$_2$][BF$_4$]. The rhodium catalyst proved competence in the hydrogenation of dehydroamino acid derivative **20** providing comparable chemical outputs (conv = 100%, ee = 80%) to the monomeric precursor **18** (Figure 4.10).

ICP-analysis of the washing phases resulted in no detection of leached rhodium (Table 4.8).

The same reduction of substituted C–C double bonds was also investigated by Kamer and co-workers, who made several L/Rh combinations by mixing polystyrene supported ligands and non-immobilized chiral species in a 2:1 L:Rh ratio.[31] However, the binding site offered by the resin chiral phosphite did not guarantee a proper selective formation of the immobilized heteroligand rhodium complexes.

4.2.2.2 NCPS-supported Oxygen-based Ligands

Desymmetrization of bis(salicylidene)ethylenediamine (salen) allowed for the incorporation of this key molecular motif also onto non-cross-linked

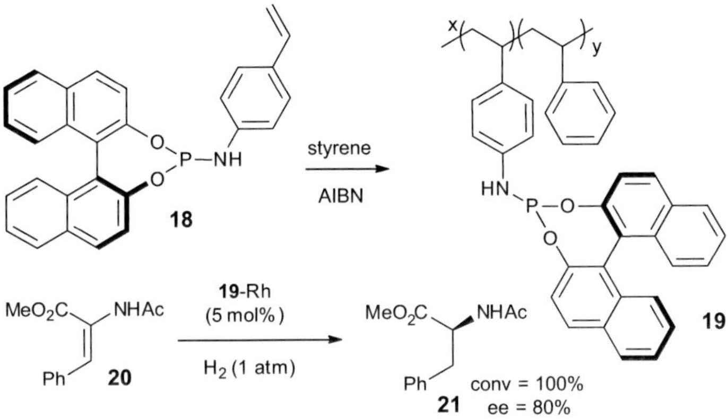

Figure 4.10 Chiral phosphoramidites grafted onto soluble polystyrene.

Table 4.8

Catalyst	Number of cycles	Method recovering	Reusability	Ref.
19	4	Precipitation from MeOH (>95% catalyst recovered each run)	Generally high, depending on substrate	30

polystyrene systems. In particular, despite the numerous efforts devoted to the heterogenization of salens through (1) grafting onto insoluble polymers and (2) polymerization of salen monomers (see below), the immobilization of salen-type ligands as flexible pendants onto soluble organic polymers has been less investigated. In this direction, Jones, Weck and co-workers utilized the readily available ammonium salt **22** for the preparation of C_1-symmetric ligand **23** bearing a styryl capping unit.[32] Free-radical co-polymerization with styrene (AIBN as the initiator) produced a range of co-polymeric species **24** with different salen/styrene compositions (i.e. 50:50, 20:80, 10:90). After complexation with Co(OAc)$_2$ the corresponding Co-polymeric structures **28** displayed great solubility in CH$_2$Cl$_2$ and were employed as the pre-catalytic entities in the asymmetric hydrolytic kinetic resolution of epichlorohydrin **26** (Figure 4.11).

Excellent levels of stereoinduction were generally recorded, with optimal reaction performances obtained when more diluted salen units were present in the macromolecular catalyst (higher flexibility of the structure and better accessibility of the reagents to the catalytically active sites, see Table 4.9).

TADDOL is an important structure for stereocontrolled transformations[33] and its heterogenization has been extensively studied through dendrimeric functionalization and immobilization on Merrifield resins. Reuse and recycling of soluble polymeric TADDOL ligand was investigated by Rosling and co-workers, by means of NCPS-supported species **27**, obtained through

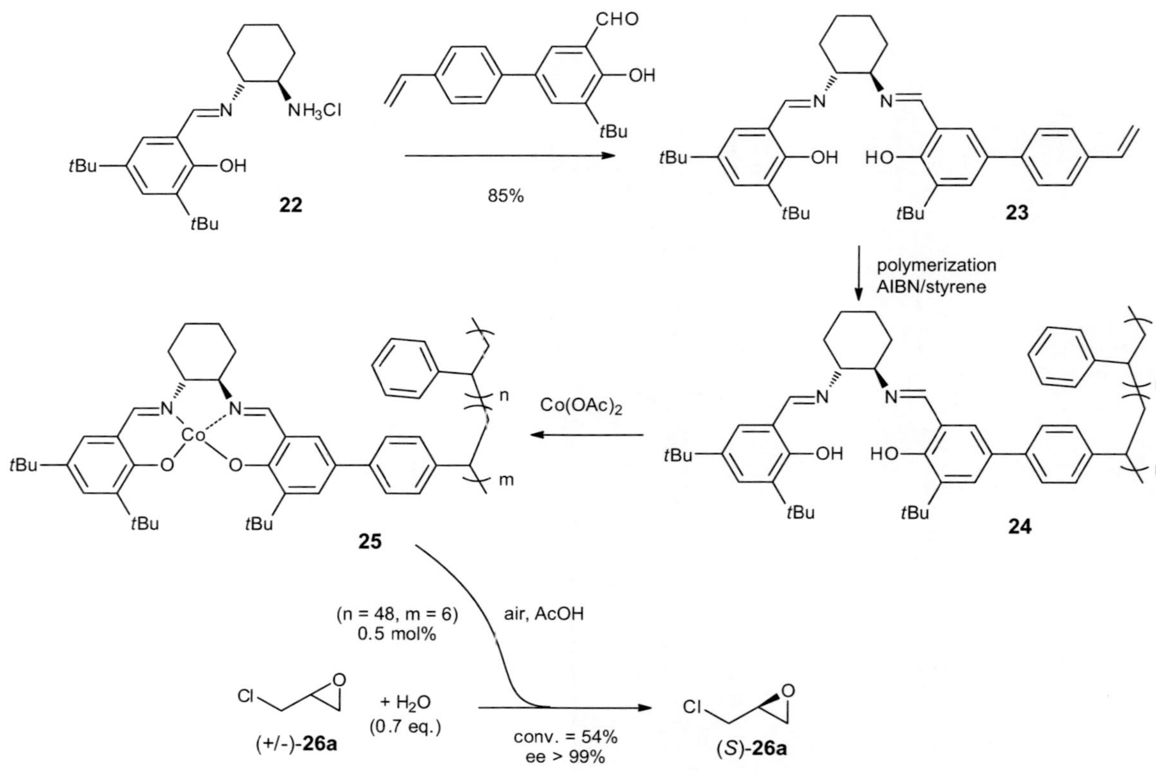

Figure 4.11 Synthesis of polystyrene supported salen ligands and their application in the enantioselective HKR of epoxides.

Table 4.9

Catalyst	Number of cycles	Method recovering	Reusability	Ref.
25	4	Precipitation from Et$_2$O, reactivation with AcOH	Very high	31

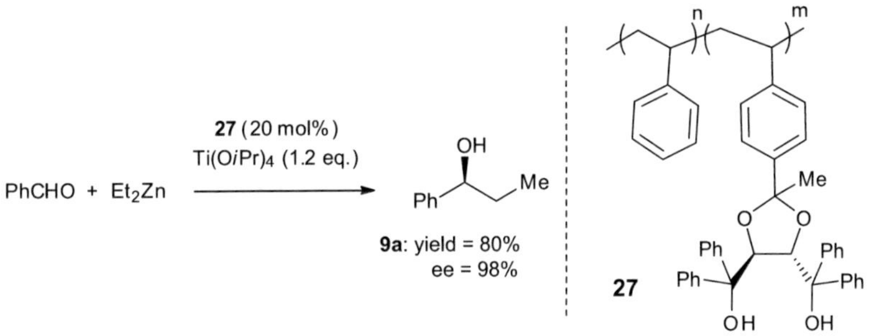

PhCHO + Et$_2$Zn $\xrightarrow{\text{27 (20 mol%)}\atop\text{Ti(O}i\text{Pr)}_4\text{ (1.2 eq.)}}$

9a: yield = 80%
ee = 98%

27

Figure 4.12 Polystyrene supported TADDOL for the enantioselective addition of Et$_2$Zn to PhCHO.

Table 4.10

Catalyst	Number of cycles	Method recovering	Reusability	Ref.
27	3	Precipitation from MeOH	Very high	34

co-polymerization with styrene (styrene/TADDOL 97.8:2.2, benzoyl peroxide as the initiator).[34] The immobilized chiral ligand (0.273 mmol/g, 20 mol% with respect to benzaldehyde) was utilized in the addition of Et$_2$Zn to PhCHO, mediated by Ti(OiPr)$_4$, and the corresponding secondary alcohol **9a** was isolated in 98% ee (Figure 4.12, Table 4.10).

4.2.3 ROMP-supported Ligands

Ring-opening polymerization of norbornene is a well-established synthetic procedure to obtain soluble and densely functionalized organic polymers.[35] Focusing on the anchoring of chiral organic ligands, Weck's team addressed many efforts in developing numerous poly-norbornene salen-metal complexes through tailor-designed four key variables: (1) type of polymeric support; (2) the nature of the linker; (3) catalyst density; and (4) catalyst-support connectivity.[36] Homo- or co-poly(norbornene)-supported salen-metal complexes (i.e. Mn^{II}, Co^{III}, Al^{III}) were obtained in a highly controlled manner and proved to be efficient in both monometallic and bimetallic transition-state stereoselective transformations. In particular, a stepwise condensation of differently

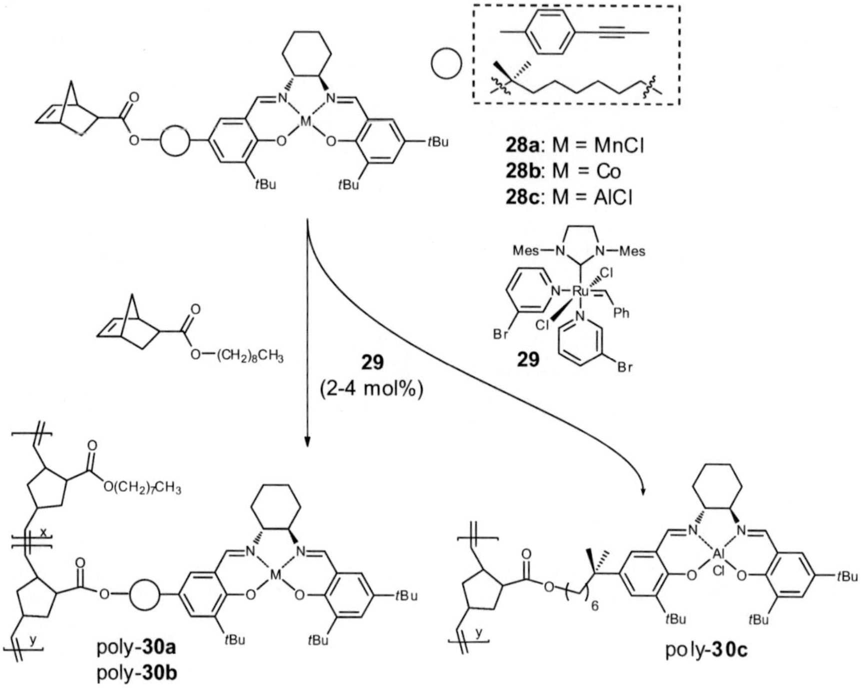

Figure 4.13 ROMP of norbornene for the immobilization of salen-metal complexes.

functionalized salicylaldehydes with monoprotected 1,2-cyclohexanediamine provided a C_1-symmetric ligand that was transformed into the corresponding metal species **28a–c** under conventional reaction conditions. Finally, ROMP reactions were performed on **28a–c** as (co)monomeric units in the presence of third generation Ru Grubbs catalyst **29** (Figure 4.13).

In details, the manganese complex **30a** (4 mol%, *y*:*x* = 9:1) promoted the stereoselective epoxidation of 1,2-dihydronaphthalene **31a** with enantioselectivity up to 88%.[37a] The corresponding cobalt species poly-**30b** (0.5 mol%, *y*:*x* = 3:1) was utilized in the hydrolytic kinetic resolution of epichlorohydrin **26a** that was isolated in 99% ee and 55% conversion (*S*-isomer).[37a] Finally, aluminium-catalyzed enantioselective conjugate addition of TMSCN to imide **32** was performed with high catalyst density (homopolymer, **30c**, 5 mol%), in accordance with the postulated bimetallic transition state of the 1,4-addition (Figure 4.14).[37b]

Attempts to recycle the polymeric metal complexes were carried out through selective precipitation of the catalytic species (Et$_2$O, MeOH) or evaporation of the reaction products with consequent reactivation of the poly(norbornene)-supported salen-metal complexes. Alternating results were obtained, with marked decreasing of activity in the Mn and Co catalysis and retention of performances for the Michael addition reaction. However, in the latter case, the large volume of organic solvents required for the complete recovering of the

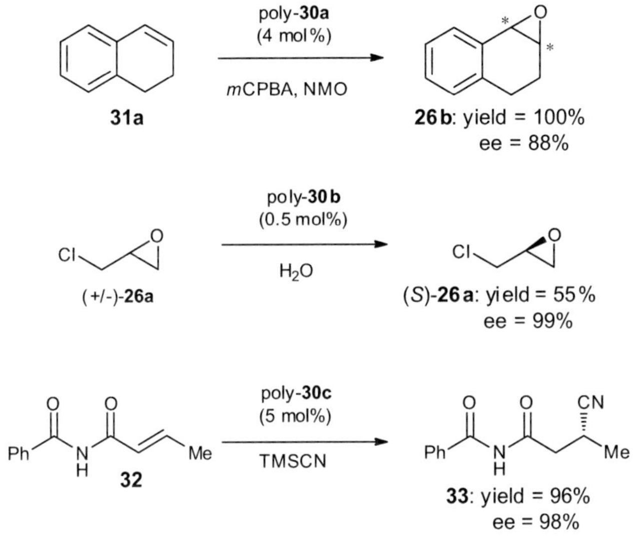

Figure 4.14 Proving the flexibility of ROMP-supported salen catalyst in asymmetric transformations.

Table 4.11

Catalyst	Number of cycles	Method recovering	Reusability	Ref.
30a	3	Precipitation from $Et_2O/$ MeOH, centrifugation	Drop of activity at the third run	37a
30b	3	Distillation of products and re-oxidation of the catalyst	Small decrease in activity	37a
30c	5	Precipitation from EtOAc	Very high	37b

salen-Al complex (overall 75 mL of EtOAc with respect to 0.26 mmol reaction scale), combined with the long procedure times (36 h), constitute important drawbacks frequently encountered in polymer-supported catalysis (Table 4.11).

4.2.4 Polyester-supported Catalysts

Anchoring of the active unit to preformed polymeric structures is a widely employed methodology to produce tailor-designed supported ligands, with defined molecular weights. The approach was addressed to the preparation of supported norephedrine ligands onto poly(alkyl methacrylate). In particular, the resulting hydroxyl groups of the easily obtainable co-polymer **34** were utilized as tethering sites for the chiral ligand *via* conventional S_N2 reactions (tosylate chemistry).[38] Wills and co-workers designed this system in order to achieve high loading of catalyst **35** (30%) that was highly efficient in controlling

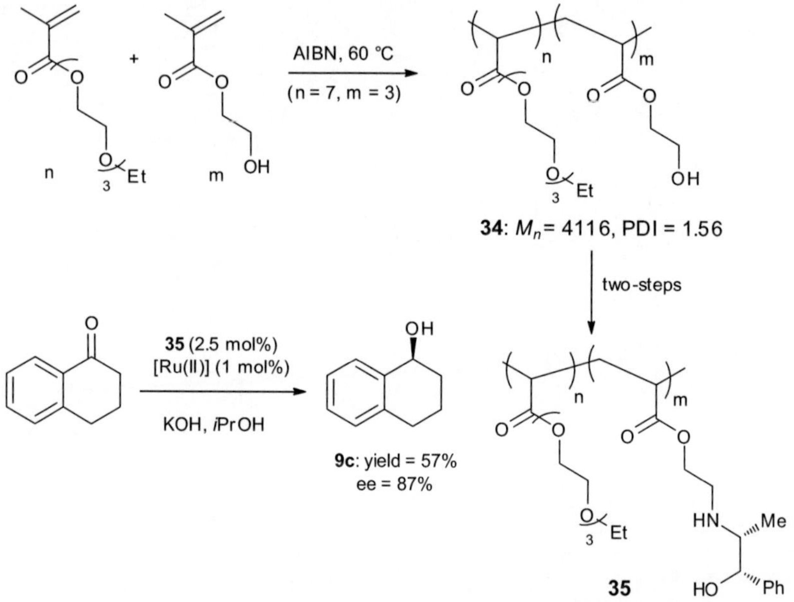

Figure 4.15 Polystyrene-supported norephedrines: valuable chiral ligands for asymmetric Ru-catalyzed hydrogen transfer to ketones.

both chemical and optical outcomes in the enantioselective Ru-mediated transfer of hydrogen to prochiral ketones. The corresponding secondary alcohol **9c** was isolated in 87% ee, with the possibility to reuse the active species simply by adding fresh reagents to the unpurified polymeric catalyst (Figure 4.15).

4.3 Inner-type Chiral Metal Polymers

An alternative approach for the immobilization of chiral ligands onto soluble polymeric structures consists in the incorporation of chiral monomeric units in the so called 'inner-sphere' type polymers in which the ligand is not part of pendants but constitutes itself as one of the exclusive repeating units of the macromolecule. The methodology has been mainly utilized for aromatic ligands such as BINAP, BINOL and salen. Mono- and bifunctional polymeric chiral ligands have been developed with the possibility to intercalate π-spacers between the repeating chiral blocks.

4.3.1 BINAP and BINAP-BINOL Co-polymers

The rigidity of polymeric chiral ligands is considered to be beneficial for the catalytic activity of the whole system, avoiding the alteration of the catalytic site frequently occurring by moving from monomeric species to flexible supported ones. Rigid poly-BINAPs and poly-BINOLs were first described by Pu and co-workers, by developing a convenient Suzuki cross-coupling approach for their preparation.[39] From a synthetic viewpoint, the polymerization event involved different combinations of functionalized 1,4-dihaloarenes and boronic ester cores, leading to the well-defined structures **36a,b** (^{1}H-NMR, ^{31}P-NMR for BINAP). Polymer **36a** ($M_w = 25\,800$ and $M_n = 14\,300$, PDI = 1.8) displayed dual catalytic function with particular regard to the enantioselective addition of Et_2Zn/Me_2Zn to aldehydes and reduction of acetophenone with catecholborane, while **36b** ($M_w = 5800$ and $M_n = 4300$, PDI = 1.35) was effectively utilized in the asymmetric hydrogenation of C=C (Rh) and prochiral aromatic ketones (Ru, Figure 4.16).

Recycling and reuse of the polymeric **36a,b** were verified over three consecutive runs (**36a** for the reduction of acetophenone and **36b** for the Ru-catalyzed hydrogenation of aromatic ketones) by precipitation with addition of methanol. The recovered ligands showed comparable reactivity/selectivity to that of the original catalyst. Having in hand these encouraging results, the team envisioned the possibility to realize a chiral co-polymer based on alternating BINAP and BINOL repeating units, with the final aim to enlarge the portfolio of chemical transformations assisted by polymeric catalysts. In particular, Suzuki cross-coupling reactions, combined with usual protecting group chemistry, allowed rigid poly(BINOL-BINAP) **37** ($M_w = 11\,600$ and $M_n = 7500$, PDI = 1.55) to be synthesized with randomly distributed BINOL and BINAP frameworks.[40] *p*-Acetylbenzaldehyde was elected as the model in a tandem stereoselective process involving the addition of Et_2Zn to the aldehydic function (BINOL catalysis), followed by the hydrogenation of the acetyl moiety after quenching of the alkylating process. Very high conversion (99%) and excellent ee/dr were recorded (92% for Et_2Zn addition, 86% for hydrogenation reaction, respectively, Figure 4.17). It is worth mentioning that the poly(BINOL-BINAP) ligand can be also utilized in the Zn catalysis or Ru-based reaction, alternatively (Table 4.12).

In the same direction, Chan and co-workers described two new bifunctional polymeric ligands, namely (poly(BINOL-BINAP) and poly(BINOL-BINAPO)) featuring a regular chiral chain polymer based on alternating chiral units. The connectivity was realized through imine bonds that offered diagnostic strong absorptions in the IR spectra to prove the existence of the co-polymer. Catalytic efficiency and reusability (precipitation with MeOH and filtration) were investigated in the hydrogenation of C–C double bonds (Ru^{II}) and addition of Et_2Zn to PhCHO as bench test processes. In some cases, the co-polymeric catalyst offered improved enantioselectivity with respect to the monomeric precursor.[41]

4.3.2 Soluble Salen Polymers

From a synthetic viewpoint, three main types of *inner-type* salen (co)polymers have been introduced. Firstly, the Zheng's type homopolymers **40a–e** originated from the condensation of C_2-symmetric bis-aldehydes **39** with enantiomerically pure 1,2-*trans*-diamines (Figure 4.18).[42]

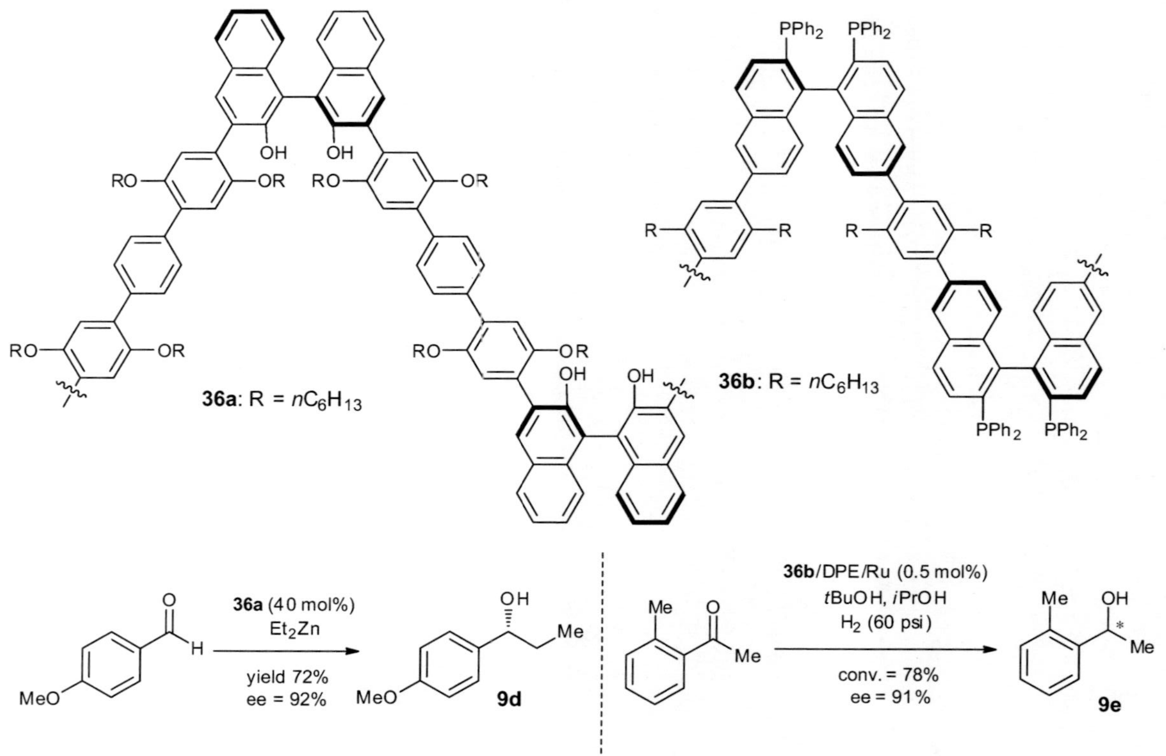

Figure 4.16 Structure of poly(BINOL) and poly(BINAP) ligands and applications in addition of Et_2Zn to aldehydes and reduction of ketones, respectively.

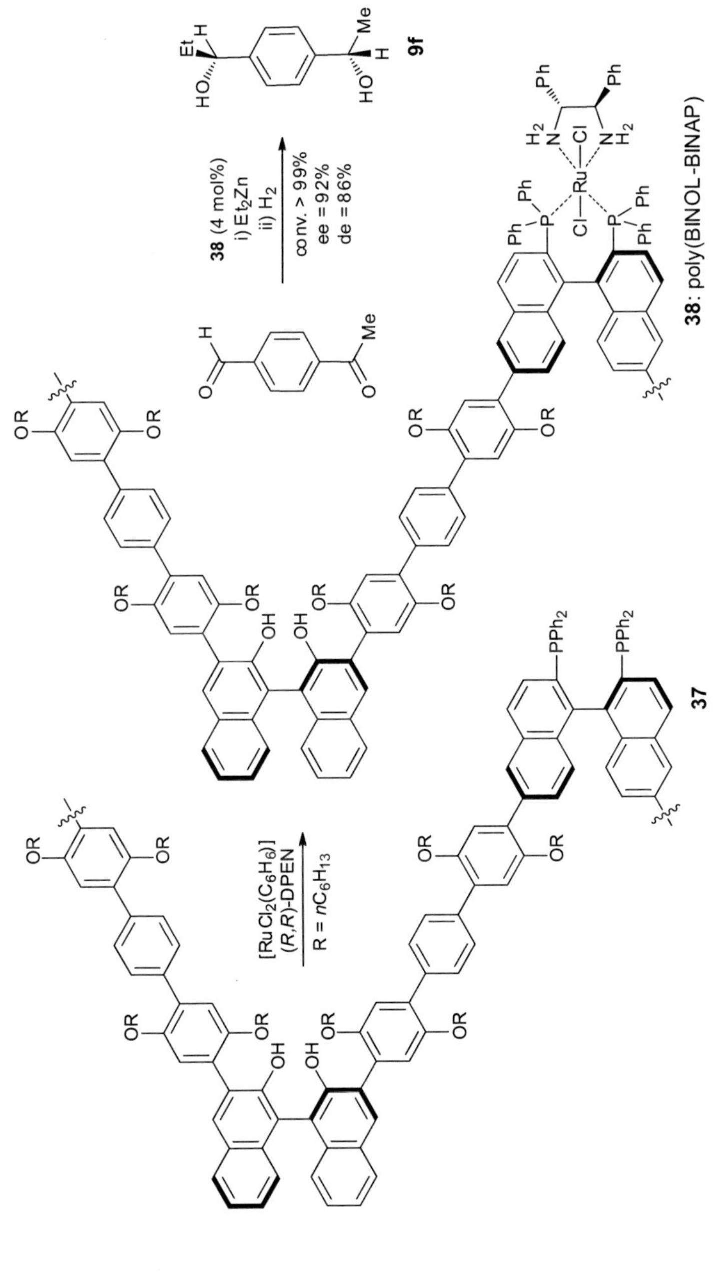

Figure 4.17   Rigid poly(BINOL-BINAP) with chiral dual catalytic function.

Table 4.12

Catalyst	Number of cycles	Method recovering	Reusability	Ref.
38	2	Precipitation from MeOH	Slight drop in diastereoselection	40

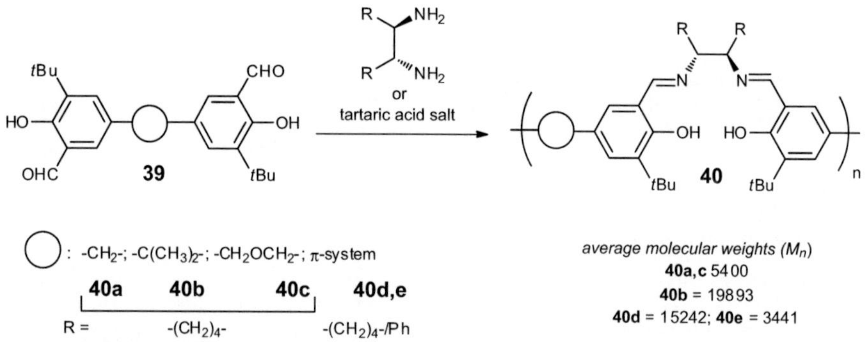

Figure 4.18   Synthesis of inner sphere type poly-salen systems.

Several spacers were considered and the corresponding poly-salen-metal complexes were utilized in a range of enantioselective reactions such as epoxidation of unfunctionalized olefins (Mn^{III}),[43] O-acetyl cyanation of aldehydes (Ti^{IV}, V^V),[44] O-trimethylsilyl cyanation of aldehydes (V^V),[45] aminolytic kinetic resolution of epoxides (Cr^{III}),[46] alkynylation and addition of Et_2Zn to aldehydes (Zn^{II}),[47] and oxidative kinetic resolutions of secondary alcohols (Mn^{III}).[48] A representative collection of stereoselective transformations promoted by poly-salen metal complexes is depicted in Figure 4.19.

In all cases, comparable results in term of catalytic activity of the polymeric salen species with respect to the monomeric unit was observed and the great insolubility of the corresponding poly-salen metal species in hydrocarbons and alcoholic solvents (i.e. MeOH) was exploited for recovering and reusing the active adducts (Table 4.13).

A second strategy for the realization of soluble poly-salen systems was reported by Yin and co-workers that underlined the efficiency of *trans*-1,2-diaminocyclohexane, combined with properly functionalized salicyl aldehydes, to obtain a linear polymeric metal catalyst precursor.[49] The corresponding Mn^{III} complex (**41**, $n \approx 14$) was employed in the enantioselective epoxidation of styrene **31c**; however, epoxide **26d** was isolated in poor ee $= 35\%$ (yield $= 99\%$, Figure 4.20). Analogously, attempts to recycle the active species (5 runs, precipitation from hexane) displayed a substantial decreasing of performance in both chemical and optical yields (yield $= 99\%$ down to 52%, ee $= 35\%$ down to 29%). A partial oxidative decomposition of the chiral

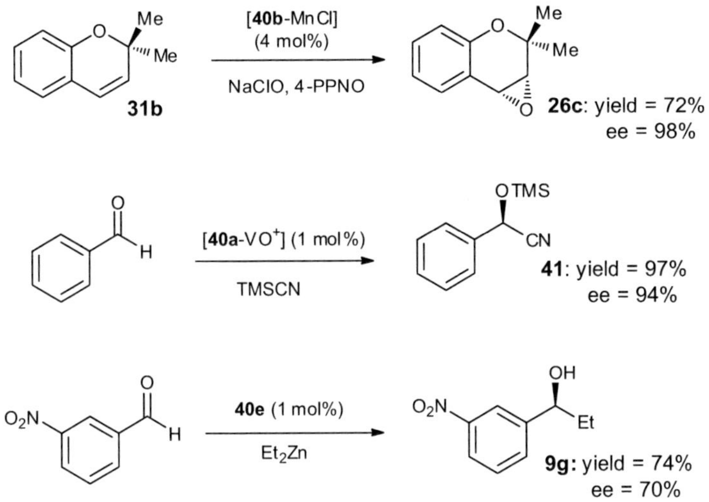

Figure 4.19 Bench-text reactions of the poly-salen metal system described by Zheng.

Table 4.13

Catalyst	Number of cycles	Method recovering	Reusability	Ref.
40b-MnCl	5	Precipitation from *n*-hexane	Very high	43
40a-VO$^+$	4	Precipitation from *n*-hexane	Small decrease in activity	44
40a-CrCl	5	Precipitation from *n*-hexane	Very high	46b
40e-Zn	2	Precipitation from MeOH	Very high, fresh Et$_2$Zn was added in the second run	47
40a-MnCl	5	Precipitation from *n*-hexane	Retention of ee, gradual decrease in activity	48a
40b-MnCl	5	Precipitation from *n*-hexane	Very high	48b

polymeric structure (m-CPBA,UV-vis spectra) was observed from the third cycle on.

The third approach for the preparation of soluble polymeric structures, containing salen ligands as repeating units, has been elegantly described by Weck and co-workers through ruthenium-catalyzed ring-expanding olefin metathesis.[50a] In particular, the C_1-symmetric salen-Co^{II} complex **42**, carrying a cyclooct-4-enecarboxylic ester group, proved efficiency as a metathesis partner for the multigram scale preparation of a mixture of macrocyclic salen-Co species **43** (quantitative yield, $M_n \approx 2460$), in the presence of the third-generation Ru Grubbs catalyst **29**. The metal complex that derived from the ring-opening metathesis polymerization (ROMP, **43**) was utilized in low catalytic loading (0.01 mol%) for the hydrolytic kinetic resolution (HKR) of

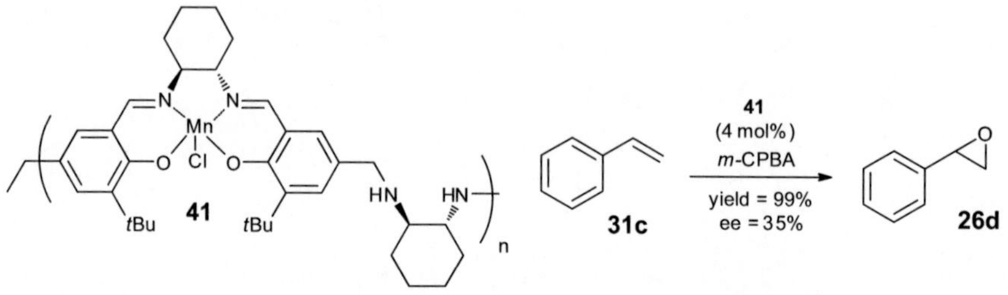

Figure 4.20   The poly-salen manganese complex described by Yin.

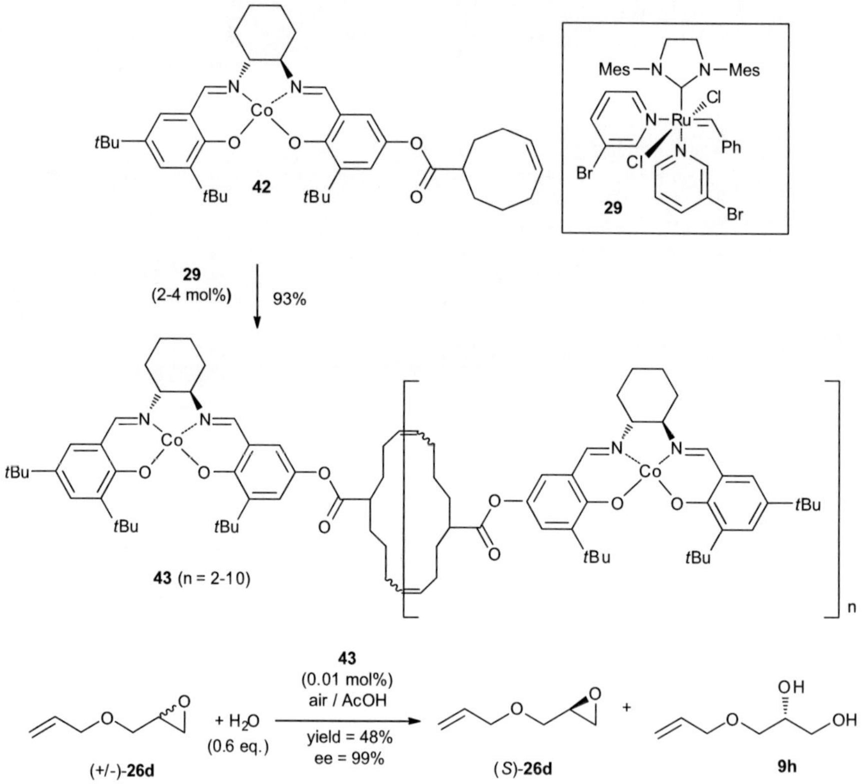

Figure 4.21   ROMP-approach to the synthesis of 'inner-sphere' poly-salen Co complex **43**.

terminal racemic epoxides (the example of allyl glycidyl ether **26d** is described in Figure 4.21). Excellent yields and enantiomeric excesses are generally obtained. Finally, the oligo(cyclooctene)-supported Co-salen catalyst **43** was recovered and reused several times (evaporation of the reaction products) with no significant loss of efficiency.[50b]

4.3.3 Polyester and Polysiloxane-supported Cinchona Alkaloid Ligands

Bis-cinchona phthalazine (DHQD)$_2$PHAL was coupled with terephthalic acid chloride to produce a chiral soluble organic co-polymer comprising PEG$_{4000}$ building blocks **44**. The polymer-bound alkaloid was utilized as the ligand in the enantioselective osmium-catalyzed (OsO$_4$) dihydroxylation of alkenes, in the presence of K$_3$Fe(CN)$_6$ as the stoichiometric oxidant.[51] Excellent catalytic activity (ee up to 99%), accompanied by the efficient reuse for five cycles, were highlighted by the authors even in absence of a detailed monitoring of the Os-leaching in the reaction product (Figure 4.22).

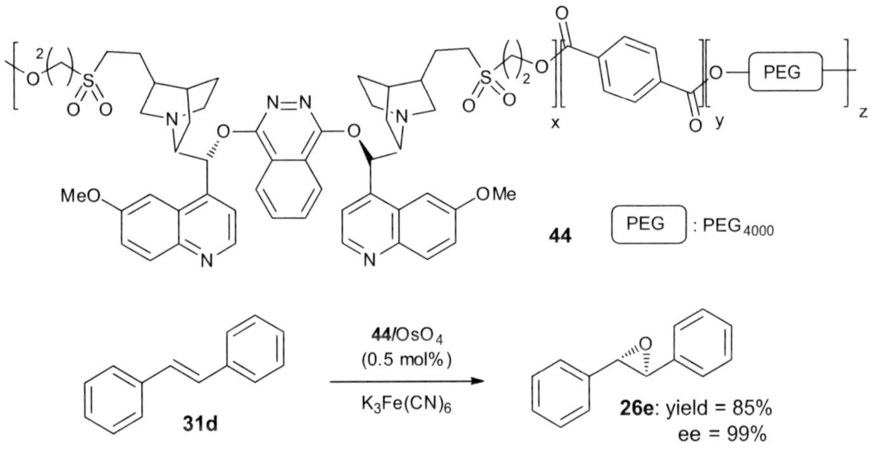

Figure 4.22 Synthesis and application of polymer-bound alkaloid **44**.

Shortly after this, the (DHQD)$_2$PHAL ligand was also investigated by Siegel and co-workers, with the final aim to synthesize and utilize in asymmetric dihydroxylation reaction (Os catalysis) a new class of soluble polysiloxane scaffolds.[52,53] The solubility was modulated by a proper mass balance of the partial hydrogenated Sharpless ligand **45** with a solubilizing group **46**, that was subjected to co-polymerization under platinum-catalyzed hydrosilylation reaction (Figure 4.23).

Two different methodologies for the recovering and reuse of **45** were proposed, namely precipitation by adding an excess of water or ultrafiltration. Interestingly, while in the first case the addition of extra OsO$_4$ and stoichiometric oxidant (K$_3$Fe(CN)$_6$) was necessary, in the second cycle, the use of the filtering unit of the centrifugal concentrator as the reaction vessel allowed the facile separation of the resulting diol from the catalytic species that was effectively reutilized with any significant loss of activity (Table 4.14).

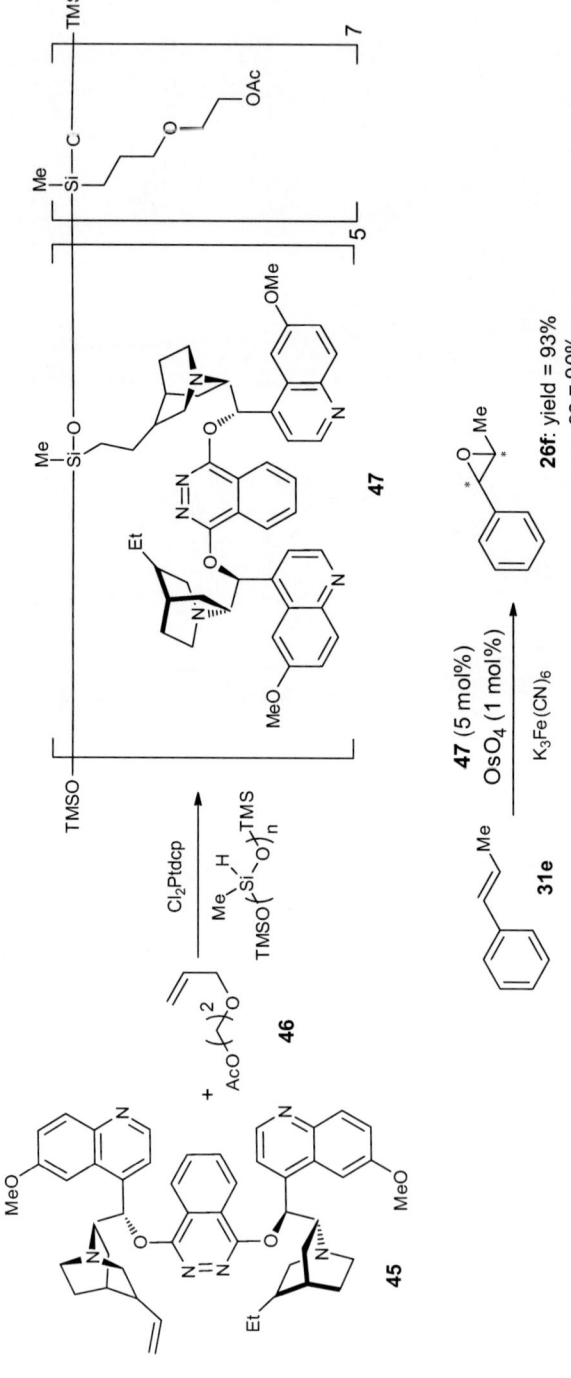

Figure 4.23 Soluble polysilane-bound (DHQD)₂PHAL ligands for asymmetric epoxidation of olefins.

Table 4.14

Catalyst	Number of cycles	Method recovering	Reusability	Ref.
47	2	Precipitation from water; ultrafiltration	addition of OsO_4; very high	52 52

4.4 Conclusions

In conclusion, the selected studies described in this chapter shown clearly the large volume of efforts recently devoted towards the realization of efficient, selective and sustainable polymer-supported metal catalytic systems, featuring high solubility in common organic reaction media. Theoretically, this approach allows to combine the peculiar activity of homogeneous catalysts with recoverability/reusability of the heterogeneous catalysis. As a matter of fact, comparable results in term of activity, with respect to the non-immobilized species (minimization of mass transport problems) were frequently highlighted. Last but not least, the inert supports can also beneficially affect the overall catalytic performances, disfavouring undesired aggregating phenomenon of the catalytic units.

Recently, great improvements in the chemical separation of the soluble polymeric species have been achieved, partially overcoming one of the main drawbacks of this approach. Selective precipitation of the catalyst and ultrafiltration are some of the most exploited methodologies that frequently allowed the recovering of the intact species in high yield. The leaching of metal into the reaction mixture can, in some instances, represent a challenging issue (especially in pharma products), and therefore careful monitoring of the released metal content in solution and further purification steps (i.e. distillation, crystallization, chromatographic separation) of the reaction crudes are routinely required. In the same direction, the reuse of the catalytic species is a mandatory requirement for polymeric-bound catalysts in order to increase their productivity (TON). In this direction, robustness and resistance toward deactivation phenomenon are key features to be satisfied by organometallic species during the immobilization on inert matrixes. All the examples reported faced the reusability of the catalytically active species. Generally, high repeatability in chemical/optical outputs is demonstrated over the whole cycles; however, in a few cases the addition of fresh metal source every run was necessary.

Focusing on reactivity platform, the remarkable chemical diversity realized with soluble polymer-bound metal catalysts must be emphasized. Stereoselective oxidations and reductions are still predominant, but remarkable advances have been achieved in numerous asymmetric C–C forming processes such as addition of R_2Zn of carbonyls, nitroaldol condensation and nucleophilic allylic alkylation.

The combination of all these aspects makes the use of homogeneous covalently bound organometallic catalysts a reliable synthetic tool to efficiently address one of the demanding challenges of the modern organic synthetic chemistry: sustainability.[54]

Abbreviations

BINAP: 2,2′-bis(diphenylphosphino)-1,1′-binaphthalene
BINOL: 1,1′-bi-2-naphthol
dcp: dichlorodi(cyclopentadienyl) platinum(II)
DIOP: 4,5-bis(diphenylphosphinomethyl)-2,2-dimethyl-1,3-dioxolane
DPEN: diphenylethylenediamine
IPA: *i*-propyl alcohol
H_2salen: bis(salicylidene)ethylenediamine
HKR: hydrolytic kinetic resolution
MeO-PEG: poly(ethylene glycol) monomethyl ether
NCPS: non-cross-linked polystyrene
PDI: polydispersity index
PEG: poly(ethylene glycol)
PHOX: diphenylphosphinooxazolines
PyBOX: pyridine bis-oxazoline
ROMP: ring-opening metathesis polymerization
TADDOL: $\alpha,\alpha,\alpha',\alpha'$-tetraaryl-1,3-dioxolane-4,5-dimethanol
TBME: *t*-butyl methyl ether
TON: turn-over number

References

1. *Comprehensive Asymmetric Catalysis*, ed. E. N. Jacobean, T. Hayashi and A. Pfaltz, Springer, Berlin, 1999.
2. E. Framery, B. Andrioletti and M. Lemaire, *Tetrahedron: Asymmetry*, 2010, **21**, 1110.
3. R. Noyori, *Nat. Chem.*, 2009, **1**, 5.
4. W. Dumont, J. C. Poulin, T.-P. Dang and H. B. Kagan, *J. Am. Chem. Soc.*, 1973, **95**, 8295.
5. (a) H. U. Blaser, B. Pugin, F. Spindler and A. C. R. Togni, *Chim.*, 2002, **5**, 379; (b) W. J. Tang and X. M. Zhang, *Chem. Rev.*, 2003, **103**, 3029.
6. D. E. Bergbreiter, J. Tian and C. Hongfa, *Chem. Rev.*, 2009, **109**, 530.
7. For some comprehensive reviews covering early discovery see: (a) S. V. Ley, I. R. Baxendale, R. N. Bream, P. S. Jackson, A. G. Leach, D. A. Longbottom, M. Nesi, J. S. Scott, R. I. Storer and S. J. Taylor, *J. Chem. Soc., Perkin Trans. 1*, 2000, 3815; (b) D. E. Bergbreiter, *Chem. Rev.*, 2002, **102**, 3345; (c) C. A. McNamara, M. J. Dixon and M. Bradley, *Chem. Rev.*, 2002, **102**, 3275; (d) S. Bräse, F. Lauterwasser and R. E. Ziegert, *Adv. Synth. Catal.*, 2003, **345**, 869.
8. T. J. Dickerson, N. N. Reed and K. D. Janda, *Chem. Rev.*, 2002, **102**, 3325.
9. M. Benaglia, A. Puglisi and F. Cozzi, *Chem. Rev.*, 2003, **103**, 3401.

10. T. Chinnusamy, P. Hilgers and O. Reiser, in *Recoverable and Recycling Catalysts*, ed. M. Benaglia, John Wiley & Sons, Chichester, 2009, Chapter 4, p. 77.
11. (a) G. Desimoni, G. Faita and K. A. Jørgensen, *Chem. Rev.*, 2006, **106**, 3561; (b) R. Rasappan, D. Laventine and O. Reise, *Coord. Chem. Rev.*, 2008, **252**, 702.
12. H. A. McManus and P. J. Guiry, *Chem. Rev.*, 2004, **104**, 4151.
13. R. Annunziata, M. Benaglia, M. Cinquini, F. Cozzi and M. Pitillo, *J. Org. Chem.*, 2001, **66**, 3160.
14. M. Benaglia, M. Cinquini, F. Cozzi and G. Celentano, *Org. Biomol. Chem.*, 2004, **2**, 3401.
15. (a) C. Geiger, P. Kreitmeier and O. Reiser, *Adv. Synth. Catal.*, 2005, **347**, 249; (b) J. M. Fraile, J. I. García, C. I. Herrerías, J. A. Mayoral, O. Reiser and M. Vaultier, *Tetrahedron Lett.*, 2004, **45**, 6765.
16. M. Glos and O. Reiser, *Org. Lett.*, 2000, **2**, 2045.
17. (a) A. Gissibl, M. G. Finn and O. Reiser, *Org. Lett.*, 2005, **7**, 2325; (b) H. Werner, C. I. Herrerías, M. Glos, A. Gissibl, J. M. Fraile, I. Pérez, J. A. Mayoral and O. Reiser, *Adv. Synth. Catal.*, 2006, **348**, 125.
18. (a) E. N. Jacobsen, W. Zhang, A. R. Muci, J. R. Ecker and L. Deng, *J. Am. Chem. Soc.*, 1991, **113**, 7063; (b) T. P. Yoon and E. N. Jacobsen, *Science*, 2003, **299**, 1691.
19. T. S. Reger and K. D. Janda, *J. Am. Chem. Soc.*, 2000, **122**, 6929.
20. U. K. Anyanwu and D. Venkataraman, *Tetrahedron Lett.*, 2003, **44**, 6445.
21. U. K. Anyanwu and D. Venkataraman, *Green Chem.*, 2005, **7**, 424.
22. *Catalytic Asymmetric Synthesis*, ed. I. Ojima, 2nd edition, Wiley-VCH, New York, 2000.
23. (a) X. Li, W. Chen, W. Hems, F. King and J. Xiao, *Org. Lett.*, 2003, **5**, 4559; (b) X. Li, W. Chen, W. Hems, F. King and J. Xiao, *Tetrahedron Lett.*, 2004, **45**, 951.
24. (a) M. Bandini, M. Benaglia, T. Quinto, S. Tommasi and A. Umani-Ronchi, *Adv. Synth. Catal.*, 2006, **348**, 1521; (b) M. Bandini, M. Benaglia, R. Sinisi, S. Tommasi and A. Umani-Ronchi, *Org. Lett.*, 2007, **9**, 2151.
25. D. Zhao and K. Ding, *Org. Lett.*, 2003, **5**, 1349.
26. D. Zhao, J. Sun and K. Ding, *Chem. Eur. J.*, 2004, **10**, 5952.
27. (a) L.-T.Chai, W.-W. Wang, Q.-R. Wang and F.-G. Tao, *J. Mol. Catal. A: Chem.*, 2007, **270**, 83; (b) L. Chai, H. Chen, Z. Li, Q. Wang and F. Tao, *Synlett*, 2006, 2395.
28. J. Chen, G. Yang, H. Zhang and Z. Chen, *React. Funct. Polym.*, 2006, **66**, 1434.
29. I. D. Kostas, *Curr. Org. Chem.*, 2008, **5**, 227, and references therein.
30. S. Doherty, E. G. Robins, I. Pál, C. R. Newman, C. Hardacre, D. Rooney and D. A. Mooney, *Tetrahedron: Asymmetry*, 2003, **14**, 1517.
31. B. H. G. Swennenhuis, R. Chen, P. W. N. M. van Leeuwen, J. G. de Vries and P. C. J. Kamer, *Eur. J. Org. Chem.*, 2009, 5796.
32. X. Zheng, C. W. Jones and M. Weck, *Chem. Eur. J.*, 2006, **12**, 576.

33. D. Seebach, A. K. Beck and A. Heckel, *Angew. Chem., Int. Ed. Engl.*, 2001, **40**, 92.
34. S. Degni, S. Strandman, P. Laari, M. Nuopponen, C.-E. Wilén, H. Tenhu and A. Rosling, *React. Funct. Polym.*, 2005, **62**, 231.
35. J. Lu and P. H. Toy, *Chem. Rev.*, 2009, **109**, 815, and references therein.
36. N. Madhavan, C. W. Jones and M. Weck, *Acc. Chem. Res.*, 2008, **41**, 1153.
37. (a) M. Holbach and M. Weck, *J. Org. Chem.*, 2006, **71**, 1825; (b) N. Madhavan and M. Weck, *Adv. Synth. Catal.*, 2008, **350**, 419.
38. S. Bastin, R. J. Eaves, C. W. Edwards, O. Ichihara, M. Whittaker and M. Wills, *J. Org. Chem.*, 2004, **69**, 5405.
39. (a) W.-S. Huang, Q.-S. Hu and L. Pu, *J. Org. Chem.*, 1999, **64**, 7940; (b) H.-B. Yu, Q.-S. Hu and L, Pu, *Tetrahedron Lett.*, 2000, **41**, 1681.
40. H.-B. Yu, Q.-S. Hu and L. Pu, *J. Am. Chem. Soc.*, 2000, **122**, 6500.
41. Q.-H. Fan, G.-H. Liu, G.-J. Deng, X.-M. Chena and A. S. C. Chan, *Tetrahedron Lett.*, 2001, **42**, 9047.
42. X. Yao, H. Chen, W. Lü, G. Pan, X. Hu and Z. Zheng, *Tetrahedron Lett.*, 2000, **41**, 10267.
43. Y. Song, X. Yao, H. Chen, G. Pan, X. Hu and Z. Zheng, *J. Chem. Soc., Perkin Trans 1*, 2002, 870.
44. W. Huang, Y. Song, J. Wang, G. Cao and Z. Zheng, *Tetrahedron*, 2004, **60**, 10469.
45. N. U. H. Khan, S. Agrawal, R. I. Kureshy, S. H. R. Abdi, V. J. Mayani and R. V. Jasra, *Tetrahedron: Asymmetry*, 2006, **17**, 2659.
46. (a) R. I. Kureshy, S. Singh, N. H. Khan, S. H. R. Abdi, S. Agrawal and R. V. Jasra, *Tetrahedron: Asymmetry*, 2006, **17**, 1638; (b) R. I. Kureshy, K. J. Prathap, S. Singh, S. Agrawal, N. U. H. Khan, S. H. R. Abdi and R. V. Jasra, *Chirality*, 2007, **19**, 809.
47. (a) K. Pathak, A. P. Bhatt, S. H. R. Abdi, R. I. Kureshy, N. U. H. Khan, I. Ahmad and R. V. Jasra, *Chirality*, 2007, **19**, 82; (b) S. Jammi, L. Rout and T. Punniyamurthy, *Tetrahedron: Asymmetry*, 2007, **18**, 2016.
48. (a) R. I. Kureshy, I. Ahmad, K. Pathak, N. U. H. Khan, S. H. R. Abdi, J. K. Prathap and R. V. Jasra, *Chirality*, 2007, **19**, 352; (b) W. Sun, X. Wu and C. Xia, *Helv. Chim. Acta*, 2007, **90**, 623.
49. R. Tan, D. Yin, N. Yu, L. Tao, Z. Fu and D. Yin, *J. Mol. Catal. A: Chem.*, 2006, **259**, 125.
50. (a) X. Zheng, C. W. Jones and M. Weck, *J. Am. Chem. Soc.*, 2007, **129**, 1105; (b) S. Jain, X. Zheng, C. W. Jones, M. Weck and R. J. Davis, *Inorg. Chem.*, 2007, **46**, 8887.
51. Y.-Q. Kuang, S.-Y. Zhang and L.-L. Wei, *Synth. Commun.*, 2003, **33**, 3545.
52. M. S. DeClue and J. S. Siegel, *Org. Biomol. Chem.*, 2004, **2**, 2287.
53. For the sake of clarity, Siegel's study was described in this section even though the related polymer should be more properly addressed as an 'outer-sphere' type polymer.
54. B. Pugin and H.-U. Blaser, *Top. Catal.*, 2010, **53**, 953.

CHAPTER 5
Enantioselective Catalytic Dendrimers

ROBERTUS J. M. KLEIN GEBBINK AND
MORGANE A. N. VIRBOUL

Utrecht University, Faculty of Science, Department of Chemistry,
Organic Chemistry & Catalysis, Universiteitsweg 99, 3584 CG Utrecht,
The Netherlands

5.1   Introduction

Homogeneous catalysis is a very active area of research because of its applications in organic synthesis, bulk and fine chemicals production. In this area, efforts in ligand design and fine-tuning are being pursued to develop catalysts with improved catalytic performance, stability and selectivity. This phenomenon is even more marked in asymmetric catalysis where the use of efficient enantiose-lective catalysts is still increasing. Unfortunately, these optimized enantioselective catalysts are often expensive due to their sophisticated ligand and precious metal components. It is for this reason that the recovery and/or reuse of enantioselective catalysts are required in many cases to make them industrially attractive. In order to overcome the difficulties of recovering homogeneous catalysts, several methods have been developed, among which aqueous and fluorous biphasic catalysis, the use of ionic liquids and supercritical carbon dioxide as reaction medium, and catalyst immobilization on insoluble and soluble supports like dendrimers.[1-4]

Dendrimers are large macromolecules with well-defined spherical or globular architectures that offer the advantages of being recoverable by precipitation, nanofiltration or ultrafiltration. In addition, dendrimers display enhanced

RSC Green Chemistry No. 15
Enantioselective Homogeneous Supported Catalysis
Edited by Radovan Šebesta
© Royal Society of Chemistry 2012
Published by the Royal Society of Chemistry, www.rsc.org

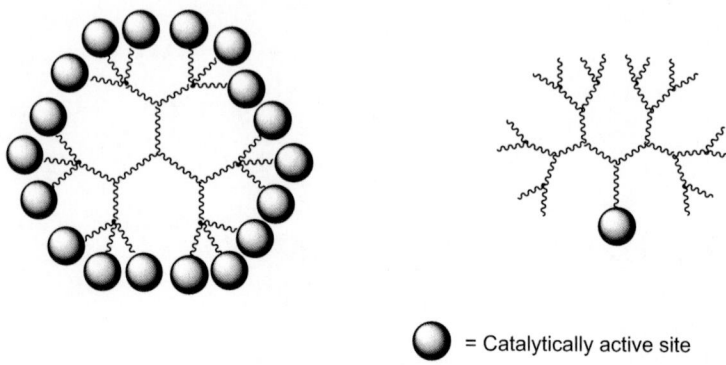

= Catalytically active site

Figure 5.1   Attachment of catalysts at the periphery or at the core of the dendrimer.

solubility profiles as compared to other polymeric supporting materials. Figure 5.1 provides a schematic representation of the two main types of metal complex attachment to dendrimers. Core-functionalized metallodendrimers (right) have a metal complex encapsulated at the centre of the dendrimer. Conversely, peripherally substituted systems (left) incorporate multiple metal species on the outer dendritic surface.

Anchoring a catalyst on a dendritic support is expected to leave the catalytic properties of the homogeneous catalysts unaltered and to potentially improve the activity by modulating the catalyst microenvironment. In this context, the type of dendrimer used for immobilization plays an important role as it can, by its intrinsic nature or its molecular geometry, have an impact on the activity of the catalyst.

In this chapter, we present an overview of the work on enantioselective homogeneous catalysts immobilized on dendrimers, where we compile all the typical examples from the literature.[5–7] The overview is organized in distinctive sections based on the position of the metal centre, either inside the dendrimer or on its surface. Next, the examples are classified according to the type of ligand that is used, where a difference is made between P-based, N-based and O-based ligands, as well as between monodentate and bidentate ligands. All examples are discussed in terms of catalytic activity, enantioselectivity and the comparison between non-immobilized and dendrimer-immobilized catalyst performance. Finally, the possibility to recover and reuse the dendrimer catalysts is discussed in terms of activity and stereoselectivity.

5.2   Dendrimer Functionalization at the Core

5.2.1   Phosphorus-based Ligands

5.2.1.1   Diphosphine Ligands

In 1994 Brunner and co-workers were the first to report the synthesis of a chiral core-functionalized dendrimer.[8] Brunner proposed that the structure of such expanded phosphines (Figure 5.2) would allow chirality to be induced to the catalyst's pocket thanks to the space-filling nature of the molecule and named these new molecules dendrizymes, as this concept is based on the resemblance to enzymatic systems.[9]

The activity of the phosphines was tested in the rhodium-catalyzed asymmetric hydrogenation of (α)-*N*-acetamidocinnamic acid (see equation in Figure 5.2). The substrate was efficiently reduced but no significant enantiomeric excess was observed. The use of these ligands in the hydrosilylation of acetophenone and the cyclopropanation of styrene with ethyl diazoacetate did not show chiral induction either. Even though the chirality-inducing effect of the 'dendrizymes' was disappointing, these studies represent the onset of many studies in the field of enantioselective dendrimer catalysis.

In a communication by Fan and Chan *et al.*,[10] the authors report the synthesis of (*R*)-BINAP ligand derivatives decorated with polyether dendrons (so-called Fréchet dendrons) of different generations. The different generations of Fréchet dendrons are thought to increase the steric bulk around the metal centre with increasing dendritic generation. The ligands 1–3 were used in the ruthenium-catalyzed asymmetric hydrogenation of 2-[*p*-(2-methylpropyl)phenyl]acrylic acid (Figure 5.3) and showed complete conversion after 24 h with good enantioselectivities. Remarkably the enantiomeric excess increased while going from the first to the second generation (ligand 1 to 2, 91.8 to 92.6% ee) but showed a slightly lower enantioselectivity when the third generation was used (ligand 3, 91.6% ee). The rate of the reaction increased as well with the size of the wedges, indicative of a positive dendritic effect. The catalysts were recycled by precipitation and reused without showing any decrease in activity or selectivity.

In a later report the same authors presented the synthesis of similar BINAP ligands 4–7 bearing an alkyl chain at the periphery to provide specific solubility properties to the catalyst.[11] The ligands were used in the ruthenium-catalyzed asymmetric hydrogenation of 2-phenylacrylic acid in an ethanol/hexane mixture that enabled an easy recovery of the catalyst by phase separation upon addition of water. The different catalytic systems displayed full substrate conversions after 4 h and good product enantioselectivities (84–91% ee). However, upon recycling the catalyst with ligand 5 showed a decreased reactivity already after the first run.

Ligands 1–3 were also used in the asymmetric hydrogenation of quinoline derivatives catalyzed by an iridium catalyst in a recent paper by Fan *et al.*[12] Full conversion and good enantiomeric excesses (85–90% ee) were obtained after 1.5 h. The rate of the reaction increased with the dendrimer generation, reaching a TOF for the most hindered ligand 3 never achieved before for this reaction (1580 h^{-1}). This strong dendritic effect could not be explained by the authors; however, they suggest that a shielding effect of the dendritic structure around the metal centre might contribute to the rate enhancement.

In 2006, Fan *et al.* reported the synthesis of axially chiral dendritic bisphosphines derived from the bridged biphenyl phosphine ligand BIPHEP (Figure 5.4).[13] The ligands 8–10 were tested in the ruthenium-catalyzed

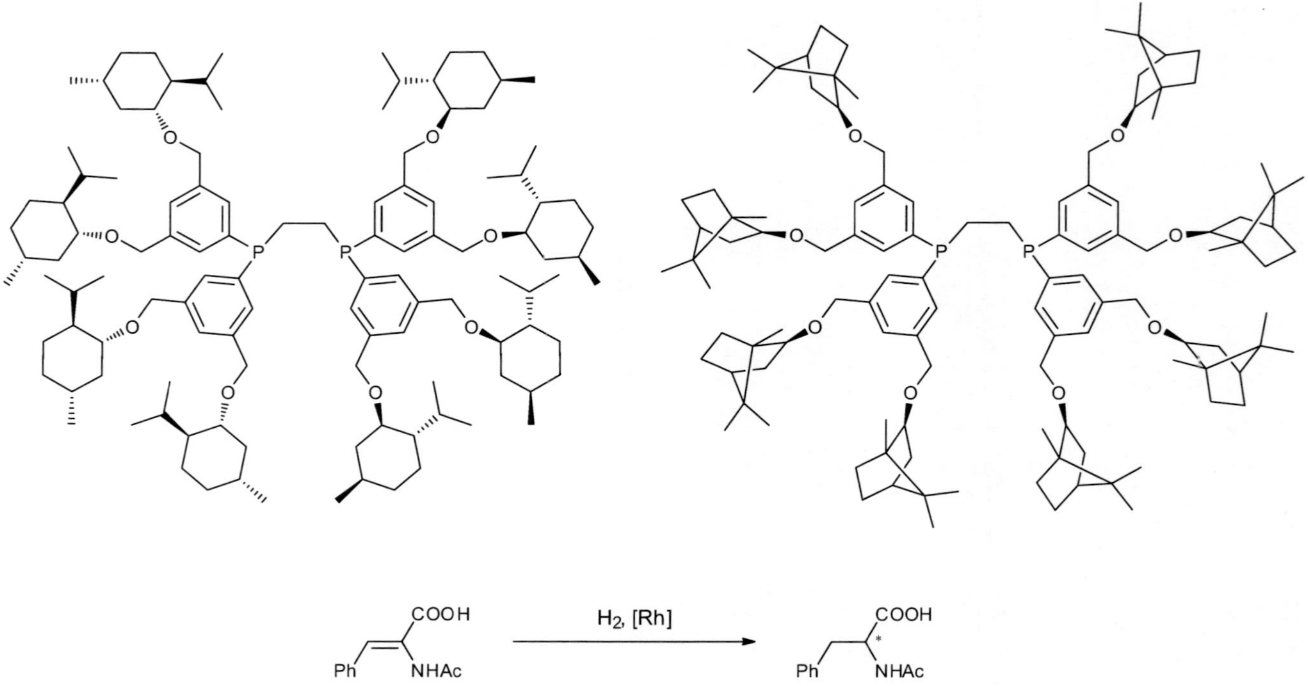

Figure 5.2 Brunner's landmark example of dendritic phosphines.

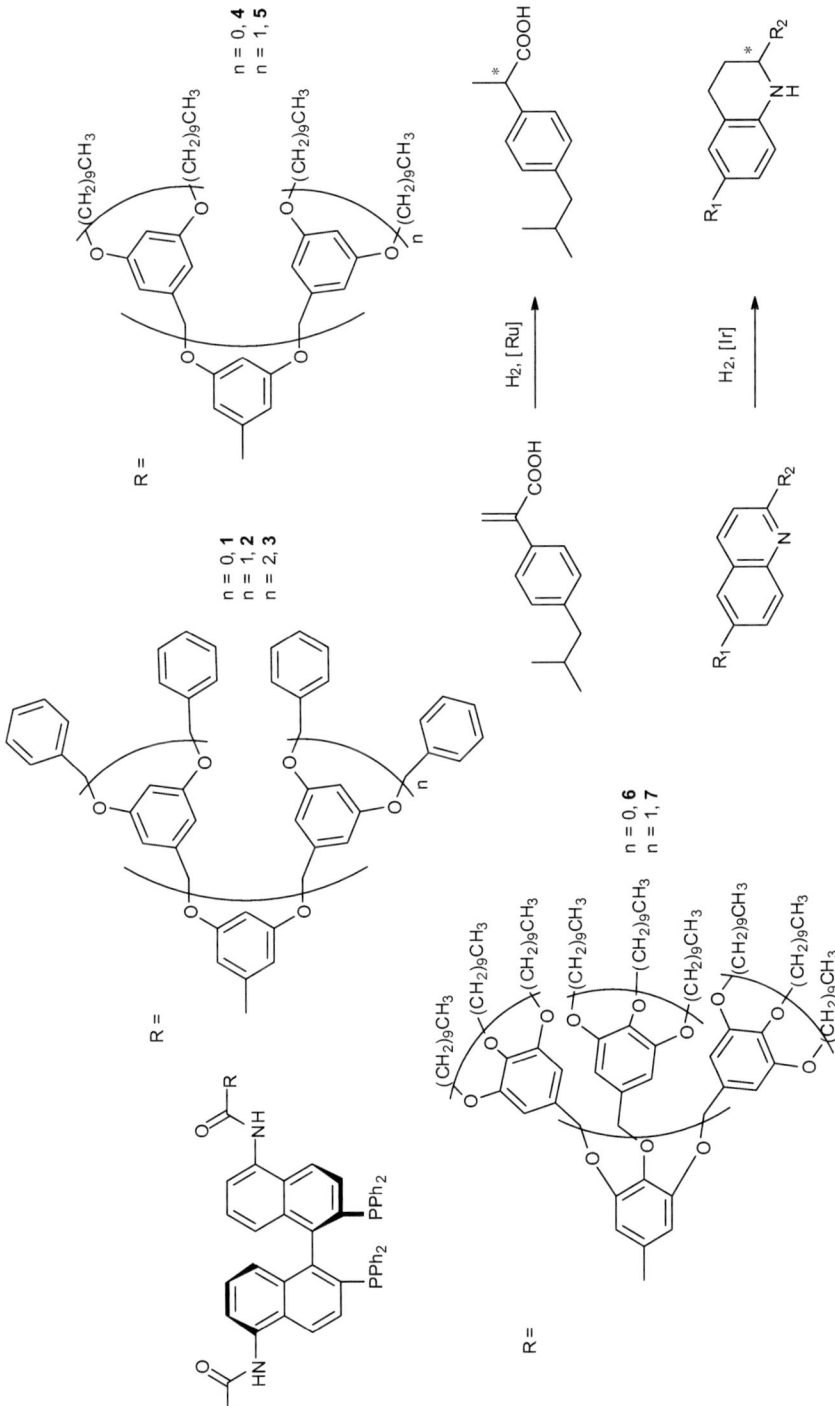

Figure 5.3 Dendritic BINAP ligands **1–7** developed by Fan's group.

hydrogenation of β-keto esters, in order to check the influence of the dendrimer generation on the catalyst activity. It was found that the size of the dendrimer had a major impact on the enantioselectivity, with the enantiomeric excess decreasing with dendrimer generation. At the same time the catalytic activity did not seem to suffer from the increased steric bulk. The authors could correlate this 'dendritic effect' with the dihedral angle of the ligands, as it is known that a larger dihedral angle tends to give reduced enantioselectivity.

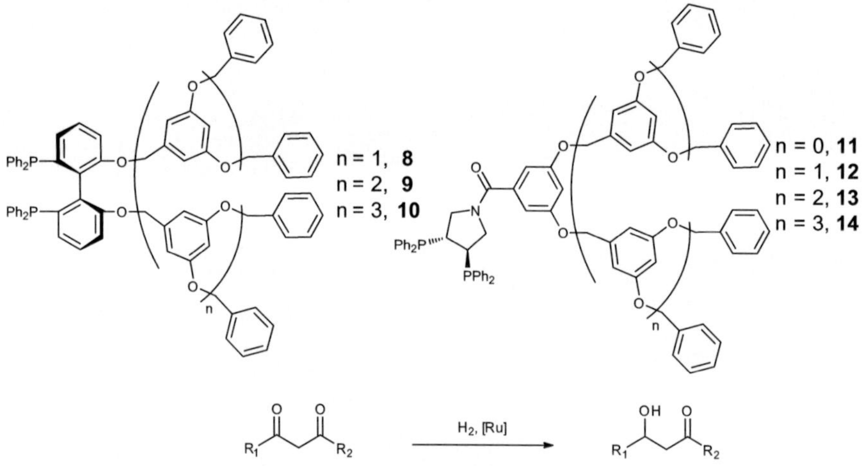

Figure 5.4 Other chiral bisphosphine dendritic ligands.

Fan and Chan also reported the synthesis of pyrphos ligands **11–14** modified at the focal point with dendritic Fréchet wedges (Figure 5.4).[14] The activity of the different generations of these chiral bisphosphine ligands was investigated in the rhodium-catalyzed asymmetric hydrogenation of α-acetamidocinnamic acid (reaction shown in Figure 5.2). When the reaction was performed in methanol/toluene, a clear decrease in reactivity was observed upon increasing the dendrimer generation (from 91 to 79% conversion), however without loss of enantioselectivity (96.9% ee). This effect was particularly flagrant when ligand **14** was used (only 20% conversion was observed), suggesting a possible encapsulation of the active site and thus a more difficult diffusion of the substrate to the active site. To further assess this hypothesis, a series of dendrimers was synthesized with a more congested catalytic centre, which displayed an even more decreased reactivity as well as a decreased enantioselectivity (from 96.9 to 94.6% ee). The recyclability of ligand **13** showed a rapid decrease of reactivity without loss of enantioselectivity. In an extension of this work, Yi and co-workers reported the synthesis of dendritic ligands similar to **11–14**, decorated with alkyl chains at the periphery.[15] The recovery of these compounds was greatly improved compared to ligand **13** and the activity started to decrease only after the fourth cycle.

5.2.1.2 Monophosphine Ligands

In 2008, Seco *et al.* published a report on the synthesis of P-stereogenic dendritic phosphines and their catalytic application in the palladium-catalyzed asymmetric hydrovinylation of styrene (Figure 5.5).[16] The authors were expecting an influence of the specific catalytic environment on the activity and stereoselectivity of the catalyst depending on the dendrimer generation. The steric congestion around the metal centre induced by the carbosilane dendrons is indeed believed to enhance the chiral induction by restricting the access to the metal centre. The steric bulk around the metal centre was first evidenced by the formation of an allyl palladium complex displaying diastereotopic protons in ^{1}H NMR, indicative of a high steric hindrance. The catalytic results reflect that increasing the dendrimer generation has a negative influence on the stereoselectivity, the best ee values being obtained with the least hindered ligand **15** (83%), ligand **16** yielding up to 82% ee, and only 73% ee for **17**. The TOF steadily decreased while increasing the steric bulk around the metal centre, suggesting a negative dendritic effect on the activity of the catalyst.

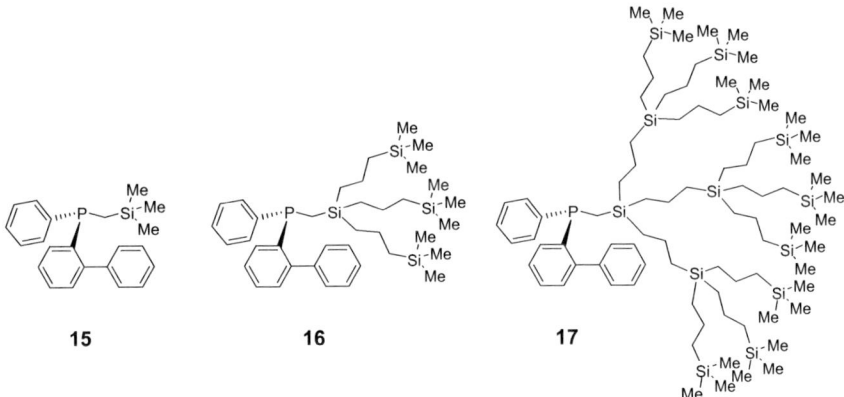

Figure 5.5 Chiral monophosphines **15–17** developed by Seco.

In a report by Klein Gebbink and co-workers, a new approach was investigated for the synthesis of chiral dendritic phosphine ligands.[17] In this case the chirality was induced by the presence of chiral Δ-TRISPHAT anions acting as counter ions for the six permanent positive charges present in the structure of the phosphine ligand (Figure 5.6). The influence of the steric bulk of ligand **18** on the activity and regioselectivity in the rhodium-catalyzed hydroformylation of styrene was investigated with different dendrimer generations, evidencing a decrease in activity and regioselectivity in comparison to PPh$_3$ (*b:l* ratio is 10 versus 21 for PPh$_3$). The stereoselectivity of **18** was also studied in the hydrogenation of dimethyl itaconate. In both reactions no significant enantiomeric excess was observed. The authors suggested that the chiral auxiliaries are

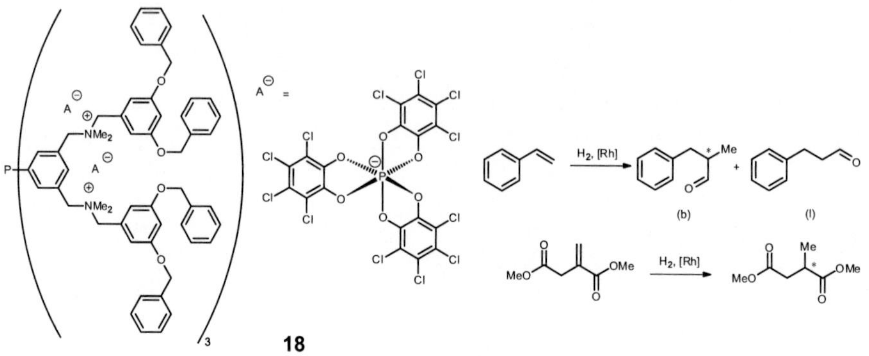

Figure 5.6   Chiral Dendriphos ligand **18** developed by Klein Gebbink.

located too far away from the metal centre and that long-range effects are not occurring even when tight ion pairing is favoured in CH_2Cl_2 solution.

5.2.1.3   Phosphite and Phosphoramidite-based Ligands

In 2008, Fan *et al.* reported on the synthesis of chiral monophosphite ligands **19–24** assembled by means of complementary hydrogen bonding between a dendritic Hamilton receptor and a barbituric acid derivative functionalized with a monophosphite moiety (Figure 5.7).[18]

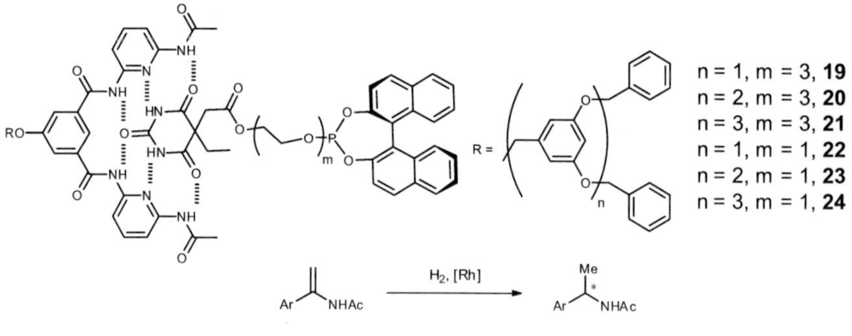

Figure 5.7   Fan's chiral dendritic phosphite ligand.

These dendritic ligands were employed in the rhodium-catalyzed asymmetric hydrogenation of α-phenylenamide and α-dehydroamino acid esters (see reaction in Figure 5.2). Interestingly, the enantioselectivity in these reactions improved with dendrimers of higher generation (from **19** (80% ee) to **20** (82% ee) and from **20** to **21** (86% ee)). On the other hand, ligands **22–24** bearing a shorter linkage displayed a reduced reactivity and enantioselectivity (88% conversion and 64% ee) compared to ligands **19–21** and the free ligand (100% yield, 93% ee) in the hydrogenation of α-phenylenamide. It is thought that this negative dendritic effect is induced by the increased steric bulk. The catalyst formed with ligand **21** was efficiently recycled and did not show deactivation even after five times.

Since the discovery that chiral monodentate ligands based on the BINOL backbone can be as efficient as the renowned bidentate BINAP ligands, a lot of efforts have been pursued in order to fine tune the selectivity and activity of this type of ligand, in particular in the development of chiral phosphoramidite ligands.

Ligands **25** and **26**, developed by Reek *et al.*,[19] were the first examples of chiral dendritic phosphoramidites ligands synthesized starting from the BICOL backbone (BICOL = bicarbazolediol, carbazole equivalent of BINOL), which can be readily functionalized with carbosilanes dendritic wedges. These two ligands were tested in the rhodium-catalyzed hydrogenation of methyl 2-acetamido cinnamate (see reaction in Figure 5.2) and compared with the very active and enantioselective MonoPhos ligand.[20] Ligand **25** showed a comparable reactivity to MonoPhos and the non-dendritic ligand **26**, indicating that the steric bulk has no negative effect on the reactivity of the catalysts. Likewise, the enantioselectivity in the reaction was not affected by the attachment of the ligand to the dendritic carbosilane scaffold (ee: **25**, 95%; **26**, 93%; MonoPhos, 95%). Shortly after the report by Reek, Fan *et al.* published several reports on the synthesis of chiral dendritic phosphoramidites ligands (compounds **27–37**, Figure 5.8) and their use in the rhodium-catalyzed asymmetric hydrogenation of prochiral alkenes (methyl 2-acetamido cinnamate, enamides and dimethyl itaconate, see reactions in Figures 5.2, 5.7 and 5.6, respectively) and palladium-catalyzed hydrosilylation of styrene (Figure 5.8).[21–23] The approach developed involves a functionalization of the phosphoramidites with Fréchet wedges on the nitrogen atoms, despite the established fact that the substituents on the nitrogen play an important role in the enantioselectivity. In the hydrogenation of methyl 2-acetamido cinnamate, ligands **27–29** exhibited a good reactivity and an even higher enantioselectivity than with the MonoPhos ligand (97.5% versus 95% ee). On the other hand, a prolonged reaction time and higher H_2 pressure than with MonoPhos or **25** were required for **27–29**. Hydrogenation of α-dehydroamino acid esters and dimethyl itaconate gave satisfying results with similar or better enantioselectivities than MonoPhos (ee: 97.0–97.7% versus 93.6%), with no evidence of the influence of the dendritic wedges on the activity of the catalyst. The modification of the chiral backbone in ligands **30–37** had no significant influence on the enantioselectivity in the hydrogenation of methyl 2-acetamido cinnamate with a rhodium catalyst. The authors could demonstrate though that a higher dendrimer generation also improved the stereoselectivity of the catalyst with other α-dehydroamino acid esters and enamides. Recycling experiments with ligand **32** showed a good recyclability of the catalyst by precipitation without loss of reactivity up to the fifth run. Ligands **30**, **32**, **35** and 3,3′-substituted derivatives were also tested in the hydrosilylation of styrene.[23] Ligand **30** exhibited a moderate reactivity and enantioselectivity (20% conversion, 11% ee), while the introduction of steric bulk by substitution on the 3,3′ position, as in **35**, increased the reactivity (>95%

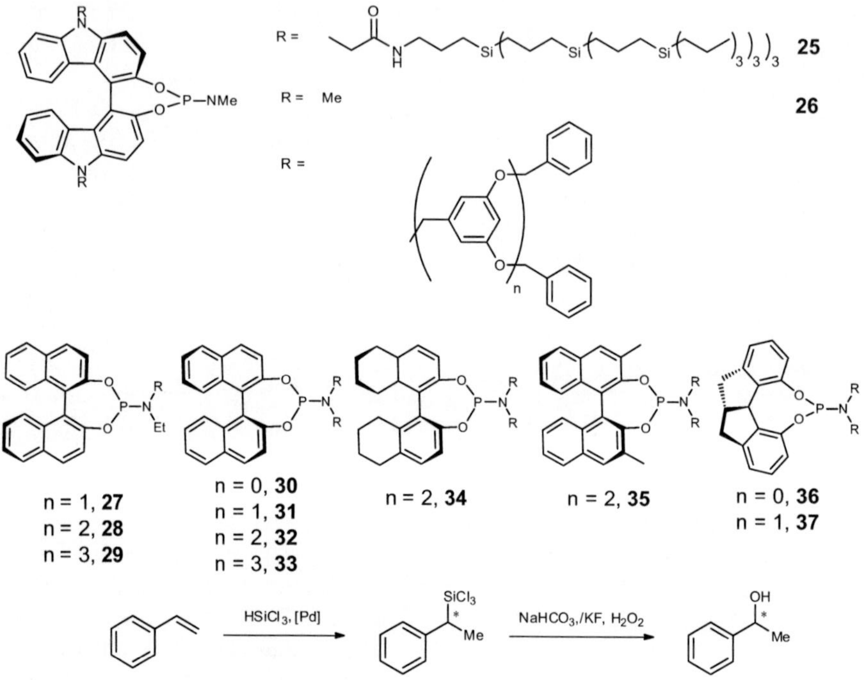

Figure 5.8 Dendritic phosphoramidite ligands developed by Reek and Fan.

conversion, 43% ee), in particular with bulkier substituents like phenyl, naphthyl or phenanthryl.

5.2.2 Nitrogen-based Ligands

5.2.2.1 Diamine Ligands

After the discovery by Noyori *et al.* of the outstanding properties of (S,S)-N-(p-tolylsulfonyl)-1,2-diphenylethylenediamine, (S,S)-TsDPEN, as ligand in the ruthenium-catalyzed asymmetric transfer hydrogenation (ATH),[24] efforts have been pursued towards the synthesis of recyclable dendritic analogues for application in asymmetric catalysis. In 2001, Deng and co-workers reported on the synthesis and application of chiral dendritic ligands **38–41**, which were synthesized by a three-step procedure involving the condensation of Fréchet type dendrons with amine-functionalized (S,S)-TsDPEN (Figure 5.9).[25] The activity of these different ligand generations was tested in the asymmetric transfer hydrogenation of acetophenone with [RuCl(p-cymene)]₂ as the metal precursor. No significant influence of the dendrimer generation on the activity and the stereoselectivity of the catalyst was observed compared to those of the monomeric ligand (*i.e.* 97% conversion in 20 h with 97.2% ee versus 99% conversion in 20 h and 97.2% ee). The catalysts formed with ligands **40** and **41** were recovered at the end of the reaction by precipitation and reused up to six times showing a decreasing reactivity, however, without loss of enantioselectivity up to the fifth run for **40** and to the sixth run for **41**. In an extension of this work, the same group reported on the synthesis of similar ligand structures and their use in ATH with prochiral ketones, imines and activated alkenes.[26] A good activity and enantioselectivity was exhibited by these dendritic ligands, again very similar to the monomeric ligand.

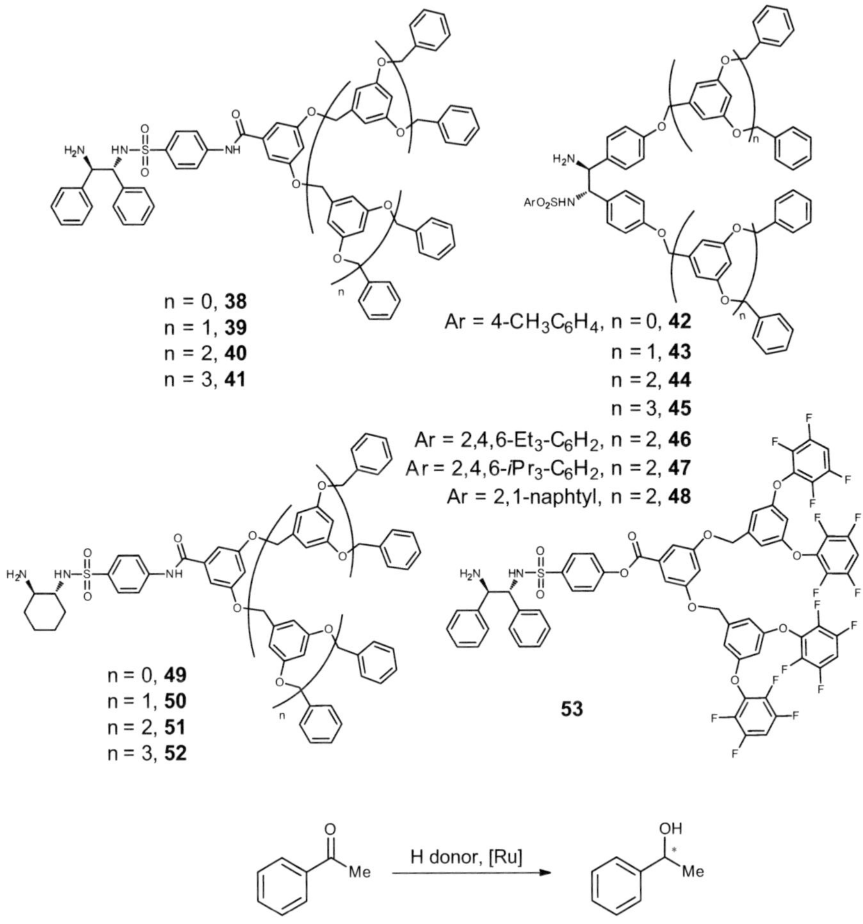

n = 0, **38**
n = 1, **39**
n = 2, **40**
n = 3, **41**

Ar = 4-CH$_3$C$_6$H$_4$, n = 0, **42**
n = 1, **43**
n = 2, **44**
n = 3, **45**
Ar = 2,4,6-Et$_3$-C$_6$H$_2$, n = 2, **46**
Ar = 2,4,6-iPr$_3$-C$_6$H$_2$, n = 2, **47**
Ar = 2,1-naphtyl, n = 2, **48**

n = 0, **49**
n = 1, **50**
n = 2, **51**
n = 3, **52**

53

Figure 5.9 Chiral dendritic diamine ligands.

The Deng group published two further consecutive reports on the development of dendritic ligands for catalytic ATH applications.[27,28] Ligands **42–48** were synthesized by introduction of the dendritic functionality on the phenyl rings of the 1,2-ethylenediamine as opposed to the amino-functionalized vicinal diamines ligands **38–41**.[25,26] The ligands were used in the ATH of the benchmark substrate acetophenone as well as of more challenging substrates like imines and alkenes. The authors noticed a slight influence of the dendrimer generation on the activity of the catalyst in the ATH of acetophenone when the reaction was performed in the presence of [RuCl(*p*-cymene)]$_2$. Ligand **44** showed a first drop in activity, which enhanced upon further increase of the steric bulk, *i.e.* the dendritic generation, however, without loss of enantios-electivity (96.1% ee).[28] The same observation was made when the reaction was performed with (*S*)-BINAP-RuCl$_2$, along with a loss of stereoselectivity though (82% ee).[27] The authors attribute this loss of reactivity to the structure of the ligands, which changes from an extended to a more globular conformation as the steric bulk induced by the dendritic wedge generation increases. In both reports, recyclability studies of the catalysts by precipitation gave satisfactory results and the catalysts could be recycled without loss of stereoselectivity up to five times (up to 94% ee), however, with a decrease in activity that can be ascribed to metal leaching as determined by ICP analysis.[28]

In their search for chiral catalysts with enhanced recyclability, the group of Deng also developed dendritic ligands **49–52** comprising a chiral 1,2-diaminocyclohexane core, which were synthesized in a similar fashion as ligands **38–41**.[29] The ligands were tested in the ruthenium- and rhodium-catalyzed ATH of prochiral ketones and showed a decreased reactivity under various reaction conditions when the steric ligand **52** was used. No influence of the dendritic generation was observed; the ligands **49–51** performing equally well as the monomeric ligand both in terms of activity and enantioselectivity, with conversions superior to 99% and ee values ranging from 85 to 96% depending on the conditions used for the reaction. The ligands also performed well when the reaction was performed in water with either a ruthenium or rhodium metal precursor. The enantioselectivity using the rhodium metal precursor was higher under these conditions than with ruthenium (96% ee versus 88% ee). The recyclability of ligand **50** by precipitation from water with hexane showed remarkable results as the catalyst could be reused up to six times in the rhodium-catalyzed ATH without loss of stereoselectivity, while only a slight decrease in reactivity was observed. Interestingly, the authors mention that the recyclability of **50** in organic solvents compared to that of **40** or **41** is very limited, the authors noted that the reactivity dropped drastically after the second use with a conversion of 46% (the first use gave a conversion of 99%) and only 7% after the third use.

In 2010, Wang and co-workers reported on the synthesis of a fluorinated TsDPEN-derived ligand (**53**) and its application in the ruthenium-catalyzed ATH of prochiral ketones in an aqueous medium.[30] Ligand **53**, which was synthesized in a six-step procedure with an overall yield of 65%, exhibited a good catalytic activity and enantioselectivity (93% ee) in water. The presence of tetrabutylammonium iodide further improved the enantioselectivity of the reaction to 97% ee. By a precipitation method, the catalyst could be recovered at the end of the reaction and reused up to an unprecedented 26 times without loss of activity and stereoselectivity. The authors suggested that this exceptional stability of the catalyst (no metal leaching was observed by ICP analysis) can be ascribed to the introduction of fluorine atoms on the ligand, thus conferring robustness to the ligand and to the catalyst.

5.2.2.2 Proline-based Ligands

Proline-based ligands represent another important type of easily accessible chiral ligands that have been well studied for their application in asymmetric catalysis. Their derivatization with dendritic wedges has been investigated by several groups, in particular by Zhao and co-workers. In 2005, they reported on the synthesis of ligands **54–57** and their application in the enantioselective addition of organozinc reagents to aldehydes (Figure 5.10).[31] This topic had already been studied by Bolm *et al.* in 1996 using a chiral pyridyl alcohol ligand decorated with Fréchet dendrons.[32] However, no effect of the dendrimer on the activity of the catalyst was demonstrated and a slight decrease of stereo-selectivity of the dendritic ligands compared to the monomeric pyridyl alcohol was found. With the proline-based ligands, very good activities and stereo-selectivities were obtained for the reaction of *p*-chlorobenzaldehyde with Et_2Zn in the presence of 20 mol% ligand. Ligands **54–56** gave better ee values than the monomeric species (98% ee versus 94% ee), except for the highest generation dendrimer **57** that exhibited a slightly reduced enantioselectivity (91% ee). Ligand **56** was recovered by precipitation at the end of the reaction and was recycled at least five times without a decrease in activity and stereoselectivity. Substituted benzaldehydes were efficiently transformed into the corresponding diaryl alcohol with the use of the dendritic proline ligands. Aliphatic aldehydes, on the other hand, were converted with lower efficiency and lower selectivity (77% conversion and 65% ee).

The use of ligands **54–57** and **58–59** as asymmetric organocatalysts was fur-ther investigated in the enantioselective reduction of ketones,[33] the enantiose-lective epoxidation of enones,[34] and the asymmetric reduction of indolones and tetralones.[35] The modified counterparts **60–63** were used in the organocatalytic asymmetric Michael addition of aldehydes to nitrostyrenes[36] and in a tandem cyclopropanation/Wittig reaction of α,β-unsaturated aldehydes with arsonium ylides.[37] These ligands showed a great versatility in their application and in all cases exhibited good activities and stereoselectivities (78–99% ee). Recovery and reuse of these ligands by means of precipitation was possible without loss of activity in up to five consecutive runs. The synthesis of pyrrolidine-derived ligands **64–66** decorated with Fréchet dendrons *via* click chemistry was reported by Gao and co-workers.[38] These organocatalysts were tested in the Michael addition of ketones to nitroolefins and exhibited a good activity (up to 99% conversion) and stereoselectivity (up to 95% ee) as well as a good recyclability with a little loss of reactivity after six runs (80% conversion and 90% ee).

In 2006, Zhao *et al.* reported the synthesis of proline-derived ligands **67–69** functionalized with dendritic wedges and their application as organocatalysts in the asymmetric direct aldol reaction in water (Figure 5.10).[39] The authors hypothesized that these chiral amphiphilic ligands would assemble in water

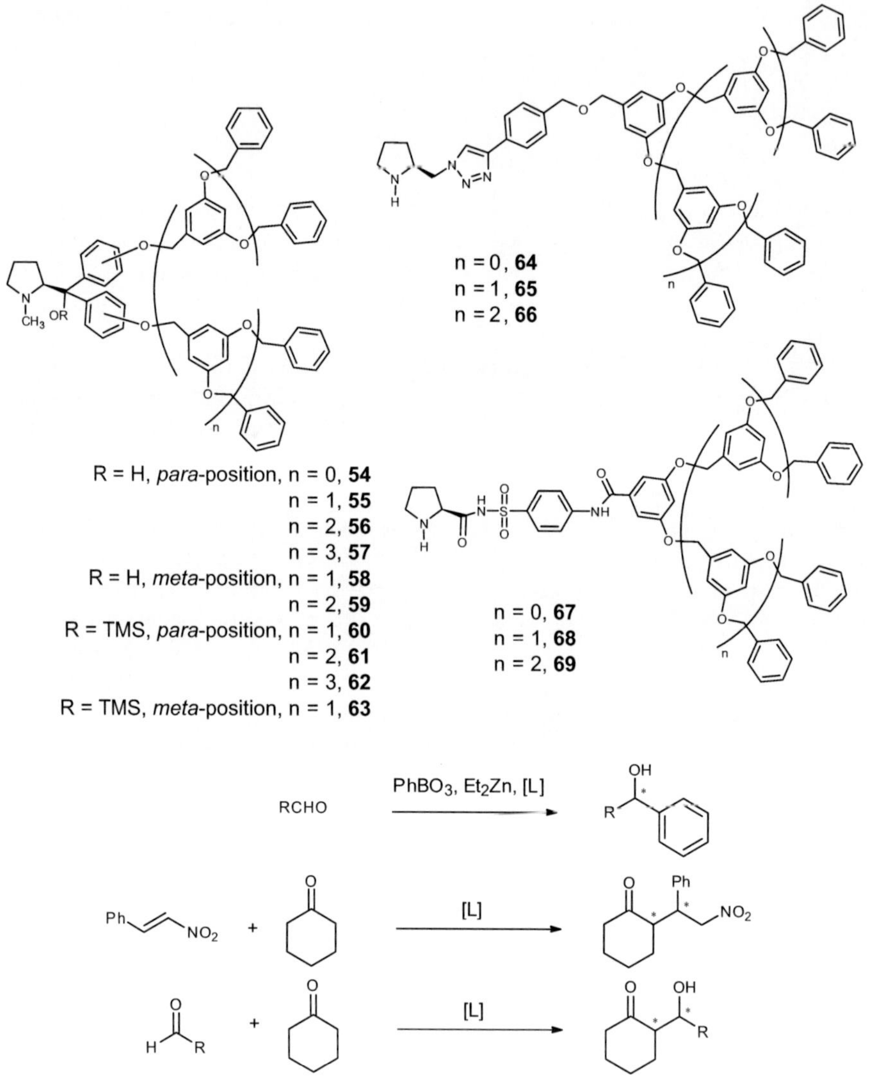

Figure 5.10 Chiral proline-derived dendritic ligands **54–69**.

with the hydrophobic reagents, keeping the reaction site away from water and thus enabling a high asymmetric induction. The aldol reaction proceeded as expected, yielding the product in good yields with high stereo- and enantios-electivity, especially with ligand **68** (*anti/syn* 99:1, 99% ee). This same ligand was used in a recycling test and could be recycled by precipitation up to four times without a decrease in activity and stereoselectivity.

In a similar approach, Chow *et al.* synthesized three series of proline-derived chiral dendritic organocatalysts for the application in aqueous asymmetric catalysis (Figure 5.11).[40] The functionalization of the proline moiety with hydrophobic hydrocarbon dendrons was expected to induce the formation of emulsions in water and to enhance the reactivity and selectivity during the catalytic reaction. The authors found that the properties of these compounds are indeed mainly due to their ability to form emulsions in water and that compounds **71**, **74** and **77** are best suited to catalyze asymmetric aldol reactions and nitro-Michael additions, for which they exhibited a good reactivity (78–87% conversion), diastereoselectivity (*syn*/*anti* > 90:10) and enantioselectivity (80–84% ee). The authors also showed that the catalysts could be recovered by solvent partitioning with heptane/methanol in which little decrease in reactivity and selectivity was found until the fifth run.

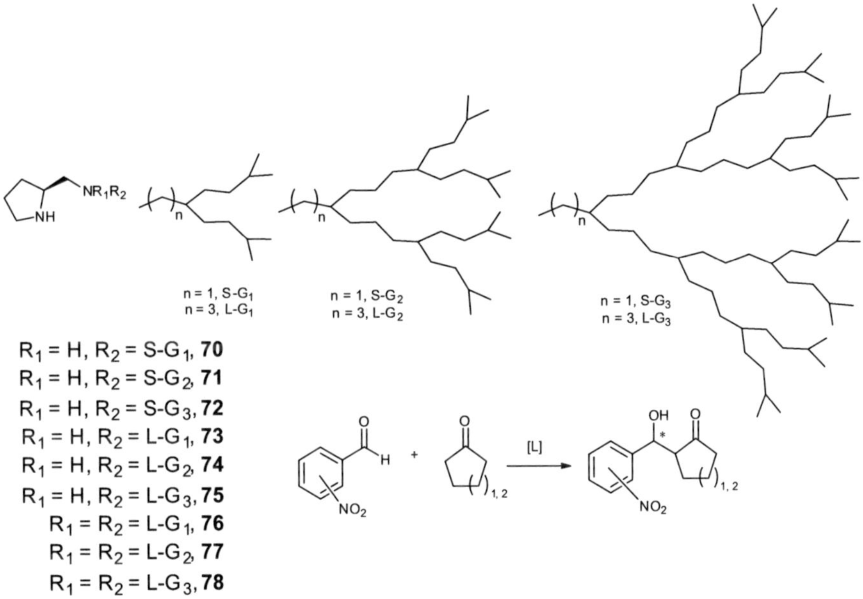
NR1R2
NH
n
n = 1, S-G1
n = 3, L-G1
n = 1, S-G2
n = 3, L-G2
n = 1, S-G3
n = 3, L-G3
R1 = H, R2 = S-G1, 70
R1 = H, R2 = S-G2, 71
R1 = H, R2 = S-G3, 72
R1 = H, R2 = L-G1, 73
R1 = H, R2 = L-G2, 74
R1 = H, R2 = L-G3, 75
R1 = R2 = L-G1, 76
R1 = R2 = L-G2, 77
R1 = R2 = L-G3, 78
O
H
+
NO2
1, 2
[L]
OH
*

Figure 5.11 Amphiphilic dendritic organocatalysts.

5.2.2.3 *Oxazoline-based Ligands*

In 2002, Moberg and co-workers investigated the functionalization of oxazolines with dendritic wedges.[41] Pyridinooxazolines and bisoxazolines were functionalized with achiral and chiral polyester dendrons to yield the ligands **79**–**89**, which were used in a palladium-catalyzed allylic alkylation (Figure 5.12). Ligands **84**–**89** showed a enantioselectivity similar to the parent ligand (76–80% versus 79% ee) and no specific influence of the dendritic wedges on the reaction was observed. On the other hand, ligands **79**–**83** showed a better enantioselectivity than their monomeric counterpart (94% versus 79% ee). The catalytic activity of the bulkier ligand **80** was quite low; *i.e.* only 10% of product was obtained after prolonged reaction time. Furthermore, the introduction of a chiral dendron on the oxazoline core moiety (ligands **81–82** and **87–88**) had no beneficial influence on the enantioselectivity (79% ee, as for the parent ligand).

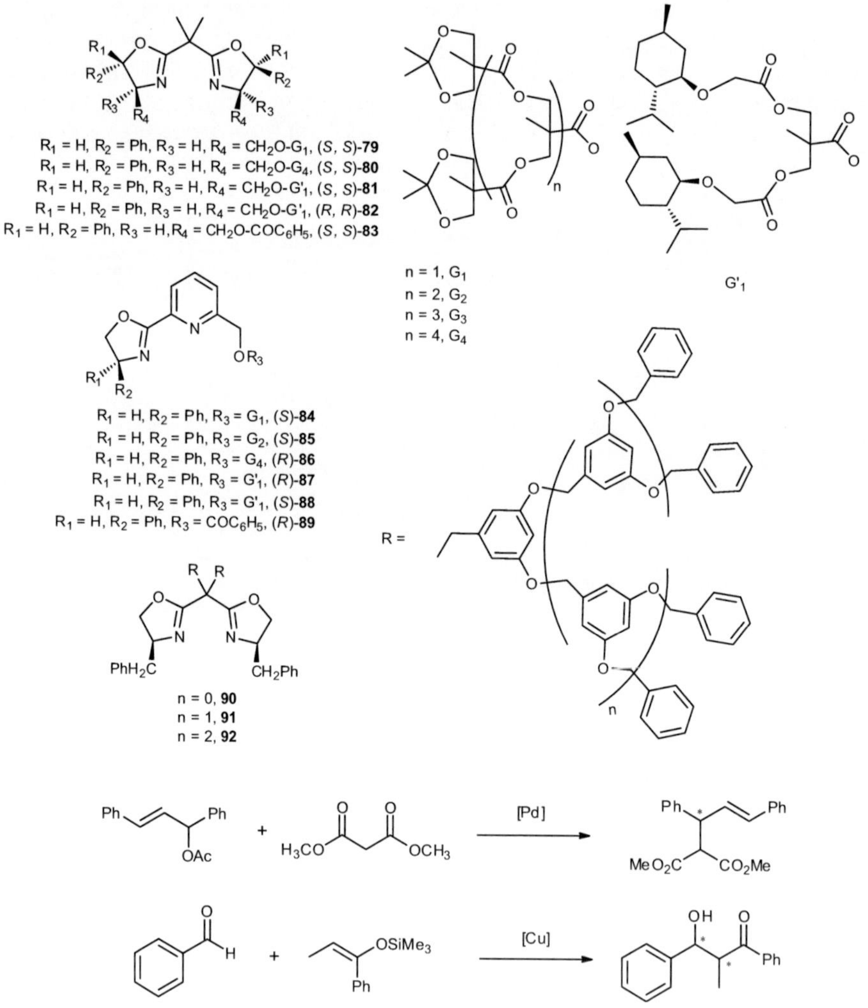

R$_1$ = H, R$_2$ = Ph, R$_3$ = H, R$_4$ = CH$_2$O-G$_1$, (S, S)-**79**
R$_1$ = H, R$_2$ = Ph, R$_3$ = H, R$_4$ = CH$_2$O-G$_4$, (S, S)-**80**
R$_1$ = H, R$_2$ = Ph, R$_3$ = H, R$_4$ = CH$_2$O-G'$_1$, (S, S)-**81**
R$_1$ = H, R$_2$ = Ph, R$_3$ = H, R$_4$ = CH$_2$O-G'$_1$, (R, R)-**82**
R$_1$ = H, R$_2$ = Ph, R$_3$ = H, R$_4$ = CH$_2$O-COC$_6$H$_5$, (S, S)-**83**

n = 1, G$_1$
n = 2, G$_2$
n = 3, G$_3$
n = 4, G$_4$

G'$_1$

R$_1$ = H, R$_2$ = Ph, R$_3$ = G$_1$, (S)-**84**
R$_1$ = H, R$_2$ = Ph, R$_3$ = G$_2$, (S)-**85**
R$_1$ = H, R$_2$ = Ph, R$_3$ = G$_4$, (R)-**86**
R$_1$ = H, R$_2$ = Ph, R$_3$ = G'$_1$, (R)-**87**
R$_1$ = H, R$_2$ = Ph, R$_3$ = G'$_1$, (S)-**88**
R$_1$ = H, R$_2$ = Ph, R$_3$ = COC$_6$H$_5$, (R)-**89**

R =

n = 0, **90**
n = 1, **91**
n = 2, **92**

Figure 5.12   Oxazoline-based dendritic ligands.

Fan *et al.* reported on the synthesis of the chiral bisoxazolines ligands **90–92** functionalized with different generations of polyether dendrons (Figure 5.12).[42] In combination with Cu(OTf)$_2$, these ligands were used as Lewis acid catalysts in the enantioselective aldol reaction of benzaldehyde with a silyl enol ether in aqueous solvent. Ligands **90–92** exhibited a good reactivity (75% conversion) but moderate stereoselectivity (*syn/anti* = 2.1/1 and 60% ee). A slight increase in product enantioselectivity was observed for a higher dendritic generation.

The catalysts were recovered at the end of the reaction by precipitation; however, the recycled catalyst gave lower yields (40%) and ee values (30% ee) compared to the freshly prepared catalysts. The same approach was pursued by Du and co-workers by the functionalization of a slightly different bisoxazoline core with Fréchet dendrons (not shown).[43] These ligands were applied in the asymmetric alkylation of indoles with nitroalkenes and showed an activity similar to the monomeric ligand (93–99% versus 96% conversion). The dendritic generation had no significant influence on the enantioselectivity but a slightly decreased reactivity was observed when the steric bulk was increased (93% conversion for the bulkier dendritic ligand versus 99% for the less hindered dendritic ligand).

5.2.3 Oxygen-based Ligands

5.2.3.1 TADDOL-derived Ligands

In two consecutive reports from 1999, Seebach *et al.* presented the synthesis of core-functionalized TADDOL dendrimers.[44,45] The TADDOL centre was decorated by four Fréchet dendrons in **93–97** and by four chiral polyether dendrons in **98** and **99** (Figure 5.13).

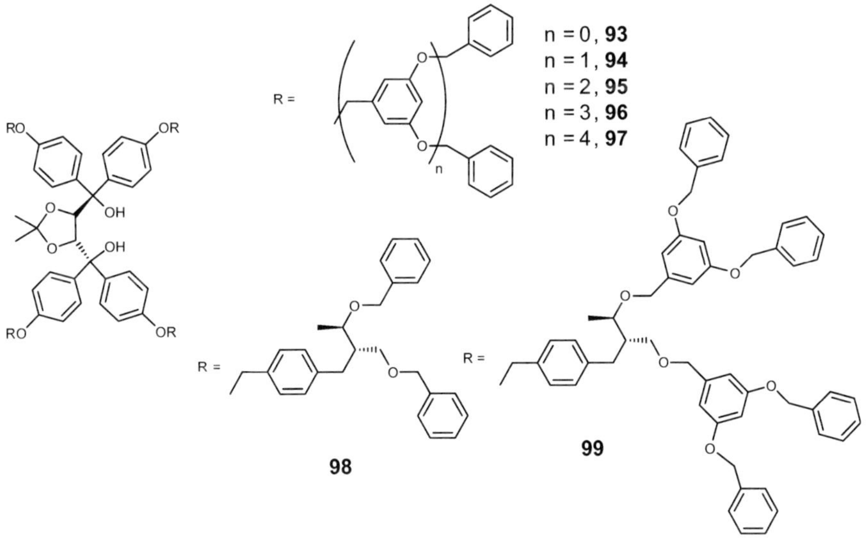

Figure 5.13   Seebach's TADDOL-derived dendrimers.

These dendrimers were used as ligands in the synthesis of titanium taddolates, which in turn were employed as catalysts in the asymmetric addition of Et$_2$Zn to benzaldehydes (see reaction in Figure 5.10). Compounds **93–99** all showed a good stereoselectivity (89–97% ee) for this reaction, which was found to be comparable to the performance of the monomeric TADDOL

ligand (98% ee). Much like in several of the previous examples, the enantioselectivity dropped slightly upon increase of the dendrimer generation. This
observation was also true for the activity of the catalyst, for which a marked
decrease between the activity of **96** and **97** was observed (94% versus 47%
conversion, respectively). No influence of the chiral dendrons **98–99** on the
stereoselectivity was demonstrated.

5.2.3.2 BINOL-derived Ligands

In 1998, Yamago and co-workers reported the synthesis of BINOL derivatives decorated with dendritic polyether dendrons (Figure 5.14).[46] Ligands
100–102 were used in the titanium-catalyzed allylation of aldehydes using
allyl stannane and showed a poor activity (18–36% conversion) for this
reaction, albeit with good enantioselectivities (88–92% ee) similar to what
was reported for BINOL (89% ee). This preliminary work on BINOL derivatives paved the way for other groups to further investigate the influence of
dendritic wedges on the activity of BINOL-based catalysts.[47,48] Ligands **100–
110** were tested in the titanium-catalyzed asymmetric addition of diethyl zinc
to benzaldehyde (see reaction in Figure 5.10). All dendritic catalysts were
found to be very active for this reaction (>99% conversion) giving 77 to
87% ee. Variations in dendron branching point (R_1/R_4 position versus R_2/R_3
position) or dendrimer size had limited to no influence on the overall catalytic
performances in these cases.

Ding *et al.* described the synthesis of the NOBIN-derived ligands **111–116**
bearing Fréchet dendrons (Figure 5.14).[49] The authors employed these dendritic ligands in titanium-mediated enantioselective hetero-Diels–Alder reaction (HDA) on Danishefsky's diene with aldehydes. The reactions proceeded
with high efficiency and good enantioselectivity (94–97% ee), though the best
results were obtained with ligands carrying Fréchet wedges that are branched
on the *meta* position (ligands **114–116**). For ligands **111–113**, the enantioselectivity was influenced by the increasing dendrimer generation, with the
enantioselectivity ranging from 92% for **111** to 75% ee for **113**. The authors
also showed that catalyst **115** could be recycled by precipitation and reused in
up to three catalytic runs with a slightly decreasing reactivity (99 to 90%
conversions for the first to the third run).

5.2.4 Other Functionalizations

The groups of Van Koten and Najera described the synthesis of cinchoni-
dine-derived ammonium salts and their application as phase-transfer catalysts
(Figure 5.15).[50] The different ligands **117–123** were used as phase-transfer
catalysts in the biphasic alkylation of *N*-(diphenylmethylene)glycine isopropyl
ester with benzyl bromide. A study on the effect of the different generations
of dendrimers on the catalytic activity showed no correlation between the
dendrimer size and the enantioselectivity. Overall reaction rates were found

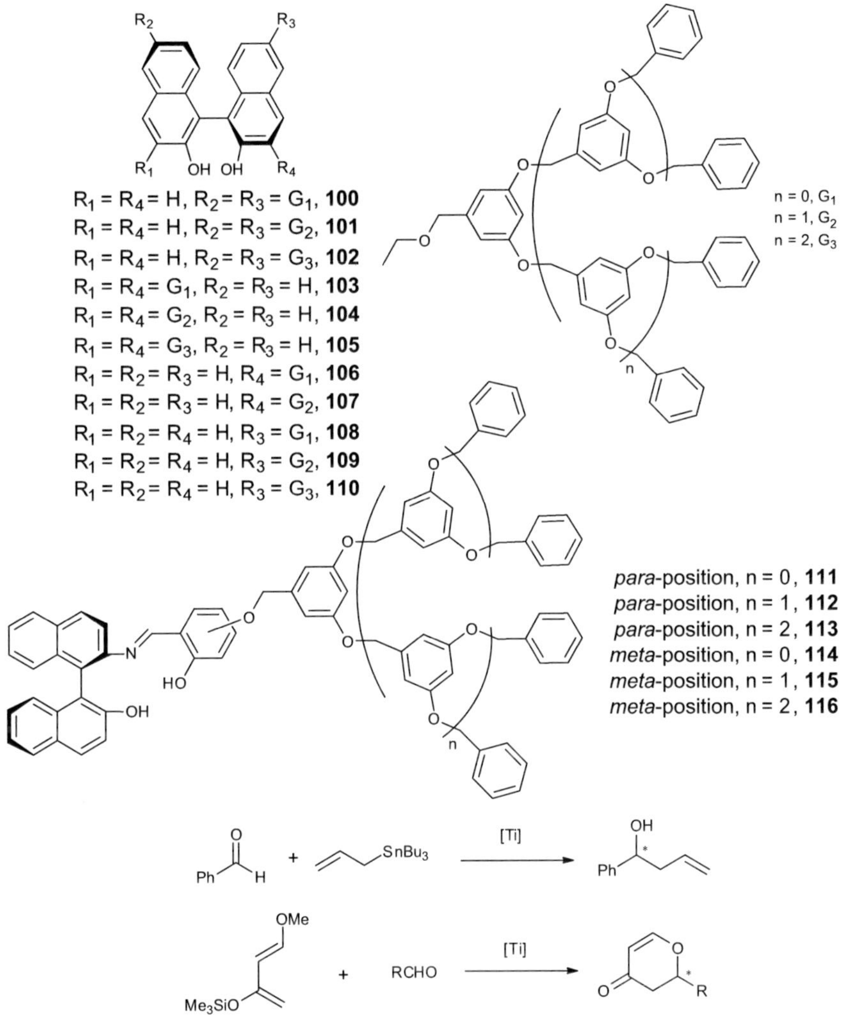

R₁ = R₄ = H, R₂ = R₃ = G₁, **100**
R₁ = R₄ = H, R₂ = R₃ = G₂, **101**
R₁ = R₄ = H, R₂ = R₃ = G₃, **102**
R₁ = R₄ = G₁, R₂ = R₃ = H, **103**
R₁ = R₄ = G₂, R₂ = R₃ = H, **104**
R₁ = R₄ = G₃, R₂ = R₃ = H, **105**
R₁ = R₂ = R₃ = H, R₄ = G₁, **106**
R₁ = R₂ = R₃ = H, R₄ = G₂, **107**
R₁ = R₂ = R₄ = H, R₃ = G₁, **108**
R₁ = R₂ = R₄ = H, R₃ = G₂, **109**
R₁ = R₂ = R₄ = H, R₃ = G₃, **110**

n = 0, G₁
n = 1, G₂
n = 2, G₃

para-position, n = 0, **111**
para-position, n = 1, **112**
para-position, n = 2, **113**
meta-position, n = 0, **114**
meta-position, n = 1, **115**
meta-position, n = 2, **116**

Figure 5.14   BINOL and NOBIN-derivatived dendritic ligands.

to range between 44 and 76% ee without an apparent trend. The recyclability of compounds **119** and **121** were tested by performing the reaction in a membrane dialysis tube, which was used as a 'tea bag' in which the reaction could take place and which could be easily transferred to a next reaction batch. Both compounds performed in a similar manner as the fresh catalyst for the first two rounds, while **119** showed a dramatic decrease in the third round with ee values going from 60 to 40% ee over a prolonged reaction time.

Recently, Nlate and co-workers reported on the synthesis of enantiopure polyoxometalates (POM) and their use as catalyst in the asymmetric sulfide

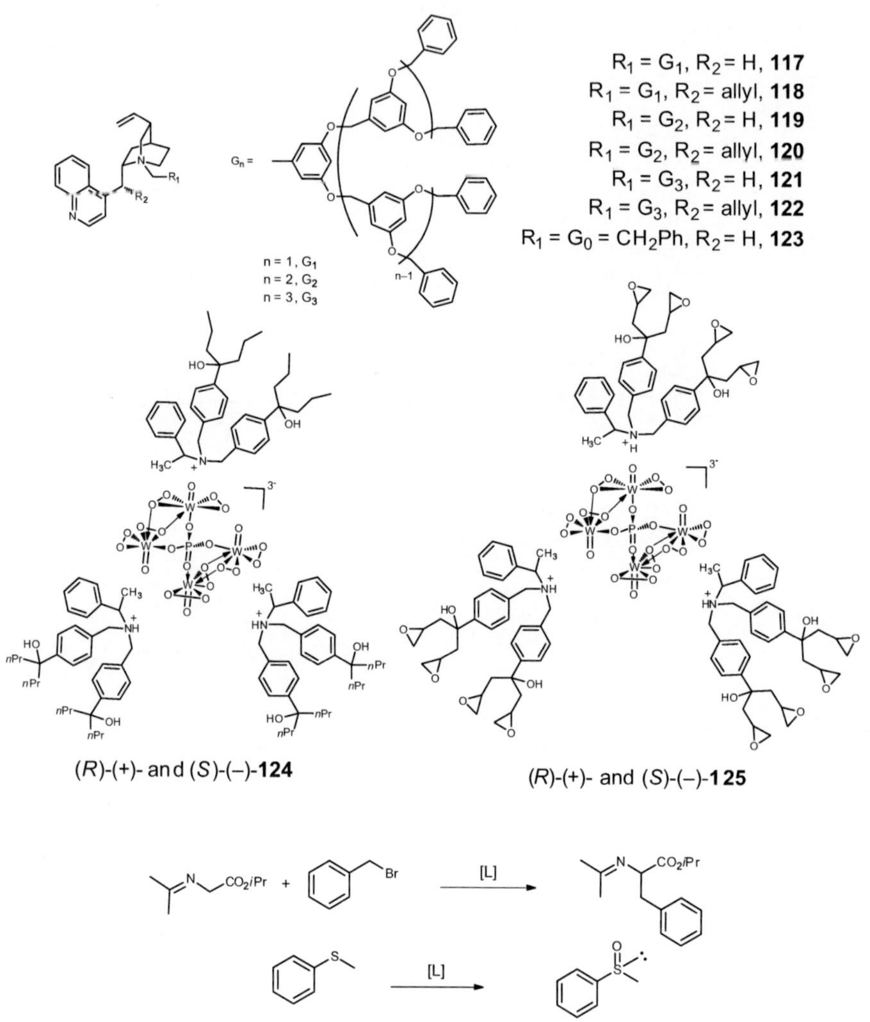

$R_1 = G_1, R_2 = H$, **117**
$R_1 = G_1, R_2 = $ allyl, **118**
$R_1 = G_2, R_2 = H$, **119**
$R_1 = G_2, R_2 = $ allyl, **120**
$R_1 = G_3, R_2 = H$, **121**
$R_1 = G_3, R_2 = $ allyl, **122**
$R_1 = G_0 = CH_2Ph, R_2 = H$, **123**

$n = 1, G_1$
$n = 2, G_2$
$n = 3, G_3$

(R)-(+)- and (S)-(−)-**124**

(R)-(+)- and (S)-(−)-**125**

Figure 5.15 Dendritic cinchonidines and dendritic POM salts.

oxidation (Figure 5.15).[51] Their original approach created chiral dendritic POMs through the interaction of three enantiopure ammonium ions with the achiral trianionic POM. The activity of compounds (R)-(+)-**125** and (S)-(+)-**125** was tested in the oxidation of thioanisole with H_2O_2, which resulted in the full oxidation to the corresponding sulfoxide with 14% ee. The POM catalysts could be recovered by precipitation with ether and were reused up to three times without any noticeable deactivation. Despite the low enantioselectivity of these catalysts, this work showed an unprecedented example of chirality transfer from an organic dendritic counter ion to the activity of a non-chiral catalyst.

5.3 Dendrimer Functionalization at the Periphery

5.3.1 Phosphorus-based Ligands

5.3.1.1 Diphosphine Ligands

In a number of papers, Togni and co-workers reported the immobilization of the chiral ferrocenyl-based diphosphine ligand Josiphos on the periphery of dendrimers (Figure 5.16).[52] This study represents one of the earlier studies on enantioselective catalysis that used peripherally functionalized dendrimers. Up to eight Josiphos ligands were immobilized on different core molecules *via* linkers that ensured sufficient flexibility of the bisphosphine moieties. These dendritic 'multi-ligands' were tested in the rhodium-catalyzed asymmetric hydrogenation of dimethyl itaconate (reaction shown in Figure 5.6). The parent Josiphos ligand is known to catalyze this reaction with good activity and enantioselectivity. The performances of compounds **126–129** were very similar to the monomeric Josiphos (98.7–98.0 versus 99% ee, respectively); the slight decrease in enantioselectivity was correlated to the increasing size of the

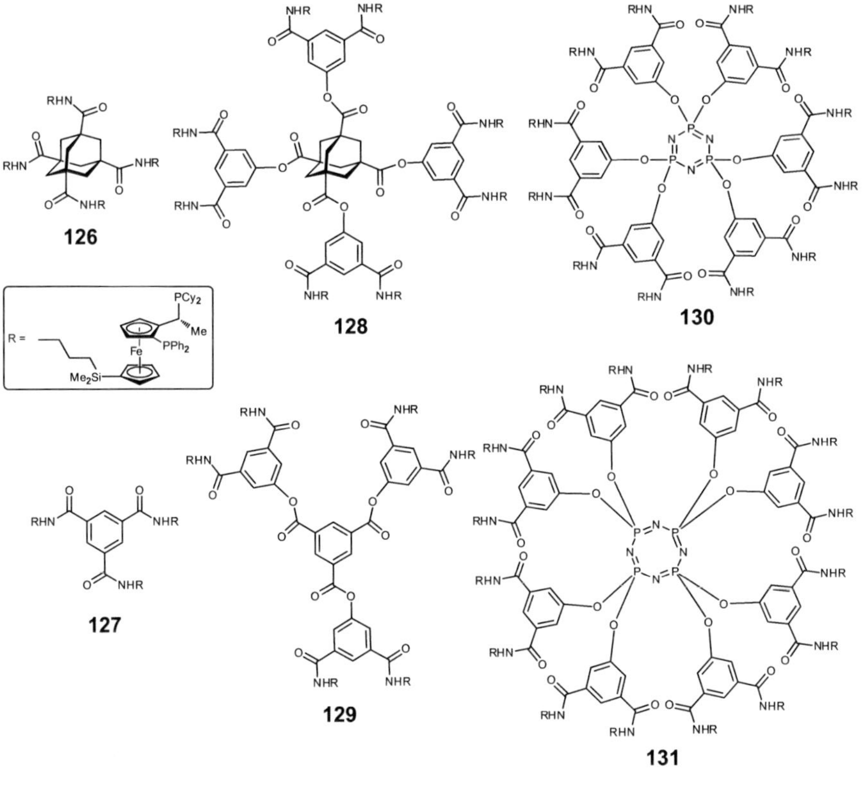

Figure 5.16 Josiphos dendrimers.

dendrimer. The recyclability of these catalysts was not investigated, although the authors demonstrated that a commercially available nano-filtration membrane was able to retain the dendrimers.

Next, the same group reported the synthesis of dendrimers **130–131** with different core molecules that could bear up to 16 Josiphos moieties.[53,54] These dendrimers were also tested in the rhodium-catalyzed asymmetric hydrogenation of dimethyl itaconate and showed a very similar performance as ligands **126–129** (the product was obtained with 98% ee), with no significant influence of the dendrimer generation on the performance of the catalyst. When the same dendrimers were used in the palladium-catalyzed substitution of allylic acetate with dimethyl malonate (85–92% yield, 85–91% ee) or the rhodium-catalyzed hydroboration of styrene (63–97% yield, 60–68% ee), no remarkable effect of the nature or generation of the dendrimer could be observed.

In a report from 2002, Gade *et al.* reported the synthesis of dendrimers bearing chiral diphosphine ligands on their peripheries (Figure 5.17).[55] Pyrphos-derived ligands were linked to the outer shell of different generations of poly-(propyleneimine) (PPI) dendrimers to form dendrimers **132–136** with up to 32 immobilized ligands. The different dendrimers were employed in the rhodium-catalyzed asymmetric hydrogenation of *Z*-methyl-α-acetamidocinnamate and dimethyl itaconate (reactions shown in Figures 5.2 and 5.6, respectively), for

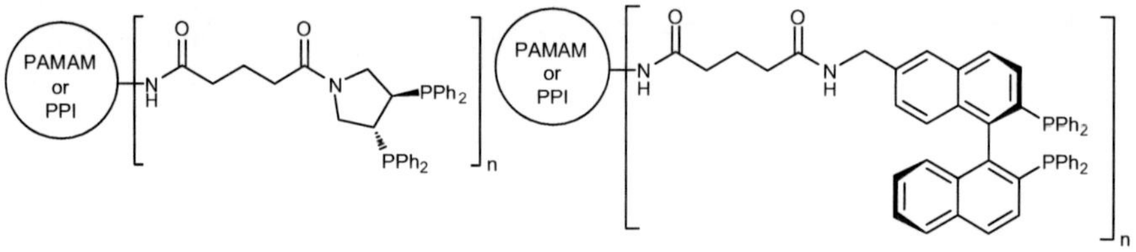

PPI: n = 4, **132**	PAMAM: n = 4, **137**	PPI: n = 4, **142**	PAMAM: n = 4, **147**
n = 8, **133**	n = 8, **138**	n = 8, **143**	n = 8, **148**
n = 16, **134**	n = 16, **139**	n = 16, **144**	n = 16, **149**
n = 32, **135**	n = 32, **140**	n = 32, **145**	n = 32, **150**
n = 64, **136**	n = 64, **141**	n = 64, **146**	n = 64, **151**

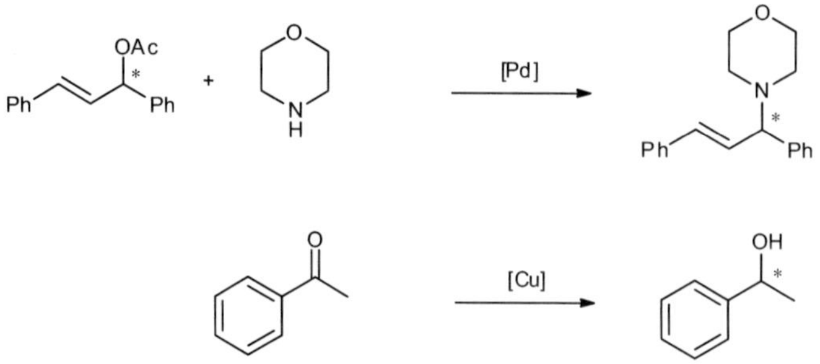

Figure 5.17 PPI and PAMAM dendrimer immobilized pyrphos (left) and BINAP (right) ligands.

which a clear relationship between the activity and the dendrimer size could be established. The stereoselectivity as well as the activity of the catalysts indeed showed a decrease with increasing dendrimer generation that was explained by the authors as the result of a potentially reduced accessibility of all metal centres. In an extension of this work, Gade *et al.* reported the synthesis of pyrphos ligands immobilized on the periphery of poly(amidoamine) dendrimers (PAMAM) (ligands 137–141) and their use in the palladium-catalyzed allylic amination of 1,3-diphenyl-1-acetoxypropene with morpholine.[56] The performance of these dendritic catalysts was compared to the monomeric pyrphos ligand and showed a dramatic improvement in the stereoselectivity, with up to 69% ee for the most selective catalyst versus 9% ee for the parent pyrphos. It was shown that the increasing stereoselectivity of the catalysts correlated with the higher dendrimer generation. Interestingly, this positive dendritic effect was also observed when PPI dendrimers 132–136 were used for this reaction, albeit that a lower extent was reached with 136 than with the higher generation 141 (40 versus 69% ee).

Despite their widespread application in homogeneous asymmetric catalysis, the first example of BINAP ligands immobilized at the periphery of a dendrimer was reported in 2008 by Gade *et al.*[57] The authors immobilized BINAP derivatives on PPI and PAMAM dendrimers of different generations in order to study the influence of the dendritic support on the catalytic performances of the compounds (Figure 5.17). The latter was studied in the copper-catalyzed hydrosilylation of acetophenone, where it was found that the compounds 142–151 performed similarly as the non-immobilized ligand with a slightly better enantioselectivity (90% ee for BINAP and 93–94% ee for ligands 142–151). No evidence for an influence of the type of support or the dendrimer generation was observed for this reaction, indicating that the conversion is controlled by the first coordination sphere around copper and that there is no mutual interaction between individual BINAP moieties that imparts the enantioselectivity.

5.3.1.2 Monophosphine Ligands

In a similar approach to the one developed by Togni,[53] Majoral and co-workers reported the synthesis of chiral ferrocenyl P,S ligands immobilized on dendrimers and their use as ligands in asymmetric catalysis (Figure 5.18).[58] The dendrimers 151–154 were used in the palladium-catalyzed allylic substitution reaction with dimethyl malonate (see reaction in Figure 5.12), where they showed high activities (87–95% yield) and enantioselectivities (81–93% ee), similar to the parent catalyst (96% yield and 93% ee) and independent of the dendrimer generation. A preliminary study on the recyclability of the dendrimer 154 by means of precipitation revealed a decrease in activity and stereoselectivity.

In 2005, the same group synthesized a new dendrimer functionalized with a P,N-iminophosphine ligand and investigated its use in the palladium-catalyzed allylic substitution, for which this ligand had previously been applied in a successful manner.[59] Dendrimer 155 showed a good activity and selectivity (87–97% yield and 84–95% ee, depending on the reaction conditions); however,

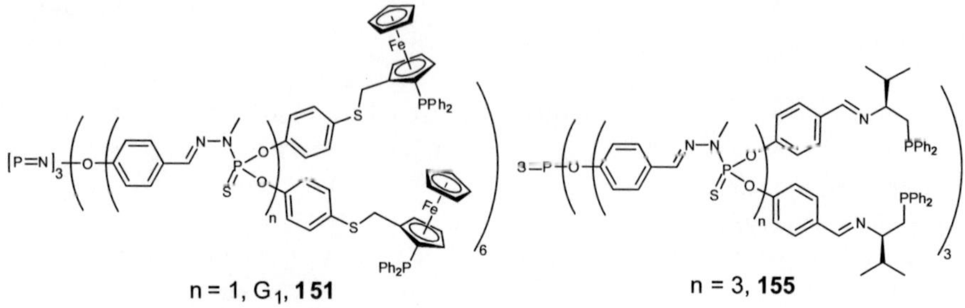

n = 1, G$_1$, **151**
n = 2, G$_2$, **152**
n = 3, G$_3$, **153**
n = 4, G$_4$, **154**

n = 3, **155**

Figure 5.18 Chiral monophosphine dendrimers.

the recycled catalyst showed a slight decrease in reactivity and enantioselectivity upon reuse.

In 2006, Rossell and co-workers reported the immobilization of chiral P-stereogenic monophosphine ligands at the periphery of different generations of carbosilane dendrimers (Figure 5.19).[60] The reaction of compounds **156–161** with [Pd(μ-Cl)(η^3-2-MeC$_3$H$_4$)]$_2$ afforded the corresponding palladodendrimers which were employed as catalysts in the hydrovinylation of styrene. The catalysts were found to have a good activity (54–95% conversion) and good enantioselectivity (up to 79% ee), which did not seem to depend on the structure of the dendrimer. The nature of the halide scavenger, and in particular its corresponding counter ion was found to have a large influence on the performances of the catalysts, in particular the use of NaBArF instead of AgBF$_4$ increased the chemoselectivity and enantioselectivity. The same reaction was also performed in supercritical carbon dioxide and the catalytic results were very similar to those obtained in organic solvents.[61]

Compounds **156–161** were also applied in the rhodium-catalyzed asymmetric hydrogenation of dimethyl itaconate and their activity was found to decrease when increasing the dendrimer generation from **158** to **160** from 94.4% to 68.6% conversion, respectively. Furthermore, the catalysts failed to induce any chirality. The ruthenium-catalyzed ATH of acetophenone was also tested with the immobilized phosphines. For this reaction, a positive effect of the dendrimer generation on the catalytic activity was observed, albeit without any enantioselectivity.[62,63]

5.3.2 Nitrogen-based Ligands

5.3.2.1 Proline-derived Ligands

Kokotos *et al.* reported in 2005 on the immobilization of *trans*-4-hydroxyproline on the periphery of different generations of PPI dendrimers (Figure 5.20).[64] The catalytic performance of dendrimers **162–166** was evaluated in the

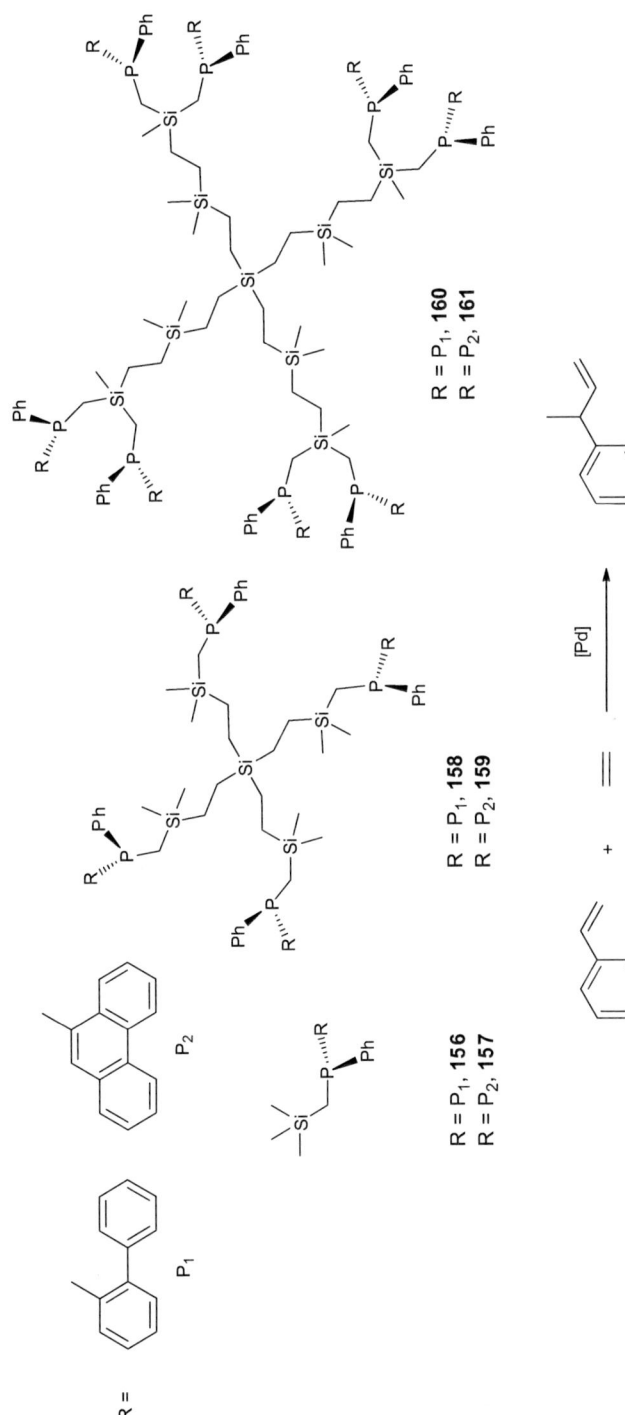

Figure 5.19 P-stereogenic ligands immobilized on carbosilane dendrimers.

asymmetric aldol reaction of 4-nitrobenzaldehyde and acetone and compared to the activity of non-immobilized (S)-proline. The activity of the second generation dendrimer **163** was found to be the most efficient with a yield and ee value comparable to the parent compound (63% yield and 69% ee versus 61% yield and 65% ee). The authors found that with this catalyst the reaction ran faster than with proline itself, however, with an actual catalyst loading of 52 mol% of proline (compared to 20 mol% for the free proline). A negative dendritic effect was observed when the higher dendrimer generations **164**–**166** were used, with both the activity and the enantioselectivity decreasing.

Diphenylprolinol ligands were used by Liang *et al.* to decorate the periphery of different types of dendrimers with varying core moieties through triazole linkers (Figure 5.20).[65] Dendrimers **167**–**170** were used as catalysts in the

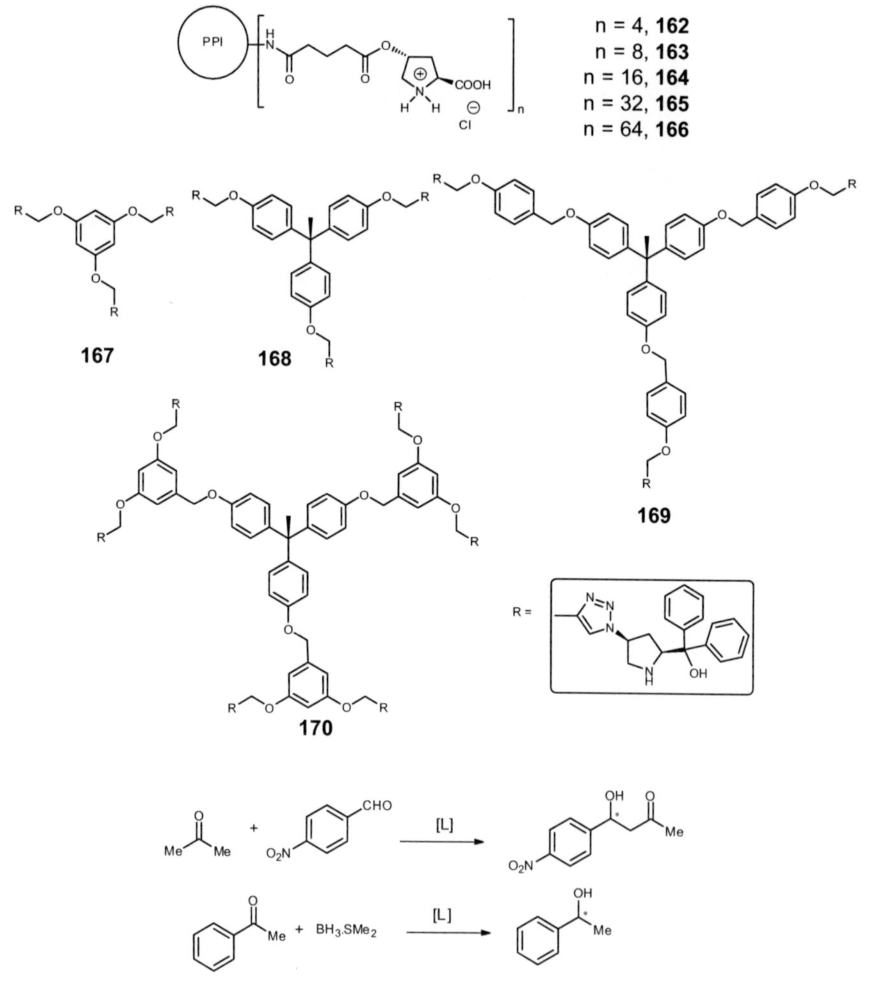
PPI
COOH
n = 4, 162
n = 8, 163
n = 16, 164
n = 32, 165
n = 64, 166
167
168
169
170
R =
OH
CHO
[L]
Me
O_2N
$BH_3.SMe_2$

Figure 5.20 Dendrimer-immobilized proline derivatives.

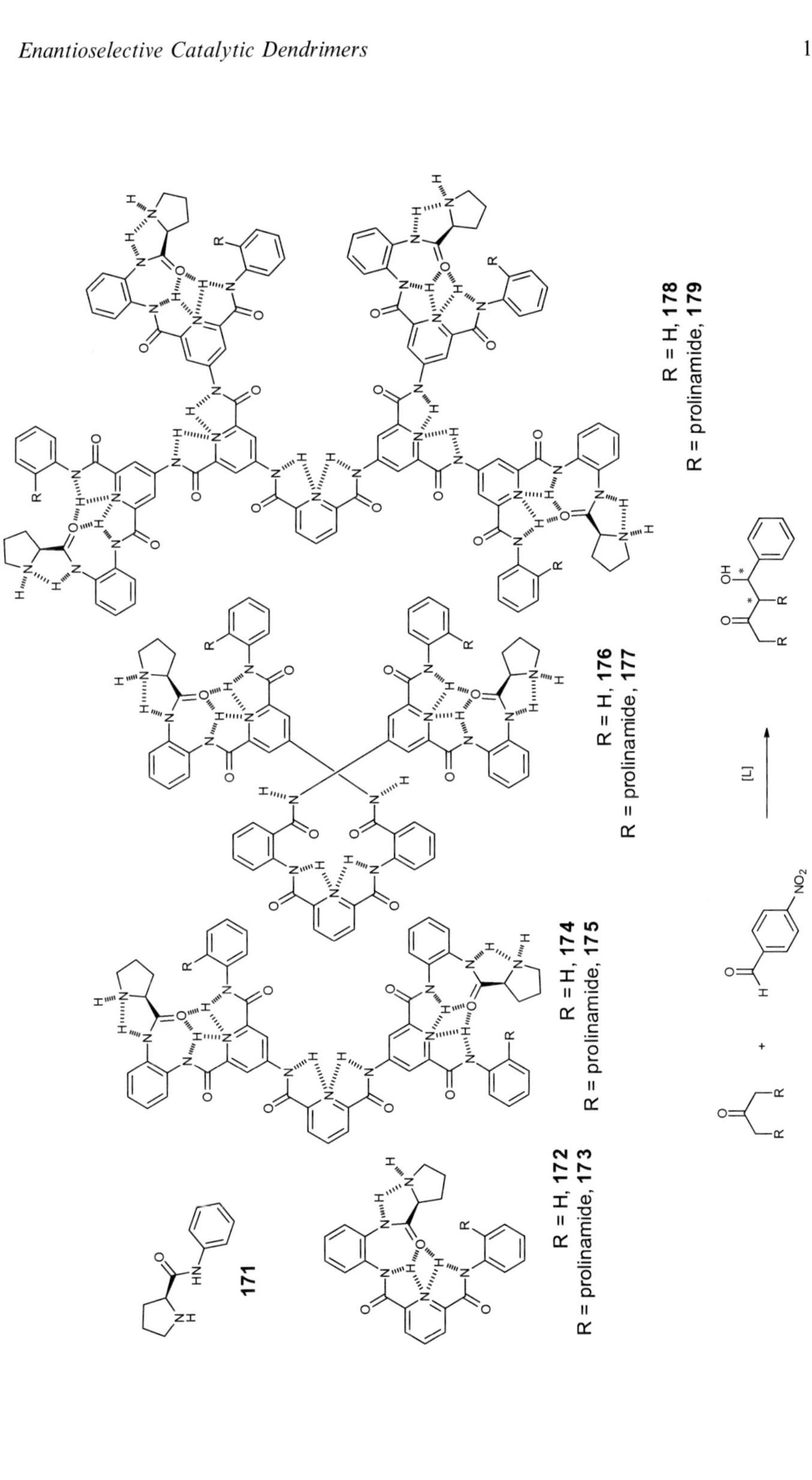

Figure 5.21 Parquette's proline-derived dendrimers.

enantioselective borane reduction of ketones where they proved to be excellent catalysts. Dendrimer **170** even showed a higher reactivity than the parent monomeric diphenylprolinol catalyst (95% ee versus 89% ee). This hexaprolinol dendrimer could be recycled by precipitation up to four times without appreciable loss of reactivity or enantioselectivity.

Parquette *et al.* prepared folded dendritic organocatalysts by attaching proline derivatives to pyridine-2,6-dicarboxyamide branching units and used dendrimers **172–179** in the asymmetric aldol reaction of 4-nitrobenzaldehyde with acyclic and cyclic ketones (Figure 5.21).[66] A significant increase in selectivity was observed when cyclic or substituted ketones were employed in these reactions (ee: 36–59% to 63–92%). Interestingly, the stereoselectivity turned out to be independent of the prolinamide density at the dendrimer periphery, i.e. no significant difference in ee was observed between compounds bearing all prolinamides or alternately (compounds with R = H or prolinamide).

5.3.2.2 Amino Alcohol Derivatives

Another early example in this field comes from Soai and co-workers, who reported the immobilization of chiral ephedrine ligands on the periphery of different dendritic supports. The so-formed dendrimers were applied in the enantioselective catalytic addition of dialkylzincs to aldehydes (Figure 5.22).[67,68] After some investigations on the addition of dialkylzincs

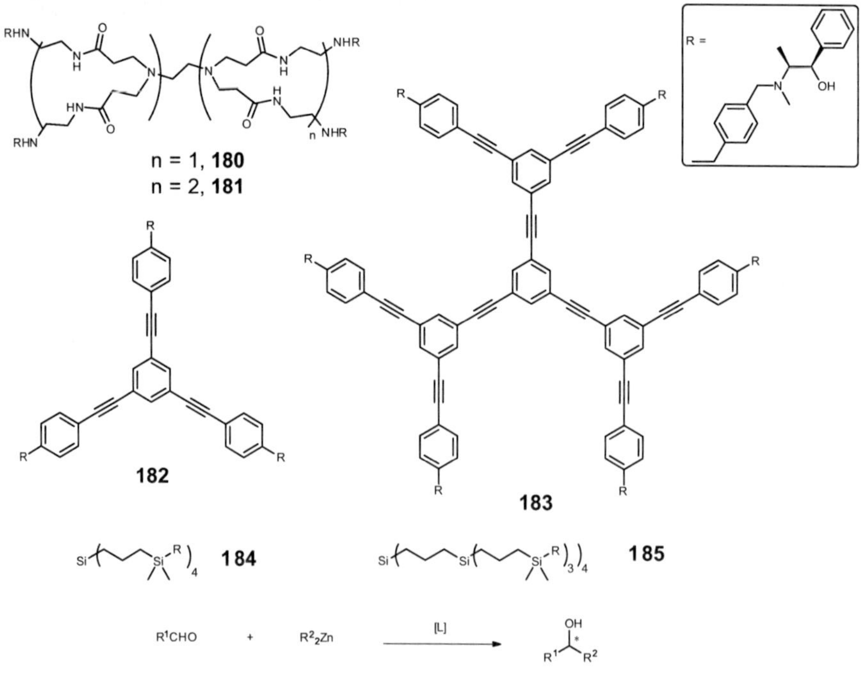

Figure 5.22 Ephedrine-derived ligand immobilized on different dendritic supports.

to *N*-diphenylphosphinylimines, the authors concluded on a negative interaction of the PAMAM support and restricted their study to the catalytic activity of dendrimers **182–185** with an inert backbone. These catalysts all exhibited a good activity for this reaction with excellent enantioselectivities (32–70% yield and 77–86% ee).

Later, Eilbracht and co-workers developed a new synthetic protocol for the preparation of polyamino alcohol dendrimers (not shown).[69] These polyamine-based dendrimers were employed in the ruthenium-catalyzed ATH reaction of acetophenone. The authors observed in general good to excellent conversions (71–86%) and moderate to good ee values (22–69% ee). Upon increase of the dendrimer generation, a negative dendritic effect on the enantioselectivity of the catalysts was observed.

5.3.2.3 Oxazoline Ligands

Recently, Gade *et al.* immobilized bis and tris(oxazoline)ligands on carbosilane dendrimers and investigated their efficiency as ligand in the copper(II)-catalyzed α-hydrazination of a β-keto ester as well as in the Henry reaction of 2-nitrobenzaldehyde with nitromethane (Figure 5.23).[70] In the first reaction, the performances of ligands **186–194** proved to be excellent with good activities and enantioselectivities (90–99% ee) obtained with a minimal catalyst loading of 1 mol%. The activity of ligands **188, 191** and **194** of the tris(oxazoline) series (type c) displayed a decreased reactivity in the Henry reaction compared to the bis(oxazoline) series (type b, Figure 5.22), albeit that the enantioselectivity was higher for the tris(oxazolines) (81–84% ee versus 52–53% ee). The latter series exhibited an improved reactivity and enantioselectivity compared to the parent compound, however without correlation with the dendrimer generation. In order to investigate the recyclability of dendrimers **190** and **191**, the catalysts were placed in a 'tea-bag', a membrane bag made of dialysis tubing, and this 'tea-bag' was placed in a fresh batch of substrates after a certain reaction time. The catalyst activity slightly decreased as well as the enantioselectivity; still displaying 77% ee after the seventh run for **190** (compared to 82% ee in the first run), but only 14% ee for **191** (69% ee in the first run).

The group of Majoral reported on the synthesis of azabis(oxazoline) ligands immobilized on phosphorus-based dendrimers *via* click chemistry (Figure 5.23).[71] The performance of ligands **199–202** was evaluated in the copper-catalyzed asymmetric benzoylation of diols and compared with the performance of ligands **195–198**, in order to determine the influence of the backbone on the catalysis, in particular the influence of the triazole ring which is known to coordinate to copper. The catalytic results indicated a moderate yield (34–41%) but a good selectivity (73–80% ee) for **195–198** and did not show any interference of the backbone on the catalytic activity, which suggested a strong affinity of the azabis(oxazoline) ligand for copper. Ligands **199–202** appeared to be less reactive (28–31% yield) than **195–198**, however with similar enantioselectivities except for the higher generation dendrimer **202** (33% ee).

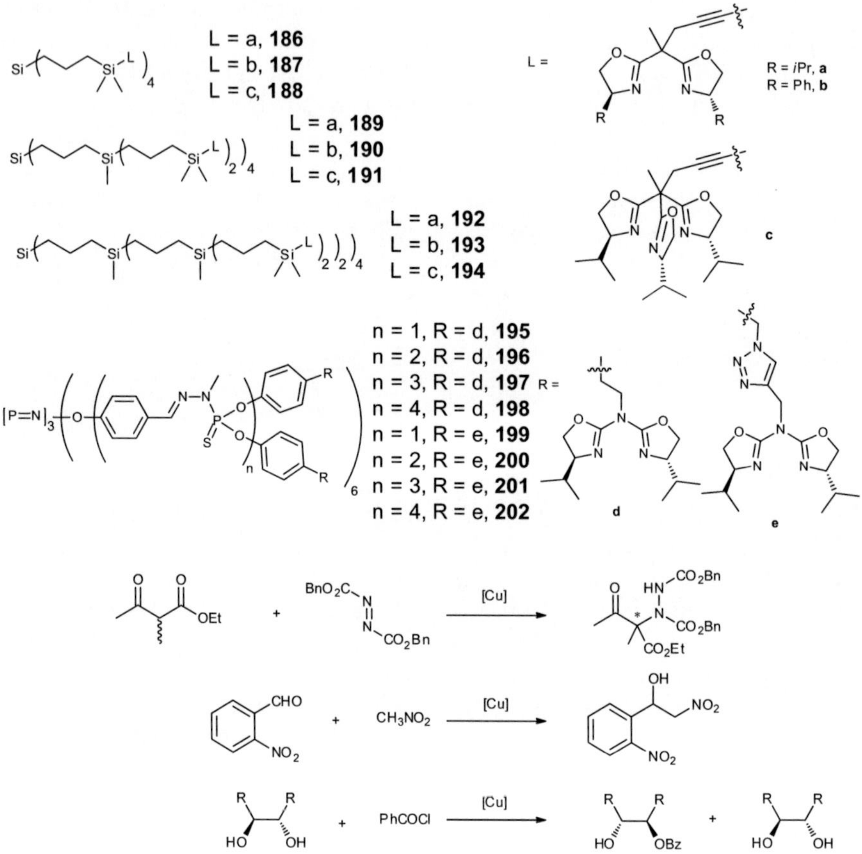

Figure 5.23 Dendrimer-immobilized bis and tris(oxazoline) ligands.

The recyclability of these catalysts by means of precipitation was also investigated with **200** and did not show any deterioration of the catalytic performances after three successive runs.

5.3.2.4 *Other Types of Ligand*

In a report from 2000, Jacobsen *et al.* reported the immobilization of [Co^{III}(salen)] complexes on the periphery of PAMAM dendrimers (Figure 5.24).[72] By doing so, the authors were expecting the proximity of the metal centres, induced by the geometry of the dendrimer, to have a positive effect on the reactivity, as it was shown earlier that the mechanism of the asymmetric ring opening (ARO) of epoxides involves cooperative bimetallic catalysis (*i.e.* a second-order kinetic dependence on [Co^{III}(salen)]). Dendrimers **203–205** were then employed in the hydrolytic kinetic resolution (HKR) of terminal epoxides and showed a dramatic improvement of the reactivity relative to the monomer: at a catalyst loading of 0.025 mol% no conversion was detected for the monomer whereas with 0.027 mol% of **204** 98% ee was obtained with 50% conversion. A further increase of the dendrimer generation, and accordingly the number of [Co^{III}(salen)] units per dendrimer, resulted in a decrease in activity/ selectivity (relative rate going from 24 to 11 and ee from 42.8 to 39.8%). According to the authors, this positive dendritic effect may be attributed to higher order productive cooperative interactions between the [Co^{III}(salen)] units that apparently were most optimal in **204**.

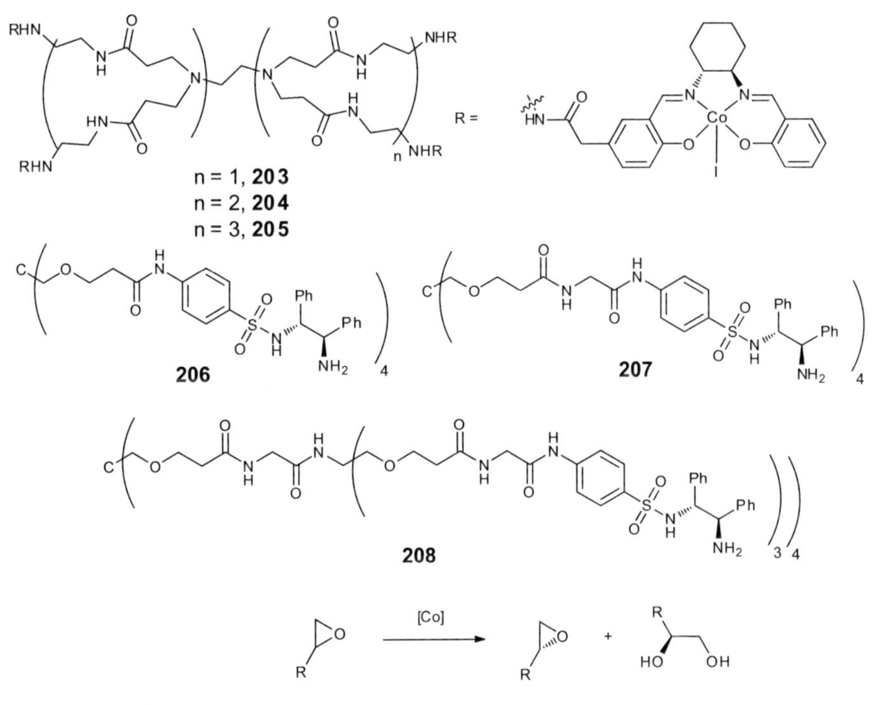

Figure 5.24 Other types of peripherally immobilized ligands.

In a continuation of their work on the influence of the dendritic backbone on the ruthenium-catalyzed ATH reaction of pro-chiral ketones with chiral diamine ligands (see Figure 5.9), Deng and co-workers reported the synthesis of dendrimers functionalized at the periphery with TsDPEN-derived ligands (Figure 5.24).[73] The performance of dendrimers **206–208** was investigated in the ATH of acetophenone and showed good activities (97% conversion) and enantioselectivities (97.6% ee), which were comparable to the monomeric TsDPEN catalyst (>99% conversion and 97.7% ee). The scope of the reaction was extended to other ketones and imines and showed in general good activities and enantioselectivities.

5.3.3   Oxygen-based Ligands

In 2002, Sasai and co-workers synthesized polyether dendrimers **209–210** functionalized with BINOL ligands at their periphery and the corresponding hetero bimetallic catalysts, synthesized by coordination of the BINOL ligands with the complex AlLibis(binaphthoxide) containing aluminium and lithium metals (Figure 5.25).[74] These dendrimers were used as catalysts in the Michael addition of dibenzylmalonate to 2-cyclohexanone and exhibited a moderate activity (57–63% yield) but excellent enantioselectivity (91–94% ee).

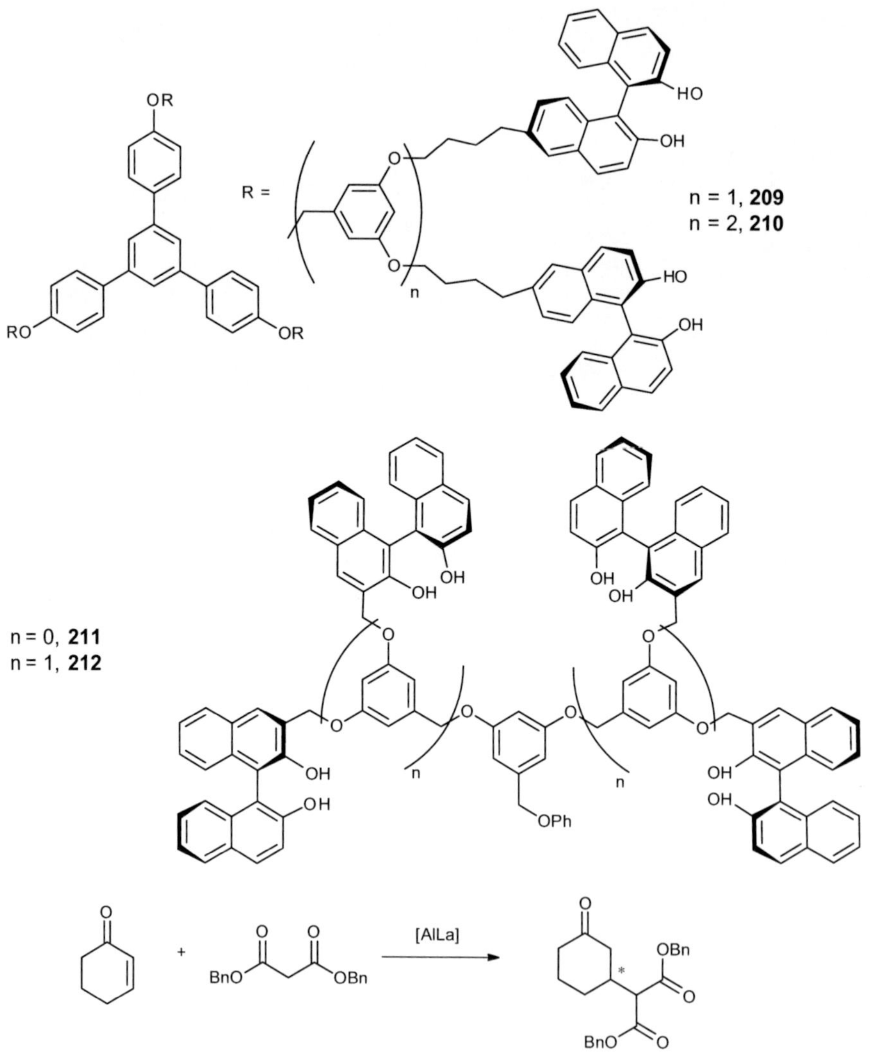

Figure 5.25   Immobilized BINOL ligands.

No evidence of the influence of the dendrimer size could be observed and the catalyst could be reused without showing a diminished activity.

In a report by Ma *et al.*, the synthesis of BINOL ligands **211**–**212** immobilized on Fréchet type dendrimers is presented.[75] Their use as catalysts in the asymmetric addition of diethyl zinc to benzaldehyde (reaction shown in Figure 5.22) was evaluated and revealed a good activity (94–96% yield) and enantioselectivity (87.1–89.2% ee) with all the catalysts in the presence of Ti(O-*i*-Pr)$_4$, similar to the reactivity of the non-immobilized BINOL ligand. Interestingly, without addition of Ti(O-*i*-Pr)$_4$ the dendrimers showed a good activity (75–78%) and moderate enantioselectivity (40% ee), though significantly higher than the BINOL ligand (17% yield and 5.2% ee). The dendritic catalyst was precipitated at the end of the reaction and reused in two extra catalytic runs without loss of reactivity.

5.4 Concluding Remarks

After about 15 years of active research, the field of catalytic metallodendrimers has seen many advances, particularly in enantioselective catalysis. In its early days, the idea of using a chiral dendrimer that would be able to induce chirality to a non-chiral reaction site over a long distance range was a tempting concept. Investigations on this concept were never successful so far and lead to a general consensus that enantioselectivity can only be induced if chirality is present in the close vicinity of the metal centre.

The success of the dendrimer immobilization approach was validated by the induction of enantiomeric product excesses observed in both types of catalyst attachments, either when the catalyst is shielded by the core of a dendrimer or dendron or when it is more exposed on the surface on a dendrimer. These investigations have also shown that the dendrimer backbone itself can be responsible for an enhancement of the catalytic activity or enantioselectivity through steric congestion induced by increasing dendrimer generation. This dendritic effect is in most cases limited to a size range above which catalytic sites are hardly accessible or where the ever closer proximity of metallic centres interferes with their activity, even though this interference can be constructive in some rare cases. The recyclability of the catalyst *via* different methods, *i.e.* precipitation or filtration, gave interesting results; nevertheless little care is taken to really 'measure' the activity of the recovered catalyst by for example determining the kinetic profiles upon reuse of the dendritic catalyst, which would be indicative of a truly unaltered catalyst performance.

These findings are now of crucial importance to make progression towards the 'Holy Grail' of homogeneous catalysis: the synthesis of highly (enantio)-selective catalysts that are recoverable and reusable without alteration of their performance. In order to reach this goal it is expected that ligand and dendrimer design have to go hand in hand with kinetic studies and advanced separation technology.

References

1. Q. H. Fan, Y. M. Li and A. S. C. Chan, *Chem. Rev.*, 2002, **102**, 3385–3465.
2. J. M. Fraile, J. I. Garcia and J. A. Mayoral, *Chem. Rev.*, 2009, **109**, 360–417.
3. J. M. Fraile, J. I. Garcia, C. I. Herrerias, J. A. Mayoral and E. Pires, *Chem. Soc. Rev.*, 2009, **38**, 695–706.
4. A. F. Trindade, P. M. P. Gois and C. A. M. Afonso, *Chem. Rev.*, 2009, **109**, 418–514.
5. Y. Ribourdouille, G. D. Engel and L. H. Gade, *C. R. Chimie*, 2003, **6**, 1087–1096.
6. J. K. Kassube and L. H. Gade, *Top. Organomet. Chem.*, 2006, **20**, 60–91.
7. A. M. Caminade, P. Servin, R. Laurent and J. P. Majoral, *Chem. Soc. Rev.*, 2008, **37**, 56–67.
8. H. Brunner and J. Furst, *Tetrahedron*, 1994, **50**, 4303–4310.
9. H. Brunner, *J. Organomet. Chem.*, 1995, **500**, 39–46.
10. Q. H. Fan, Y. M. Chen, X. M. Chen, D. Z. Jiang, F. Xi and A. S. C. Chan, *Chem. Commun.*, 2000, 789–790.
11. G. J. Deng, Q. H. Fan, X. M. Chen, D. S. Liu and A. S. C. Chan, *Chem. Commun.*, 2002, 1570–1571.
12. Z. J. Wang, G. J. Deng, Y. Li, Y. M. He, W. J. Tang and Q. H. Fan, *Org. Lett.*, 2007, **9**, 1243–1246.
13. G. J. Deng, G. R. Li, L. Y. Zhu, H. F. Zhou, Y. M. He, Q. H. Fan and Z. G. Shuai, *J. Mol. Catal. A: Chem.*, 2006, **244**, 118–123.
14. B. Yi, Q. H. Fan, G. J. Deng, Y. M. Li, L. Q. Qiu and A. S. C. Chan, *Org. Lett.*, 2004, **6**, 1361–1364.
15. B. Yi, H. P. He and Q. H. Fan, *J. Mol. Catal. A: Chem.*, 2010, **315**, 82–85.
16. L. I. Rodriguez, O. Rossell, M. Seco and G. Muller, *Organometallics*, 2008, **27**, 1328–1333.
17. D. J. M. Snelders, K. Kunna, C. Muller, D. Vogt, G. van Koten and R. J. M. Klein Gebbink, *Tetrahedron: Asymmetry*, 2010, **21**, 1411–1420.
18. Y. Li, Y. M. He, Z. W. Li, F. Zhang and Q. H. Fan, *Org. Biomol. Chem.*, 2009, **7**, 1890–1895.
19. P. N. M. Botman, A. Amore, R. van Heerbeek, J. W. Back, H. Hiemstra, J. N. H. Reek and J. H. van Maarseveen, *Tetrahedron Lett.*, 2004, **45**, 5999–6002.
20. M. van den Berg, A. J. Minnaard, E. P. Schudde, J. van Esch, A. H. M. de Vries, J. G. de Vries and B. L. Feringa, *J. Am. Chem. Soc.*, 2000, **122**, 11539–11540.
21. W. J. Tang, Y. Y. Huang, Y. M. He and Q. H. Fan, *Tetrahedron: Asymmetry*, 2006, **17**, 536–543.
22. F. Zhang, Y. Li, Z. W. Li, Y. M. He, S. F. Zhu, Q. H. Fan and Q. L. Zhou, *Chem. Commun.*, 2008, 6048–6050.
23. F. Zhang and Q. H. Fan, *Org. Biomol. Chem.*, 2009, **7**, 4470–4474.
24. R. Noyori and S. Hashiguchi, *Acc. Chem. Res.*, 1997, **30**, 97–102.
25. Y. C. Chen, T. F. Wu, J. G. Deng, H. Liu, Y. Z. Jiang, M. C. K. Choib and A. S. C. Chan, *Chem. Commun.*, 2001, 1488–1489.

26. Y. C. Chen, T. F. Wu, L. Jiang, J. G. Deng, H. Liu, J. Zhu and Y. Z. Jiang, *J. Org. Chem.*, 2005, **70**, 1006–1010.
27. W. G. Liu, X. Cui, L. F. Cun, J. Wu, J. Zhu, J. G. Deng and Q. H. Fan, *Synlett*, 2005, 1591–1595.
28. W. G. Liu, X. Cui, L. F. Cun, J. Zhu and J. Deng, *Tetrahedron: Asymmetry*, 2005, **16**, 2525–2530.
29. L. Jiang, T. F. Wu, Y. C. Chen, J. Zhu and J. G. Deng, *Org. Biomol. Chem.*, 2006, **4**, 3319–3324.
30. W. W. Wang and Q. R. Wang, *Chem. Commun.*, 2010, **46**, 4616–4618.
31. X. Y. Liu, X. Y. Wu, Z. Chai, Y. Y. Wu, G. Zhao and S. Z. Zhu, *J. Org. Chem.*, 2005, **70**, 7432–7435.
32. C. Bolm, N. Derrien and A. Seger, *Synlett*, 1996, 387–388.
33. G. Y. Wang, X. Y. Liu and G. Zhao, *Synlett*, 2006, 1150–1154.
34. X. Y. Liu, Y. W. Li, G. Y. Wang, Z. Chai, Y. Y. Wu and G. Zhao, *Tetrahedron: Asymmetry*, 2006, **17**, 750–755.
35. G. Y. Wang, C. W. Zheng and G. Zhao, *Tetrahedron: Asymmetry*, 2006, **17**, 2074–2081.
36. Y. W. Li, X. Y. Liu and G. Zhao, *Tetrahedron: Asymmetry*, 2006, **17**, 2034–2039.
37. Y. H. Zhao, C. W. Zheng, G. Zhao and W. G. Cao, *Tetrahedron: Asymmetry*, 2008, **19**, 701–708.
38. G. H. Lv, R. H. Jin, W. P. Mai and L. X. Gao, *Tetrahedron: Asymmetry*, 2008, **19**, 2568–2572.
39. Y. Y. Wu, Y. Z. Zhang, M. L. Yu, G. Zhao and S. W. Wang, *Org. Lett.*, 2006, **8**, 4417–4420.
40. C. M. Lo and H. F. Chow, *J. Org. Chem.*, 2009, **74**, 5181–5191.
41. M. Malkoch, K. Hallman, S. Lutsenko, A. Hult, E. Malmstrom and C. Moberg, *J. Org. Chem.*, 2002, **67**, 8197–8202.
42. B. Y. Yang, X. M. Chen, G. J. Deng, Y. L. Zhang and Q. H. Fan, *Tetrahedron Lett.*, 2003, **44**, 3535–3538.
43. H. Liu and D. M. Du, *Eur. J. Org. Chem.*, 2010, 2121–2131.
44. P. B. Rheiner, H. Sellner and D. Seebach, *Helv. Chim. Acta*, 1997, **80**, 2027–2032.
45. P. B. Rheiner and D. Seebach, *Chem. Eur. J.*, 1999, **5**, 3221–3236.
46. S. Yamago, M. Furukawa, A. Azuma and J. Yoshida, *Tetrahedron Lett.*, 1998, **39**, 3783–3786.
47. Q. H. Fan, G. H. Liu, X. M. Chen, G. J. Deng and A. S. C. Chan, *Tetrahedron: Asymmetry*, 2001, **12**, 1559–1565.
48. G. H. Liu, W. J. Tang and Q. H. Fan, *Tetrahedron*, 2003, **59**, 8603–8611.
49. B. M. Ji, Y. Yuan, K. L. Ding and A. B. Meng, *Chem. Eur. J.*, 2003, **9**, 5989–5996.
50. G. Guillena, R. Kreiter, R. van de Coevering, R. J. M. Klein Gebbink, G. van Koten, P. Mazon, R. Chinchilla and C. Najera, *Tetrahedron: Asymmetry*, 2003, **14**, 3705–3712.
51. C. Jahier, M. Cantuel, N. D. McClenaghan, T. Buffeteau, D. Cavagnat, F. Agbossou, M. Carraro, M. Bonchio and S. Nlate, *Chem. Eur. J.*, 2009, **15**, 8703–8708.

52. C. Kollner, B. Pugin and A. Togni, *J. Am. Chem. Soc.*, 1998, **120**, 10274–10275.

53. R. Schneider, C. Kollner, I. Weber and A. Togni, *Chem. Commun.*, 1999, 2415–2416.

54. C. Kollner and A. Togni, *Can. J. Chem.*, 2001, **79**, 1762–1774.

55. G. D. Engel and L. H. Gade, *Chem. Eur. J.*, 2002, **8**, 4319–4329.

56. Y. Ribourdouille, G. D. Engel, M. Richard-Plouet and L. H. Gade, *Chem. Commun.*, 2003, 1228–1229.

57. J. K. Kassube, H. Wadepohl and L. H. Gade, *Adv. Synth. Catal.*, 2008, **350**, 1155–1162.

58. L. Routaboul, S. Vincendeau, C. O. Turrin, A. M. Caminade, J. P. Majoral, J. C. Daran and E. Manoury, *J. Organomet. Chem.*, 2007, **692**, 1064–1073.

59. R. Laurent, A. M. Caminade and J. P. Majoral, *Tetrahedron Lett.*, 2005, **46**, 6503–6506.

60. L. I. Rodriguez, O. Rossell, M. Seco, A. Grabulosa, G. Muller and M. Rocamora, *Organometallics*, 2006, **25**, 1368–1376.

61. L. I. Rodriguez, O. Rossell, M. Seco, A. Orejon and A. M. Masdeu-Bulto, *J. Organomet. Chem.*, 2008, **693**, 1857–1860.

62. L. I. Rodriguez, O. Rossell, M. Seco and G. Muller, *J. Organomet. Chem.*, 2007, **692**, 851–858.

63. L. I. Rodriguez, O. Rossell, M. Seco and G. Muller, *J. Organomet. Chem.*, 2009, **694**, 1938–1942.

64. E. Bellis and G. Kokotos, *J. Mol. Catal. A: Chem.*, 2005, **241**, 166–174.

65. Y. N. Niu, Z. Y. Yan, G. Q. Li, H. L. Wei, G. L. Gao, L. Y. Wu and Y. M. Liang, *Tetrahedron: Asymmetry*, 2008, **19**, 912–920.

66. K. Mitsui, S. A. Hyatt, D. A. Turner, C. M. Hadad and J. R. Parquette, *Chem. Commun.*, 2009, 3261–3263.

67. I. Sato, T. Shibata, K. Ohtake, R. Kodaka, Y. Hirokawa, N. Shirai and K. Soai, *Tetrahedron Lett.*, 2000, **41**, 3123–3126.

68. K. Soai and I. Sato, *C. R. Chimie*, 2003, **6**, 1097–1104.

69. M. A. Subhani, K. S. Muller and P. Eilbracht, *Adv. Synth. Catal.*, 2009, **351**, 2113–2123.

70. M. Gaab, S. Bellemin-Laponnaz and L. H. Gade, *Chem. Eur. J.*, 2009, **15**, 5450–5462.

71. A. Gissibl, C. Padie, M. Hager, F. Jaroschik, R. Rasappan, E. Cuevas-Yanez, C. O. Turrin, A. M. Caminade, J. P. Majoral and O. Reiser, *Org. Lett.*, 2007, **9**, 2895–2898.

72. R. Breinbauer and E. N. Jacobsen, *Angew. Chem., Int. Ed. Engl.*, 2000, **39**, 3604–3607.

73. Y. C. Chen, T. F. Wu, J. G. Deng, H. Liu, X. Cui, J. Zhu, Y. Z. Jiang, M. C. K. Choi and A. S. C. Chan, *J. Org. Chem.*, 2002, **67**, 5301–5306.

74. T. Arai, T. Sekiguti, Y. Iizuka, S. Takizawa, S. Sakamoto, K. Yamaguchi and H. Sasai, *Tetrahedron: Asymmetry*, 2002, **13**, 2083–2087.

75. L. Yin, R. Li, F. S. Wang, H. L. Wang, Y. F. Zheng, C. F. Wang and J. T. Ma, *Tetrahedron: Asymmetry*, 2007, **18**, 1383–1389.

CHAPTER 6
Fluorous Catalysts

GIANLUCA POZZI

CNR – Consiglio Nazionale delle Ricerche, Istituto di Scienze e Tecnologie Molecolari, via Golgi 19, 20133 Milano, Italy

6.1 Introduction

Fluorine is the most electronegative of all elements; it has a high ionization potential and a very low polarizability. The C–F bond is one of the most stable single bonds found in organic chemistry ($\sim 485\,\mathrm{kJ\,mol^{-1}}$ as compared with $\sim 425\,\mathrm{kJ\,mol^{-1}}$ for a standard C–H bond), owing also to effective overlapping of orbitals.[1] It follows that the physicochemical characteristics of organofluorine compounds can differ quite markedly from those of their hydrogenated counterparts, with relevant effects in various science and technology domains.[2] This is well exemplified by fluorocarbons, organic molecules with carbon skeletons and fluorine 'skins', which have found a wide range of applications spanning from electronics to medicine thanks to their unique properties and chemical behaviour.[3]

According to IUPAC recommendations, only compounds consisting wholly of fluorine and carbon should be referred to as fluorocarbons,[4] although this term has been also in use as an all-embracing name for organic fluorides.[5] It is also worth noting that liquid saturated fluorocarbons (*e.g.* perfluorooctane, C_8F_{18}), ethers and amines in which all hydrogen atoms have been replaced by fluorine atoms (*e.g.* perfluoro(2-butyl-tetrahydrofurane), $C_4F_9(C_4F_7O)$; perfluoro(tripentylamine), $(C_5F_{11})_3N$) are often collectively known as perfluorocarbons (PFC) because of their close similarities.[6] PFC are apolar, colourless liquids of high density, characterized by low refraction index, very

RSC Green Chemistry No. 15
Enantioselective Homogeneous Supported Catalysis
Edited by Radovan Šebesta
© Royal Society of Chemistry 2012
Published by the Royal Society of Chemistry, www.rsc.org

low surface tension, high fluidity, low dielectric constant and high compressibility. All these compounds share many other common features including non-toxicity, non-flammability, thermal stability and chemical inertness. As an example, perfluorotrialkylamines are extremely weak Lewis bases and tend to bind protons even less than alkanes. The exceptional stability of PFC has been exploited in many technological applications, but it has also a negative effect from the environmental impact point of view: although PFC are not ozone-depleting compounds, their atmospheric lifetimes range from 400 to over 2000 years.[7] This, coupled with their infrared absorption profiles, translates to very high global warming potential. PFC are excellent solvents for O_2 and other gaseous substances, as popularized by a famous experiment showing that the mouse could survive by breathing while immersed in an oxygen saturated solution of perfluoro(2-butyl-tetrahydrofurane).[8] On the other hand, PFC poorly dissolve organic compounds and are not miscible with most common solvents (*e.g.* ethanol, acetone and ethyl acetate) or with water. The phase diagrams of mixtures of PFC and non-fluorinated solvents are characterized by strong deviations from the Raoult law, and an abrupt increase in miscibility with temperature can be often observed.[9]

The characteristics of PFC, in particular their solubility behaviour, were generally considered as incompatible with the role of reaction media until the early 1990s. At that time Zhu highlighted the benefits of using inert PFC in reactions that call for drastic conditions or involve unstable compounds, or where low boiling products must be separated from the final mixture,[10] whereas in a ground-breaking paper Horváth and Rábai introduced the concept of 'fluorous biphase system' (FBS),[11] exemplified by the efficient catalytic hydroformylation of C_8–C_{12} olefins in a toluene/perfluoro(methylcyclohexane) biphasic mixture showing temperature- and pressure-dependent miscibility (Scheme 6.1). The reaction was catalyzed by a rhodium complex designed to dissolve selectively in PFC and it was performed under homogeneous conditions at $T = 100\,^{\circ}C$ and $P = 150\,\mathrm{psi}$. When the reactor was cooled and depressurized, the two solvents quickly demixed with the organic products confined in the toluene layer and the catalyst in the bottom perfluoro(methylcyclohexane) (PFMCH) layer. The latter was decanted off and reused without further treatment in a new reaction cycle with a fresh solution of olefin in toluene. Since the PFC phase had a similar function to the aqueous phase in aqueous/organic systems where the catalyst is immobilized in the water layer, it was dubbed 'fluorous phase'. As a consequence, the tailor-made catalyst soluble in PFC was referred to as a 'fluorous soluble' catalyst or, more simply, as a 'fluorous' catalyst. The original FBS concept was soon adapted to meet the requirements of many different catalytic and stoichiometric reactions,[12] and the usage of the adjective 'fluorous' evolved accordingly. An inclusive definition that generically associates the term fluorous with the characteristics of highly fluorinated saturated organic materials, molecules or molecular fragments is now widely accepted.[13] Fluorous catalysts can be thus plainly identified by the presence of extended fluorocarbon domains in their structure, which can trigger unusual solubility properties, phase behaviour or interactions with solid supports,

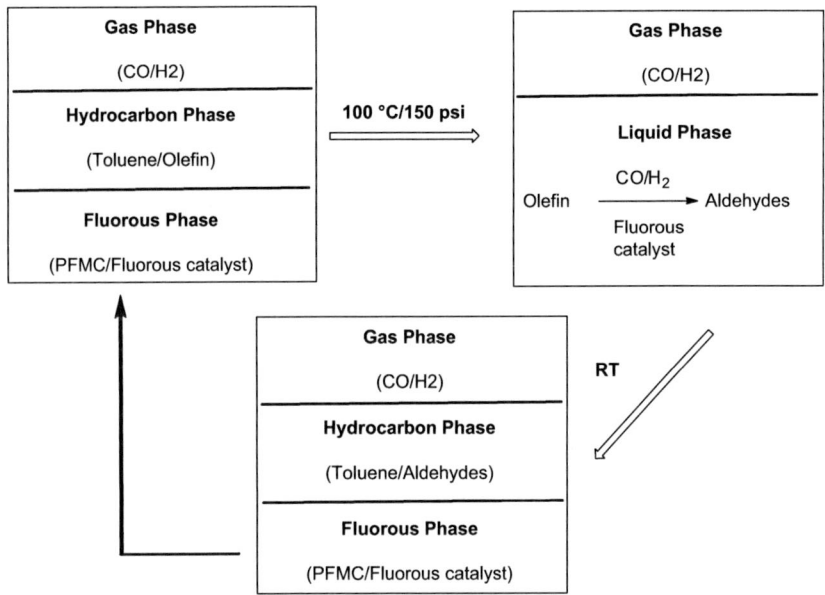

PFMC = Perfluoro(methylcyclohexane)

Fluorous catalyst = $HRh(CO)\{P[CH_2)_2(CF_2)_5CF_3]_3\}_3$ (from

$P[CH_2)_2(CF_2)_5CF_3]_3$, $Rh(CO)_2(acac)$, CO and H_2)

Scheme 6.1 Hydroformylation of olefins under original FBS conditions.

without necessarily entailing preferential solubility in PFC. These features can be then exploited in various ways to achieve a quick separation of the catalyst from reaction mixtures.

While the basic notion of fluorous catalysis is straightforward, its translation into practice can be challenging. This issue will not be discussed in the present chapter, which is instead focused on the applications of enantioselective fluorous catalysis: the interested reader can find detailed information on the design and synthesis of fluorous catalysts, including experimental procedures and practical tips, in ref. 12. Nevertheless, a few significant aspects are worth recalling briefly.

Most chiral fluorous catalysts developed so far feature linear perfluoroalkyl chains C_nF_{2n+1} (R_F, $n = 6$ to 12) as fluorocarbon modifiers. These substituents (known as 'fluorous ponytails') are strongly electron-withdrawing and their presence can significantly influence the electronic distribution and consequently the activity and selectivity of a fluorous catalyst. To avoid this, short alkyl chains, aryl rings, heteroatoms, or a combination of all these elements are used as spacers to keep fluorous ponytails and the catalyst core apart. Selecting proper spacer units is not trivial because their nature can affect both the choice of the synthetic approach and the properties of the catalyst.

Fluorous chiral catalysts are primarily meant for quick separation and recycling. In principle, a clear-cut orthogonal fluorocarbon-like behaviour typical of 'heavy fluorous' molecules would be thus helpful, which can be roughly attained by inclusion of longer or multiple medium-sized fluorous ponytails ensuring a fluorine content of the catalyst higher than 60% by weight. Reaching this goal, however, adds further complexity to the synthetic scheme. In addition, the limited solubility of heavy fluorous molecules in common organic solvents together with their quite unpredictable reactivity can make their chemical modification and manipulation difficult throughout the whole preparation process. The adoption of 'light fluorous' chiral catalysts with reduced fluorine content allows a good balance between synthetic accessibility and ease of separation and recovery.[14] Indeed, light fluorous compounds,[15,16] often bearing a single fluorous ponytail, have regular solubility in many organic solvents in which their reactivity is not too dissimilar to their non-fluorous analogues. Still, their recovery from mixtures containing organic products is feasible by means of techniques that no longer depend on the use of PFC either as reaction media or as separation auxiliaries.

Enantioselective fluorous catalysis has obviously benefited from the evolution of the original FBS concept. As an example, it is now well-known that generation of a homogeneous liquid phase upon heating is not a strict requirement for FBS catalytic reactions because even reagents poorly soluble in the fluorous catalytic phase can react at a reasonable rate at the interface of the two immiscible layers.[17] Since high temperatures are often detrimental to enantioselectivity, the opportunity to work under truly biphasic fluorous–organic conditions at low temperatures encouraged the development of the first fluorous chiral catalysts.[18] The emergence of alternative fluorous separation strategies that minimize, or even eliminate, the need for rather expensive and environmentally persistent PFC further stimulated the research efforts in this area.

The 'fluorous solid-phase extraction' (F-SPE) technique was first introduced by Curran and co-workers for the purification of reaction mixtures containing heavy fluorous by-products.[19] It soon proved useful for the separation of organic from light fluorous molecules,[20] including light fluorous chiral ligands and catalysts.[21,22] In an standard F-SPE procedure (Scheme 6.2) a mixture of fluorous and non-fluorous compounds is loaded onto the top of a cartridge filled with a fluorous solid support, typically silica gel with a fluorocarbon-bonded phase, hereafter called fluorous silica, and eluted with a solvent with poor affinity for fluorous compounds, such as aqueous MeOH or CH_3CN. During this step, the fluorous compound adsorbs onto the fluorous support, while the organic compounds are extracted off, moving with or near the front of the 'fluorophobic' solvent. A subsequent elution with a more 'fluorophilic' solvent (an ether or even dry MeOH) washes fluorous compounds off the cartridge, which can be regenerated by further washing with the fluorophilic and fluorophobic solvents. F-SPE processes are operationally similar to filtrations and they are easy to conduct in parallel, either manually or by using various automated techniques.[23] Combining light fluorous catalysis run under

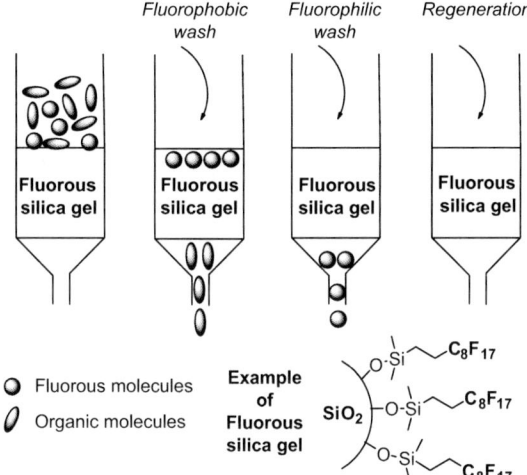

Scheme 6.2   Fluorous solid-phase extraction (F-SPE).

homogeneous conditions in a routine organic solvent and F-SPE is a good way to overcome the need for PFC in any step of the catalytic process. The same result can be achieved through other means, for instance by immobilizing a fluorous catalyst onto fluorous silica prior to the reaction stage. The resulting material can be then used as a heterogeneous catalyst,[24] even under solvent-free conditions,[25] and recovered by simple solid–liquid separation, typically filtration or centrifugation. Also the highly temperature-dependent solubilities in organic solvents exhibited by some fluorous catalysts allow for the development of PFC-free reaction and separation protocols, as independently shown by the groups of Gladysz and Yamamoto.[26,27] This thermomorphic behaviour can be exploited in various ways,[28] the most straightforward being depicted in Scheme 6.3. Here, a suspension of the solid fluorous catalyst in an organic solution of the reactants is warmed to dissolve the catalyst and reach one-phase reaction conditions. Upon completion of the reaction the catalyst is deposited by just cooling the system, and eventually recovered by solid–liquid phase separation. In order to facilitate the manipulation and recovery of very small quantities of thermomorphic fluorous catalysts in laboratory-scale operations, Gladysz and co-workers have introduced the concurrent use of insoluble fluoropolymer supports to sorb the catalyst on cooling.[29] Both Teflon® and the more porous Gore-Rastex® fibre gave interesting results.[30,31] In that case, the catalyst previously supported onto the fluoropolymer is added to the organic phase and released in solution during the reaction step carried out at high temperature. It is then seized again by the always insoluble support when the system is cooled.

The manifold reaction/separation/recycling options offered by the fluorous approach, including solubilization of fluorous catalysts in supercritical or compressed CO_2, have been now explored in several fundamental asymmetric catalytic transformations. A broad overview of the achievements and problems

Organic Solvent

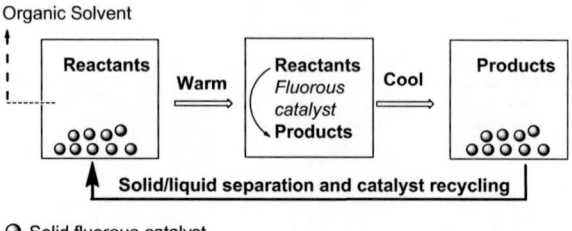

Scheme 6.3   Catalytic reaction with an ideal thermomorphic fluorous catalyst.

connected with the application of fluorous chiral catalytic systems is presented in the following sections.

6.2   Transition Metal Based Catalytic Systems

Most enantioselective fluorous catalysts developed so far have consisted of transition metal complexes, where chelation by enantiopure fluorous ligands provides a suitable chiral environment around the active metal site. Fluorous analogues of classic nitrogen, oxygen and phosphorus chiral ligands have been developed to this purpose, providing the key elements of easily recoverable catalytic systems. These ones are here classified according to the type of asymmetric process in which they were tested.

6.2.1   Oxidation Catalysts

Fluorous biphasic catalytic systems for the aerobic oxidation of hydrocarbons based on achiral transition metal complexes of polydentate fluorous nitrogen ligands were reported soon after the disclosure of the original FBS concept.[17,32] These initial studies opened the way to the development of asymmetric fluorous catalysis which was first demonstrated in 1998 through the use of PFC-soluble manganese(III) complexes derived from fluorous chiral salen ligands **1** and **2** (Figure 6.1) in the aerobic epoxidation of indene (Scheme 6.4).[18] This reaction was performed in a biphasic system CH_2Cl_2/perfluoroctane at 20 °C under an atmospheric pressure of oxygen in the presence of pivalaldehyde. Both [Mn-**1**]Cl and [Mn-**2**]Cl complexes were found to be active catalysts, affording indene oxide in isolated yields of 77% and 83%, respectively. High enantiomeric excesses (ee) were also observed (ee = 90% and 92%), and the fluorous liquid phase recovered by simple decantation could be recycled in a second run without appreciable decrease of catalytic activity and enantioselectivity. Under similar conditions, the epoxidation of other linear and cyclic alkenes proceeded smoothly, but much lower enantioselectivities (ee = 0 to 15%) were achieved. A systematic examination confirmed that the nature of the primary oxidant and the fluorous biphasic reaction conditions had only minor effects on this trend,

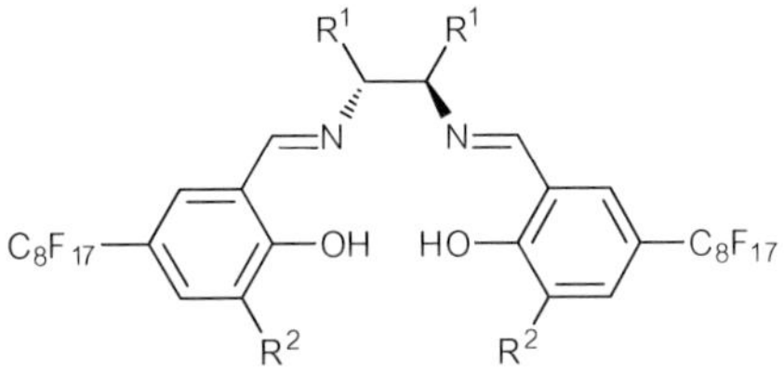

1 $R^1 = Ph$, $R^2 = C_8F_{17}$

2 $R^1, R^1 = -(CH_2)_4-$, $R^2 = C_8F_{17}$

3 $R^1 = Ph$, $R^2 = Bu^t$

4 $R^1, R^1 = -(CH_2)_4-$, $R^2 = Bu^t$

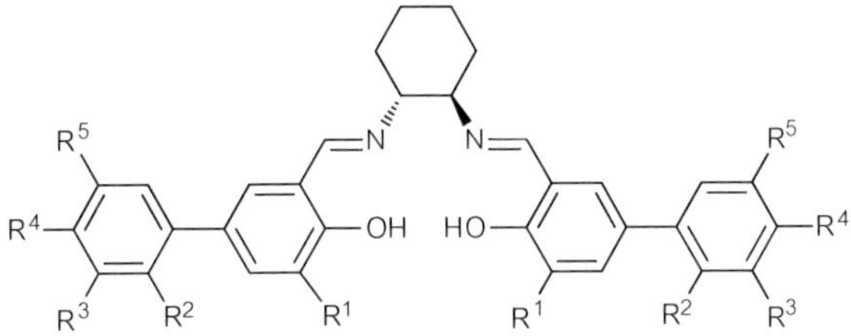

5 $R^1 = Bu^t$, $R^2 = R^3 = R^4 = C_8F_{17}(CH_2)_2O-$, $R^5 = H$

6 $R^1 = Bu^t$, $R^2 = R^4 = H$, $R^3 = R^5 = C_8F_{17}$

Figure 6.1 Fluorous salen ligands.

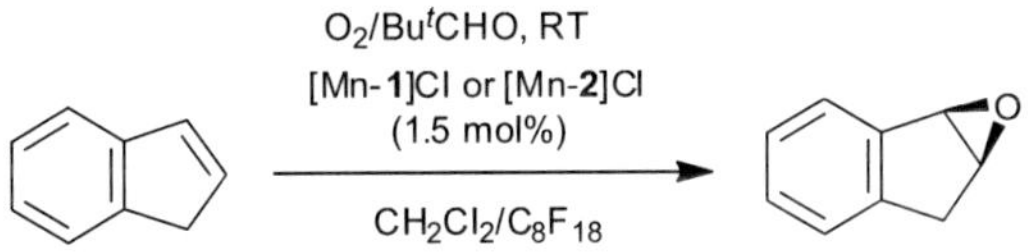

Scheme 6.4 Enantioselective epoxidation of indene under fluorous biphasic conditions.

which was governed by the intrinsic stereoelectronic features of the catalysts.[33] With most prochiral alkenes the presence of linear, rigid R_F substituents close to the binding sites of the ligand was not sufficient to differentiate the enantiotopic faces of the substrate. At the same time, the electron-withdrawing effect associated with R_F enhanced the reactivity of the high-valent Mn-oxo intermediate involved in the oxygen transfer to the substrate, thus further lowering the stereocontrol level of the process. Second-generation salen ligands **3–6** (Figure 6.1) were designed with the aim to eliminate these drawbacks.[34,35] The catalytic behaviour of the corresponding manganese(III) complexes was investigated in the presence of various oxidizing systems that are commonly applied to the asymmetric epoxidation of alkenes promoted by chiral Mn-salen complexes,[36] such as *m*-chloroperbenzoic acid in combination with *N*-methyl-morpholine *N*-oxide (MCPBA/NMO), or iodosylbenzene in combination with pyridine *N*-oxide (PhIO/PNO).

For model epoxidation reactions performed in a homogeneous solvent mixture CH_2Cl_2-BTF (BTF = α,α,α-trifluoromethyltoluene), the enantioselec-tivities attained with [Mn-**3**]Cl to [Mn-**6**]Cl were consistently higher than those attained with [Mn-**1**]Cl and [Mn-**2**]Cl. As an example, the oxidation of dihy-dronaphthalene with MCPBA/NMO at −50 °C afforded the corresponding epoxide with enantioselectivities ranging from 33% ee (catalyst = [Mn-**3**]Cl) to 63% ee (catalyst = [Mn-**5**]Cl), with a clear improvement over those observed with [Mn-**1**]Cl (ee = 12%) and [Mn-**2**]Cl (ee = 12%).[35] Comparison of all the data obtained showed that the replacement of R_F substituents in the 3,3'-positions of ligands **1** and **2** with bulky *t*-butyl groups was the single most important factor determining such a progress, but, in order to reach enantio-selection levels close to those obtained with standard Mn-salen complexes (*e.g.* Jacobsen or Katsuki catalysts),[36] the concurrent interposition of efficient insulating elements between R_F substituents and the binding O,N sites as in ligands **5** and **6** was required. It was also established that the nature of the counter anion X^- did not significantly influence the catalytic behaviour of fluorous complexes [Mn-**3**]X to [Mn-**6**]X when they were employed under homogeneous conditions. At the same time the fluorous phase affinity of these cationic complexes could be tuned by varying the nature of X^-. This allowed the investigation of their use under FBS conditions too (Scheme 6.5). Indeed, Mn^{III} complexes of ligands **5** and **6** featuring the highly fluorinated counter anion $C_7F_{15}COO^-$ showed a distinctive solubility in PFC and were thus tested as catalysts for the epoxidation of dihydronaphthalene and other cyclic and linear alkenes with PhIO/PNO in a fluorous biphasic system CH_3CN/perfluorooctane (Scheme 6.5). In this case both enantioselectivities and epoxide yields increased with temperature, the best results being obtained at 100 °C, corresponding to the boiling point of *n*-perfluorooctane. Although the biphasic mixture did not become homogeneous at 100 °C, the contact among the components of the catalytic system was obviously facilitated by the increased miscibility of the two layers. The two catalysts behaved quite similarly under these optimal condi-tions, affording epoxides in 68–98% yields and 50–92% ee, very close to the values obtained using Jacobsen catalysts in CH_3CN. The recycling of the fluorous layer containing the catalyst could be repeated up to four times, with

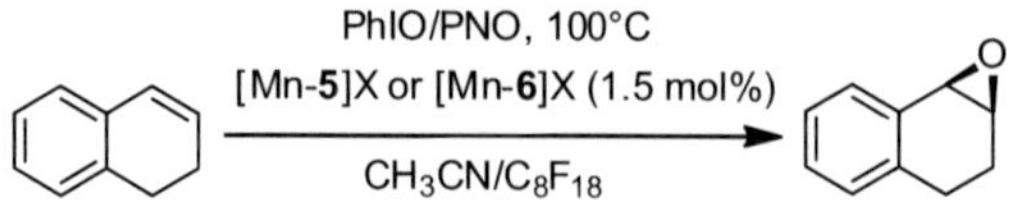

Scheme 6.5 Enantioselective epoxidation of dihydronaphthalene under fluorous biphasic conditions.

no apparent loss of activity and enantioselectivity in the first three consecutive runs. The lower activity generally observed for the fourth run was due mainly to the oxidative decomposition of the catalyst. In this respect, the fluorous approach was only marginally better than previously reported immobilization methods based on the use of soluble or insoluble polymers.[37]

Highly fluorinated counter anions can contribute to enhance not only the fluorous phase affinity, but also the catalytic activity of cationic complexes of fluorous chiral salen ligands, as next demonstrated in the case of the hydrolytic kinetic resolution (HKR) of terminal epoxides (Scheme 6.6).[38,39] Neutral cobalt(II) complexes of ligands **2–4** were oxidized by air in the presence of $C_8F_{17}COOH$ or CH_3COOH to give the corresponding fluorous cobalt(III) complexes $[Co-L]^+RCO_2^-$.[38] The latter were tested in the HKR of racemic 1-hexene oxide performed in a homogenous system where the epoxide acted as the solvent. Complexes featuring the $C_8F_{17}CO_2^-$ counter anion displayed the highest activities and enantioselectivities. In particular, $[Co-4]C_8F_{17}CO_2$ gave outstanding results, affording enantiomerically enriched (*R*)-1-hexene oxide (ee = 91.2%) and (*S*)-1,2-hexanediol (ee $\geq$ 99%) in 55% and 41% isolated yield, respectively, very close to the 50% values expected for an ideal HKR process. Kinetic resolutions of other terminal epoxides were likewise accomplished over reasonable reaction times. The catalyst could be conveniently isolated by distilling the products off, and reactivated upon treatment with $C_8F_{17}COOH$ in air. After four subsequent runs the activity of the catalyst was somewhat decreased, although the chemical yields and enantioselectivities were almost unaffected.

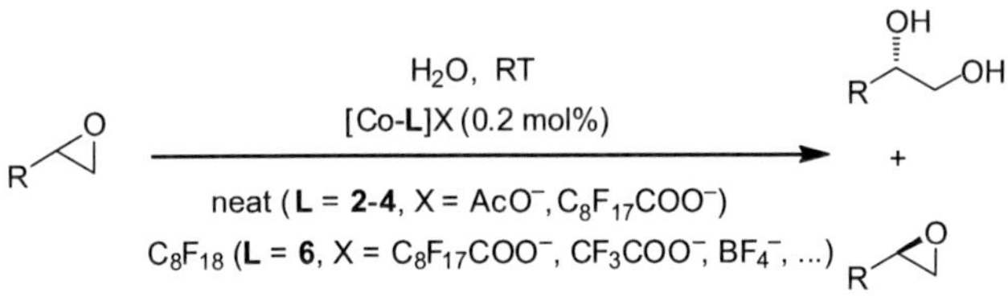

Scheme 6.6 Hydrolytic kinetic resolution of terminal epoxides.

HKR experiments carried out on 1-hexene oxide under fluorous biphasic conditions epoxide/perfluorooctane further evidenced the beneficial role of $C_8F_{17}COO^-$, even in comparison to other highly fluorinated counter anions.[39] The PFC soluble complex $[Co-6]C_8F_{17}CO_2$ showed remarkable catalytic properties, the reaction rate being higher than that observed with standard Co-salen catalysts employed under homogeneous conditions.[40] Reaction products (*S*)-1,2-hexane diol and (*R*)-1-hexene oxide were readily isolated by fractional distillation of the upper organic layer in 49% and 46% yields, respectively, with enantioselectivities as high as 99% ee. Unfortunately the bottom fluorous layer could not be used as such for a subsequent run since it contained variable amounts of Co^{II}-**6**. Regeneration of the active Co^{III} complex was attempted, but regardless of the oxidation procedure the initial catalytic activity could not be restored, although enantioselectivities were maintained.

Chiral salen ligands are known to be highly effective for a wide variety of useful asymmetric transformations catalyzed by different metals,[41] but attempts to extend the use of fluorous chiral salen ligands beyond the epoxidation of alkenes and HKR of epoxides have been less successful.[42]

Bis(oxazolines) (Box) are another versatile class of ligands being thoroughly investigated in enantioselective catalysis.[43] Just as for salen ligands, the potential of fluorous analogues of chiral Box has been investigated in various representative catalytic asymmetric processes,[44–50] among which the copper-catalyzed oxidation of cycloalkenes with peresters to give allyl esters (Scheme 6.7).[51]

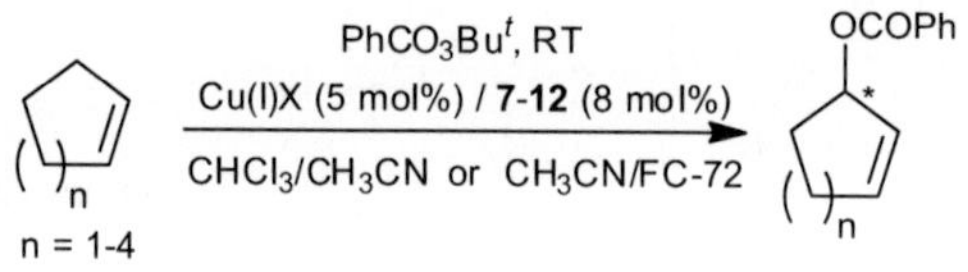

$$\text{PhCO}_3\text{Bu}^t,\ \text{RT}$$
$$\text{Cu(I)X (5 mol\%) / 7-12 (8 mol\%)}$$
$$\text{CHCl}_3/\text{CH}_3\text{CN or CH}_3\text{CN/FC-72}$$

n = 1-4

Scheme 6.7   Allylic oxidation of cycloalkenes.

Bayardon and Sinou devised a convenient access to fluorous Box with fluorine content lower than 60% by the direct attachment of two fluorous ponytails to the methylene bridge of the Box framework.[44] Ligands 7–10 (Figure 6.2) were thus prepared and combined with Cu^{I}X (X = OTf or PF$_6$) under homogeneous or fluorous biphasic conditions to give the corresponding Cu^I complexes that were used *in situ* as catalysts for the asymmetric allylic oxidation of cycloalkenes.[45] Other fluorous Box obtained through more complex synthetic pathways (*e.g.* 11 and 12, Figure 6.2) were similarly employed.[45,46] The nature of the reaction media and that of the Cu^I precursor had little or no influence on the outcome of the reactions that proceeded slowly (up to 7 days) to afford allyl esters in moderate yields and enantioselectivities,

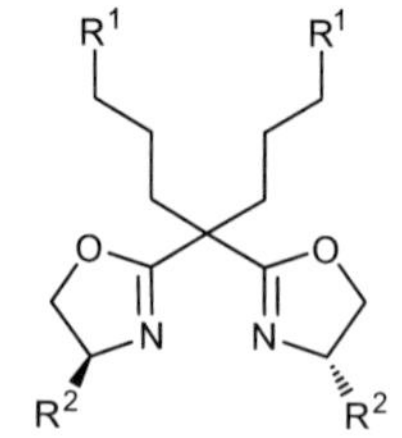

7   R^1 = C$_8$F$_{17}$, R^2 = Ph   8   R^1 = C$_{10}$F$_{21}$, R^2 = Ph

9   R^1 = C$_8$F$_{17}$, R^2 = Pr^i 10 R^1 = C$_{10}$F$_{21}$, R^2 = Pr^i

11 R^1 = C$_8$F$_{17}$, R^2 = TBDMS OCH$_2$–

12 R^1 = C$_8$F$_{17}$, R^2 = C$_7$F$_{15}$CH$_2$O–⟨benzene⟩–CH$_2$–

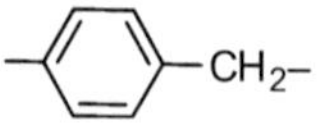

Figure 6.2   *C$_2$*-symmetric fluorous bis(oxazolines).

as high as those obtained with the non-fluorous analogues of the ligands used. Enantioselectivities ranging from 50% ee (ligand **12**) to 73% ee (ligand **7**) were obtained in the allylic oxidation of cyclohexene, depending mostly on the nature of the R^2 substituents present on the fluorous Box.

The limited fluorous phase affinity of ligands **7–12** prevented the direct recycling of the catalytic system by simple liquid–liquid phase separation. Indeed, it was observed that most of the catalyst was in the organic phase and not in the fluorous phase during reactions run under biphasic conditions CH_3CN/FC-72 (FC-72 = perfluorohexanes mixture). For this reason, reactions were better performed under homogeneous conditions in $CHCl_3/CH_3CN$. Upon completion of the reaction, the organic solvents were evaporated and the catalyst was precipitated by addition of hexane: the allylic benzoate soluble in hexane was readily separated from the solid catalyst that was reused in a next run without further treatment. The effectiveness of this procedure was directly linked to the fluorophilicity of the Box ligand, as shown by comparison of the results obtained with **9** and with the more fluorophilic ligand **12** featuring four fluorous ponytails in the oxidation of cyclohexene: cyclohexen-2-yl benzoate was isolated in 51% yield and 57% ee after 2 days in the presence of the fresh Cu^I catalyst prepared from **9**, whereas the recovered catalyst afforded the product in 53% yield and 37% ee after 3 days.[45] Under otherwise identical conditions, the catalyst prepared from **12** did not show any loss of enantio-selectivity (ee = 50%) and only a slightly decreased activity (43% versus 49% yield after 2 days) when reused.[46]

A preceding, less successful fluorous approach to the asymmetric allylic oxidation of cyclohexene with *t*-butyl perbenzoate had been reported by Fache and Piva, who studied the combined use of fluorous proline **13** (Figure 6.3, 13 mol%) and Cu_2O (5 mol%) in various fluorinated and non-fluorinated sol-vents.[52] The catalytic system was completely inactive under fluorous biphasic conditions and the highest allyl ester yield (77%) and enantioselectivity (20% ee) were achieved using hexafluoroisopropanol as a solvent. The fluorous catalyst recovered after evaporation of the solvent and extraction of the organic compounds with petroleum ether was reused in a subsequent run affording the allyl ester in 54% yield and 13% ee.

$C_8F_{17}(CH_2)_3O_{\prime\prime\prime}$

13

N
|
H
CO_2H

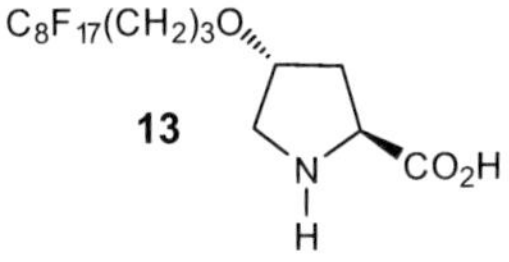

Figure 6.3 Early example of fluorous proline derivative.

6.2.2 Catalysts for C–C Bond Formation Reactions

As anticipated earlier, the peculiar solubility properties of fluorous catalysts have found convenient application in asymmetric reactions performed in a benign and safe medium such as compressed or supercritical CO_2 ($scCO_2$).[53] Indeed, organic substrates and products have adequate solubility in $scCO_2$ whilst most transition metal complexes developed as chiral catalysts for homogeneous reactions in organic solvents are only sparingly soluble in this medium. This solubility difference can be actually exploited to allow catalyst separation, but it also entails considerable mass transfer limitations during the reaction stage. The use of fluorous ligands with their improved affinity for $scCO_2$ offers the opportunity to overcome this drawback. Reactions can be thus performed in a single supercritical phase after which the catalyst can be separated out by controlled changes in pressure and/or temperature in downstream processing.[54] Franciò and Leitner pioneered the use of enantioselective fluorous catalysts in $scCO_2$ and first demonstrated this concept in the case of the rhodium-catalyzed asymmetric hydroformylation of vinyl arenes to give a mixture of linear and branched aldehydes (Scheme 6.8).[55,56]

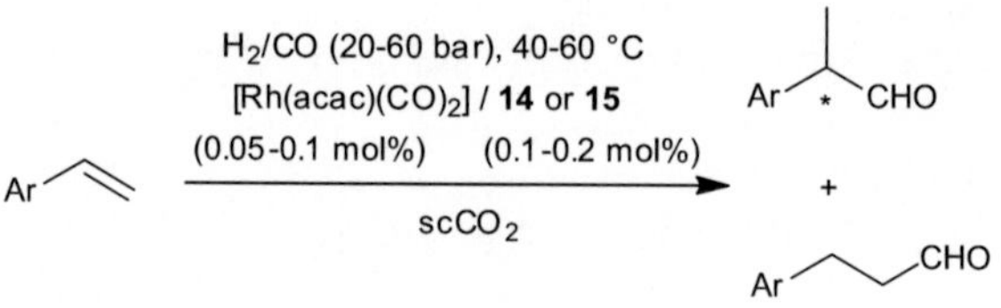

Scheme 6.8 Hydroformylation of styrene derivatives.

Preliminary experiments revealed the viability of the catalytic system based on a rhodium complex containing the chiral phosphine/phosphite ligand **14** (Figure 6.4) for reactions carried out in $scCO_2$. Higher activities and, more strikingly, regio- and enantioselectivities were observed in comparison to reactions carried out in benzene in the presence of the parent non-fluorous (*R*,*S*)-2-(diphenylphosphino)-1,1'-binaphthalen-2'-yl-1,1'-binaphthalene-2,2'-diyl (BIN-APHOS) ligand. Control experiments with **14** in benzene proved that these

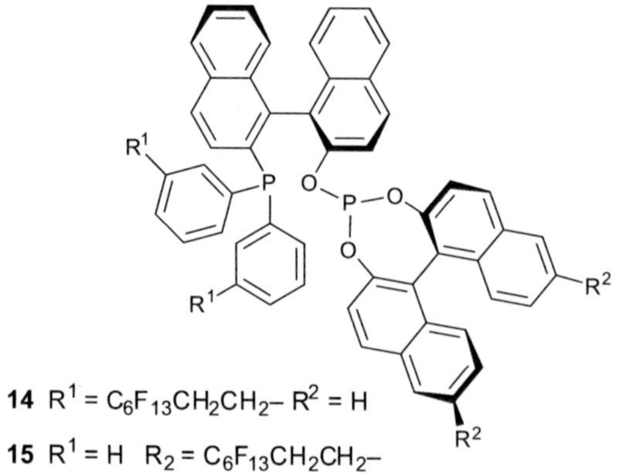

14 $R^1 = C_6F_{13}CH_2CH_2-$ $R^2 = H$

15 $R^1 = H$ $R_2 = C_6F_{13}CH_2CH_2-$

Figure 6.4 Fluorous BINAPHOS ligands for reactions in $scCO_2$.

beneficial effects arise from the presence of fluorous ponytails and not from the nature of the solvent. Also the physical state of CO_2 (liquid or supercritical) did not influence the outcome of the reaction to a large extent, at least in the case of styrene.[55] A wide range of vinyl arenes were hydroformylated with good conversions (70–99%) regioselectivities (up to 95/5 branched to linear aldehyde) and enantioselectivities (up to 95% ee). Recycling of the catalytic system was demonstrated for the enantioselective hydroformylation of styrene. The catalyst could be separated from the product by lowering the density of the scCO$_2$ in the reactor so a second liquid phase formed, in which the catalyst was preferentially soluble. The compressed gas phase containing the products was stripped away by purging the reactor with CO_2 and the residual catalyst could be reused without further treatment. The enantioselectivity of the reaction slowly decreased from an initial value of 86.6% ee to 70.2% ee in the seventh successive run. Nevertheless, the catalytic system was still active after a total turnover of more than 12 000 and rhodium leaching was usually lower than 1 ppm for each run.[56]

Ojima and co-workers developed closely related (*R,S*)- and (*S,R*)-BINA-PHOS derivatives (*e.g.* **15**, Figure 6.4) bearing fluorous ponytails at the peripheral binaphthyl phosphite moiety. The corresponding rhodium complexes generated *in situ* by reaction with [Rh(acac)(CO)$_2$] were tested in the hydroformylation of styrene in different reaction media.[57] The fluorous catalysts showed good activity (styrene conversion = 84–100%) and selectivity for aldehydes (100%), with a ratio branched to linear aldehyde = 9/1 and 88–95% ee when reactions were carried out in benzene, the standard solvent for the hydroformylation catalyzed by Rh-BINAPHOS complexes. However, when the hydroformylation was carried out in scCO$_2$, enantioselectivities dropped to 70–74% ee. This poorer performance was tentatively ascribed to the rapid racemization of the branched aldehyde under the reaction conditions. Preliminary data concerning the use of ligand **15** in fluorous media (*e.g.* biphasic systems toluene/PFMCH or neat PFMCH) showed that enantioselectivities were in the same range as those observed in benzene. Unfortunately, ligands such as **15** with their low fluorine content were preferentially soluble in the organic phase and their recovery and recycling was not attempted.

The use of fluorous separation techniques has been thoroughly explored in the case of the enantioselective 1,2-addition of Et$_2$Zn to benzaldehyde (Scheme 6.9), a popular test reaction for new chiral ligands and catalysts.[58] A variety of fluorous catalytic systems proved to be useful for such an asymmetric C–C bond formation process and the results obtained are summarized in Table 6.1.

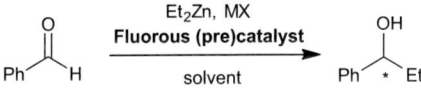

Scheme 6.9 Enantioselective 1,2-addition of Et$_2$Zn to benzaldehyde.

Table 6.1 Enantioselective 1,2-addition of Et$_2$Zn to benzaldehyde promoted by fluorous catalysts.[a]

Entry	Catalyst or Ligand (mol%)	MX	T [°C]	Solvent System[b]	Ee [%] Run 1	Ee [%] Run (n)	Ref.
1	**16** (2.5)	–	RT	Octane/PFMCH	87	37 (3)	59, 60
2	**17** (2.5)	–	RT	Octane/PFMCH	79	36 (4)	59, 60
3	**18** (2.5)	–	RT	Octane/PFMCH	92	43 (4)	59, 60
4	**19** (2.5)	–	RT	Octane/PFMCH	90	60 (9)	60
5	**20** (2.5)	–	RT	Octane/PFMCH	90	70 (13)	60
6	**21** (10)	–	RT	Hexane-toluene	83 (90)[c]	84 (10)	62
7	**22** (10)	–	RT	Hexane-toluene	83 (91)[c]	n.r.	62
8	**23** (10)	–	RT	BTF-hexane	25 (54)[c]	n.r.	62
9	**24** (3)	*n*-BuLi	40	Hexane/FC-72	92	81 (9)	63
10	**25** (10)	–	0 to RT	Hexane-toluene	18 (38)[c]	n.r.	65
11	**26** (10)	–	0 to RT	Hexane-toluene	93 (99)[c]	n.r.	65
12	**27** (10)	–	0 to RT	Hexane-toluene	89 (97)[c]	n.r.	65
13	**28** (20)	Ti(O*i*-Pr)$_4$	45	Hexane/PFMD	54	55 (9)	67, 68
14	**29** (20)	Ti(O*i*-Pr)$_4$	45	Hexane/PFMD	41	31 (3)	68
15	**30** (20)	Ti(O*i*-Pr)$_4$	45	Hexane/PFMD	70	28 (2)	68
16	**31** (20)	Ti(O*i*-Pr)$_4$	45	Hexane/PFMD	64	41 (3)	68
17	**30** (20)	Ti(O*i*-Pr)$_4$	0	CH$_2$Cl$_2$	77	n.r.	68
18	**32** (20)	Ti(O*i*-Pr)$_4$	0	Hexane-toluene/ FC-72	83 (81)[c]	80 (5)	21, 69
19	**33** (20)	Ti(O*i*-Pr)$_4$	0	BTF-hexane-toluene	84 (92)[c]	83 (4)	69
20	**33** (20)	Ti(O*i*-Pr)$_4$	0	Hexane-toluene/ FC-72	79 (82)[c]	78 (3)	69
21	**34** (20)	Ti(O*i*-Pr)$_4$	–20	CH$_2$Cl$_2$-hexane	74	n.r.	70
23	**39** (5)	Ti(O*i*-Pr)$_4$	–78 to –30	Hexane-toluene	97 (97)[c]	97 (7)	72, 73
27	**40** (3)	Ti(O*i*-Pr)$_4$	–35	Hexane	98 (99)[c]	n.r.	75

[a]Catalysts and ligands, see Figures 6.5, 6.6 and 6.7.
[b]PFMCH = perfluoro(methylcyclohexane); BTF = α,α,α,-trifluoromethyltoluene; FC-72 = perfluorohexanes mixture; PFMD = perfluoro(methyldecalin).
[c]Isolated yield in the first run in parenthesis.

Fluorous versions of enantiopure zinc *ortho*-aminoarene thiolate catalysts (Figure 6.5) were developed by van Koten and co-workers who first recognized the potential of the fluorous approach for the 1,2-addition of diorganozinc reagents to aldehydes.[59,60] Fluorous complexes **16–20** were found to be active as pre-catalysts in the asymmetric addition of Et$_2$Zn to benzaldehyde in hexane, their enantioselectivity (up to 94% ee using **18**) being even superior to that of their non-fluorous equivalents (up to 72% ee).[59] The same reaction was then performed with comparable results under fluorous biphasic conditions in octane/PFMCH (Table 6.1, entries 1–5). In this case the fluorous layer was simply decanted and reused. The effectiveness of this recovery technique heavily depended on the fluorine content and number of ponytails of the catalyst. The best results were obtained with **20** that could be used for ten cycles before the value of the ee started to diminish, whereas a considerable drop in enantioselectivity was observed already in the third consecutive run with the pre-catalysts **16–18**.[60]

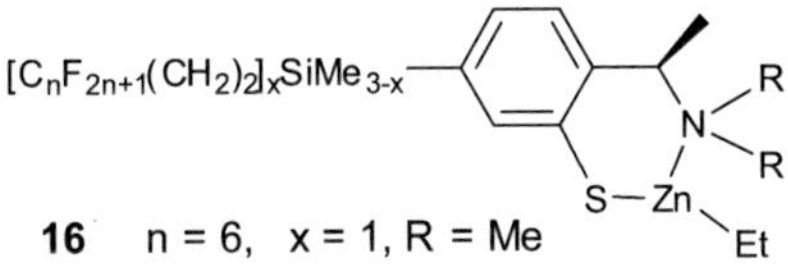

$[C_nF_{2n+1}(CH_2)_2]_xSiMe_{3-x}$

16 n = 6, x = 1, R = Me

17 n = 10, x = 1, R = Me

18 n = 10, x = 1, R,R = -(CH$_2$)$_4$-

19 n = 8, x = 2, R,R = -(CH$_2$)$_4$-

20 n = 8, x = 3, R,R = -(CH$_2$)$_4$-

Figure 6.5 Fluorous zinc *ortho*-aminoarene thiolate catalysts for the 1,2-addition of Et$_2$Zn to aldehydes.

Amino alcohols are well known to react *in situ* with dialkylzincs to generate Zn-based chiral Lewis acid complexes which can further coordinate with both the aldehyde substrates and the dialkylzinc reagents to conduct the catalytic addition.[61] Thus, Nakamura, Takeuchi and co-workers examined the addition of Et$_2$Zn to benzaldehyde in the presence of 10 mol% of ephedrine-based fluorous β-amino alcohols **21–23** (Figure 6.6).[62] Ligands **21** and **22** gave similar results in a mixture of toluene and hexane as the solvent (entries 6 and 7). The more sterically hindered and fluorophilic ligand **23** (entry 8) was insoluble in toluene and had to be tested in a homogeneous mixture of BTF and hexane, affording the secondary alcohol in lower yield and poor enantioselectivity. Recovery of ligand **21** was achieved by F-SPE of the crude reaction mixture and the ligand could be recycled up to ten times with no apparent loss in catalytic activity and enantioselectivity. The enantioselective addition of Et$_2$Zn to representative aldehydes in the presence of **21** was also demonstrated, with enantioselectivities ranging from 84% ee (4-chlorobenzaldehyde) to 70% ee (*trans*-cinnamaldehyde and 3-phenylpropanal).

Fluorous prolinol **24** (Figure 6.6) is an interesting amino alcohol, originally introduced by Bolm, Kim and co-workers as a ligand for the addition of diorganozinc to aldehydes.[63] The additional perfluoroalkyl substituents caused only slightly reduced enantioselectivities with respect to the parent non-fluorous prolinol. Moreover, for the addition of Et$_2$Zn to benzaldehyde performed under fluorous biphasic conditions hexane/FC-72 (entry 9), enantioselectivity increased with the temperature as previously observed in enantioselective FBS epoxidations.[34] The ee value of the product increased from 81% to 86% to 92% when the reactions were performed at 0, 20 and 40 °C, respectively. A further increase of the reaction temperature to 60 °C resulted in a small drop in enantioselectivity (ee = 88%). The preferential solubility of **24** in PFC allowed its recovery by separation of the fluorous layer. The addition of fresh solutions of *n*-BuLi and Et$_2$Zn in hexane restored the catalytic system. Enantioselectivities and catalytic activities were maintained for six consecutive runs performed at 40 °C, after which the ee value of the product decreased to 81% in the

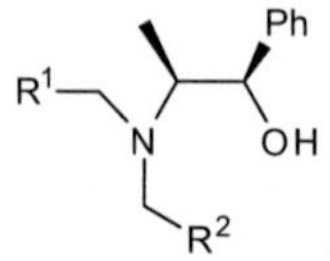

21 $R^1 = 4\text{-}[(C_6F_{13}CH_2CH_2)_3C]C_6H_4\text{-}, R^2 = H$

22 $R^1 = 4\text{-}[(C_6F_{13}CH_2CH_2)_3Si]C_6H_4\text{-}, R^2 = H$

23 $R^1 = R^2 = 4\text{-}[(C_6F_{13}CH_2CH_2)_3C]C_6H_4\text{-}$

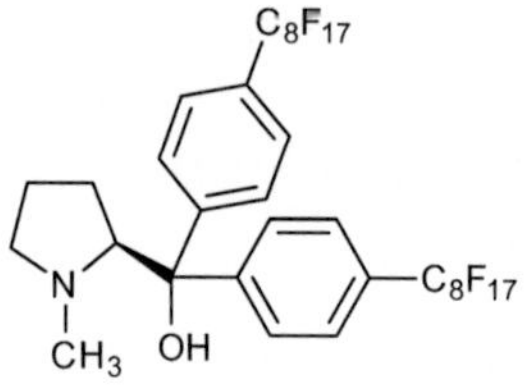

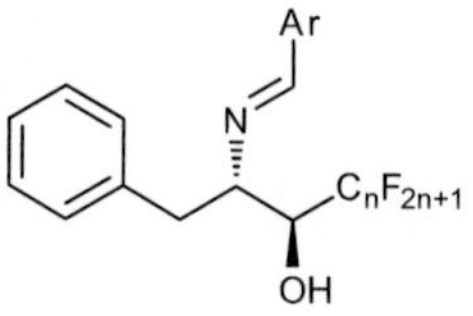

25 n = 8, Ar = 2-Hydroxyphenyl

26 n = 8, Ar = 1-Naphthyl

27 n = 4, Ar = 1-Naphthyl

Figure 6.6 Fluorous β-amino alcohols and Schiff bases tested as ligands in the 1,2-addition of diorganozinc reagents to aldehydes.

ninth run. This value was close to the one obtained using 1 mol% of fresh **24** at the same temperature, thus pointing out to a non-negligible loss of ligand (initial loading 3 mol%) throughout the whole process.

Chiral α-perfluoroalkyl-β-amino alcohols were used as starting material for the preparation of Schiff bases able to bind Et$_2$Zn *in situ* and form chiral complexes with enhanced Lewis acidity thanks to proximity of the electron-withdrawing R$_F$ substituents to the coordinated metal.[64,65] Stereochemistry of the β-amino alcohol backbone had a major impact on the outcome of the asymmetric addition of Et$_2$Zn to benzaldehyde catalyzed by such complexes, with higher enantioselectivities consistently obtained with ligands based on the *anti*-diastereoisomers (*e.g.* **25–27**, Figure 6.6). The length of the R$_F$ chain and the nature of the aromatic aldehyde moiety Ar also played a role (entries 10–12), but their effect can be hardly rationalized. Among the various Schiff bases examined, ligand **26** gave the best results, outperforming the otherwise similar ligand **25** deriving from salicylaldehyde.[65] However, when the same fluorous ligands were tested in the enantioselective alkynylation of benzaldehyde with phenylacetylene/Et$_2$Zn in toluene opposite results were observed, with **25** affording (*R*)-1,3-diphenylprop-2-yn-1-ol in 99% yield and 67% ee against 97% yield and 43% ee obtained with **26**.[66] Despite their promising catalytic behaviour, fluorous Schiff bases were highly sensitive to moisture and could not possibly stand the acidic aqueous work-up conditions. This lack of stability thwarts any effort to set up suitable recovery methods.

Titanium complexes generated *in situ* by enantiopure fluorous chiral diols (Figure 6.7) with Ti(OR)$_4$ represent a third, useful family of catalysts for the enantioselective addition of organometallic reagents to aromatic aldehydes.

The groups of Takeuchi and Chan independently reported the synthesis of enantiopure fluorous 1,1′-bi(2-naphthol) (BINOL) derivatives and studied their behaviour in the Ti-catalyzed addition of Et$_2$Zn to benzaldehyde. Chan's BINOLs (*e.g.* **28–31**, Figure 6.7) were characterized by the presence of up to

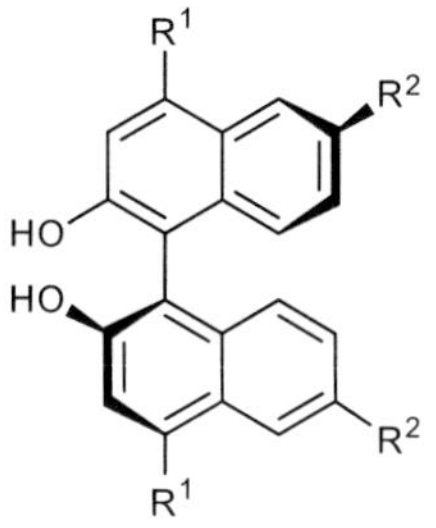

28 $R^1 = R^2 = C_8F_{17}$

29 $R^1 = R^2 = C_4F_9$

30 $R^1 = H, R^2 = C_4F_9$

31 $R^1 = H, R^2 = C_8F_{17}$

32 $R^1 = H, R^2 = (C_6F_{13}CH_2CH_2)_3Si^-$

33 $R^1 = H, R^2 = (C_8F_{17}CH_2CH_2)_3Si^-$

34 $R^1 = H, R^2 = C_6F_{13}$

35 $R^1 = H, R^2 = C_6F_{13}CH_2CH_2-$

36 $R^1 = H, R^2 = C_8F_{17}CH_2CH_2-$

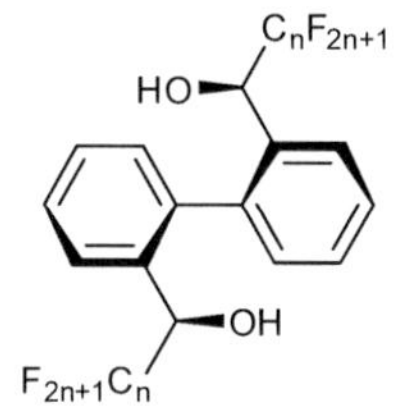

37 n = 1

38 n = 2

39 n = 7

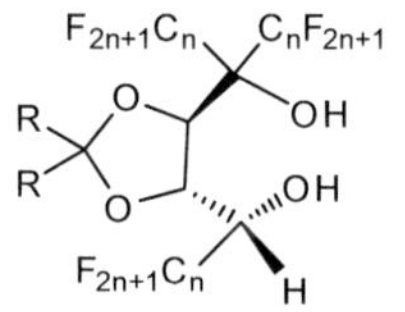

40 n = 8, R = Me

41 n = 6, R = Me

42 n = 4, R = Me

43 n = 8, R,R = - $(CH_2)_5-$

44 n = 4, R,R = - $(CH_2)_5-$

Figure 6.7 Fluorous diols tested as ligands in the 1,2-addition of diorganozinc reagents to aldehydes.

four R_F substituents of various length directly linked to the 4,4'- and 6,6'-positions of the binaphthyl scaffold,[67,68] whereas both ligands **32** and **33** developed by Takeuchi bear hefty fluorous ponytails stemming from silicon atoms at the 6,6'-positions.[21,69] This second approach provided superior catalytic systems, as shown by comparison of the results obtained with the two sets of ligands in the model reaction (entries 13–17 versus 18–20). Chan's ligands **28–31** were tested under FBS conditions of hexane/perfluoro(methyldecalin) (PFMD) at 45 °C, upon which a single phase was formed (entries 13–16). At the end of the reaction, the homogeneous system was cooled down to 0 °C and divided into two layers which were readily separated. The upper organic phase was subjected to acidic aqueous work-up to give the enantiomerically enriched secondary alcohol, whereas the bottom fluorous phase in which the fluorous ligand was immobilized was reused after addition of fresh Ti(*i*-PrO)$_4$. The highest ee value was obtained using the light fluorous ligand **30** (70%, entry 15), which, however, was not suitable for FBS recovery and reuse. The same ligand used under standard homogeneous conditions (entry 17) afforded the product with 77% ee.[68] Ligand **34**, an analogue of **30** bearing longer R_F chains, behaved similarly (entry 21),[70] whereas the parent non-fluorous (*R*)-BINOL or Takeuchi's ligand **33** (entry 19) gave the product with 88% and 84% ee, respectively.[69]

Among Chan's BINOLs, **28** gave the best results in the FBS recycling experiments, but the enantioselectivity observed was moderate (54% ee, entry 13). Similar enantioselectivities were obtained in the FBS addition of Et_2Zn to other aromatic aldehydes in the presence of **28** and $Ti(i\text{-}PrO)_4$, with the enantioselectivities remaining constant after three consecutive runs (*e.g.* 54–51% ee for 4-chlorobenzaldehyde). On the other hand, ligands **32** and **33** were suitable for repeated use in the Ti-catalyzed addition of Et_2Zn to benzaldehyde under fluorous biphasic conditions at 0 °C, affording the product with ee values of 83% and 79%, respectively in the first run (entries 18 and 20). However, when the reaction was performed in the presence of **32** in toluene/hexane/FC-72 about 10% of the fluorous ligand was recovered from the organic phase after the hydrolytic work-up procedure required for liberating the alcohol. For experiments carried out in toluene/FC-72 the amount of **32** that leached into the organic phase was less than 1% and the enantioselectivity was only slightly lowered with respect to that achieved in toluene/hexane/FC-72 (ee = 78% versus 83%). It should be noted that hexane did not enhance the leaching of **33** into the organic phase that was consistently <1%.[69] The addition of Et_2Zn to other aromatic aldehydes catalyzed by $Ti(i\text{-}OPr)_4$ in combination with **32** and **33** was studied under homogeneous conditions using BTF as the solvent. The organic products and the fluorous ligand were easily separated by F-SPE and secondary alcohols were isolated in excellent yields (>90%) with enantioselectivities ranging from 78% ee in the case of 2-naphthylbenzaldehyde to 91% ee in the case of 1-naphthylbenzaldehyde.[69]

Fluorous BINOLs **35** and **36** (Figure 6.7) where straight R_F substituents and methylene insulating units are combined were developed by Zhao and co-workers and tested in the asymmetric allylation of benzaldehyde with allyl-tributyltin in the presence of $Ti(i\text{-}PrO)_4$.[71] The highest enantioselectivities, up to 90% ee, were obtained under fluorous biphasic conditions of hexane/FC-72 at 0 °C. The two ligands gave similar results and, because of their relatively low fluorine content, they could be only partially recovered by phase separation followed by repeated extraction of the organic layer with FC-72 and evaporation of the combined fluorous layers. The same reaction performed in CH_2Cl_2 at 0 °C was investigated by Stuart and co-workers, who reported the efficient separation and recovery of ligand **34** by F-SPE followed by acidic hydrolysis of the crude fluorous material containing a non-polar Sn-**34** adduct.[70] Nevertheless, the enantioselection level achieved with **34** (ee = 66%) was lower than that obtained with **35** (ee = 76.6%) under otherwise similar conditions, thus highlighting the beneficial effect of the methylene spacers.

As a part of their ongoing effort to make the most of perfluoroalkyl groups as active components of chiral ligands and catalysts,[64,65] Ando, Kumadaki and co-workers designed fluorous chiral diols **37–44** (Figure 6.7) and proved their viability in the Ti-catalyzed enantioselective addition of organometallic reagents to aromatic aldehydes.[72–76] Enantiopure biphenyl diols **37–39** characterized by the hindered rotation around the aryl–aryl bond and by the presence of two stereogenic quaternary carbons bearing both perfluoroalkyl- and hydroxy groups efficiently interact with Ti(i-PrO)$_4$ to generate catalytically active chiral complexes.[72] Enantioselectivities up to 97% ee were observed in the addition of Et$_2$Zn to benzaldehyde under homogeneous conditions in the presence of ligand **39** (Table 6.1, entry 23). Under the same conditions, 85% ee and 91% ee were achieved using ligands **37** and **38**, respectively, thus showing a significant dependence of the level of enantioselection on the length of the R$_F$ chains. Ligand **39** was recovered almost quantitatively from the reaction mixture by repeated extraction with FC-72 and evaporation of the combined fluorous layers, and reused after addition of fresh Ti(i-PrO)$_4$. Seven subsequent runs were carried out with no apparent loss of catalytic activity and enantioselectivity.[73] The presence of electron-withdrawing R$_F$ substituents close to the binding sites of the ligand resulted in a remarkably high activity of the catalytic system based on **39** and Ti(i-PrO)$_4$ and a reduced amount of ligand (5 mol%) was required with respect to fluorous BINOLs. The increased catalytic activity was confirmed in a more challenging reaction, namely the methylation of aromatic and aliphatic aldehydes using Me$_2$Zn generated *in situ* from ZnCl$_2$ and MeMgBr.[74] 1-Phenylethanol was obtained from benzaldehyde in 89% yield with 94% ee in the presence of 20 mol% of **39**. Enantioselectivities ranging from 12% ee (*p*-anisaldehyde) to 92% ee (*p*-*i*-Pr-benzaldehyde) were observed in the methylation of other aromatic aldehydes, whereas extremely high enantioselective methylation was attained with aliphatic aldehydes such as octanal (99% ee). As in the case of Et$_2$Zn addition, the ligand could be recovered by liquid–liquid extraction with FC-72.

The same factors that ensured the success of **39** were exploited in the design of fluorous TADDOL-like ligands **40–44** (Figure 6.7).[75,76] Preliminary investigations showed that **40** was highly effective in the Ti-catalyzed addition of Et$_2$Zn to benzaldehyde under homogeneous conditions (Table 6.1, entry 27).[75] The application of **40** to the addition of Me$_2$Zn to a broad range of aromatic and aliphatic aldehydes was also successful, secondary alcohols being obtained in excellent yields and ee values ranging from 84% (benzofuran-2-ylcarbaldehyde) to >99% (4-trifluoromethylbenzaldehyde) using only 6 mol% of ligand. The fluorous diol **40** could be easily separated from the products by precipitation from cold toluene, following a procedure similar to that previously described for fluorous Box **7–12**. Recycling of **40** was demonstrated in the case of the addition of Me$_2$Zn to benzaldehyde: the ligand could be reused up to five times with a progressive erosion of the ee values that was possibly due to its incomplete recovery.[75] The activity and recyclability of fluorous TADDOL-like diols **41–44** was next studied.[76] Compared to **40**, these ligands were more soluble in organic solvents owing to the presence of shorter R$_F$ substituents and/or more extended ketal moieties. This allowed to reduce the volume of hexane, the organic solvent used in the reaction step. At the same time, recovery of these less fluorophilic ligands was not as effective as in the case of **40**. In particular, diol **42** bearing C$_4$F$_9$ substituents could not be separated by precipitation from reaction mixtures and its extraction with perfluorohexane was

required. Shortening the fluorous ponytails had little effect on enantioselectivities, in contrast with the marked influence of this factor observed with
biphenyl diols **37–39**. Finally, in the case of some substrates such as cyclo-
hexancarbaldehyde, catalytic systems based on cyclohexanone derivatives **43**
and **44** were shown to give higher ee values than those based on acetone
derivatives **40–42** (*e.g.* 93% ee with **44** versus 84% ee with **40**).

Despite the great deal of attention paid to transition metal complexes of
fluorous trialkyl- and triarylphosphines at the onset of FBS catalysis,[11,77] the few
enantioselective fluorous catalysts based on chiral phosphorous-based ligands
reported in the late 1990s were developed for use in chemical processes performed in scCO$_2$, such as the above mentioned rhodium-catalyzed asymmetric
hydroformylation of vinyl arenes.[55] Chiral phosphines specifically designed for
fluorous applications were reported a few years later, starting with examples in
the field of palladium(0)-catalyzed asymmetric transformations. Sinou, Pozzi
and co-workers synthesized the chiral monodentate phosphine **45** (Figure 6.8),
a fluorous analogue of the versatile 2-(diphenylphosphino)-2′-methoxy-1,1′-
binaphthyl (MOP) ligand, and described its use in the asymmetric allylic
alkylation of 1,3-diphenyl-2-propenyl acetate with various carbon nucleophiles
(Scheme 6.10).[78] Reactions were conveniently carried out in toluene or BTF at
room temperature or 50 °C, in the presence of a mild base such as BSA/KOAc
(BSA = bis(trimethylsilyl)acetamide). The substitution products were obtained

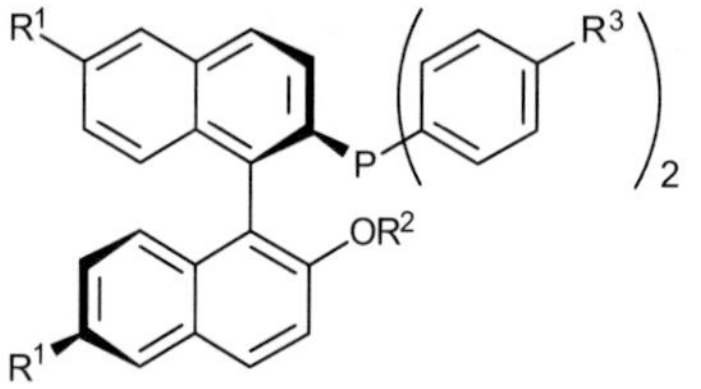

45 R^1 = H, R^2 = C$_7$F$_{15}$CH$_2$-, R^3 = C$_7$F$_{15}$CH$_2$O-

46 R^1 = C$_8$F$_{17}$, R^2 = C$_7$F$_{15}$CH$_2$-, R^3 = H

47 R^1 = (C$_6$F$_{13}$CH$_2$CH$_2$)$_3$Si-, R^2 = Me, R^3 = H (S)-configuration

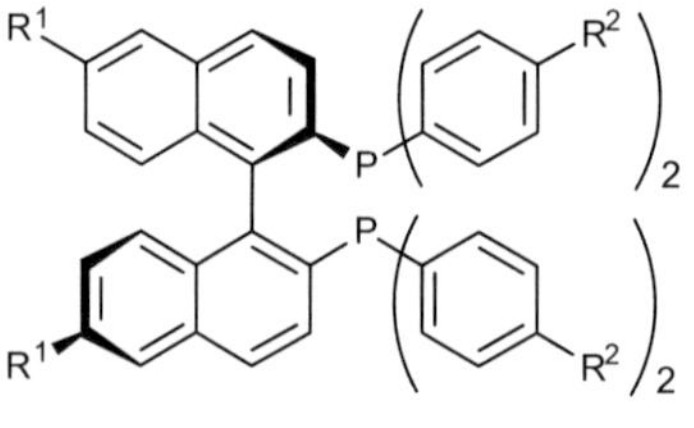

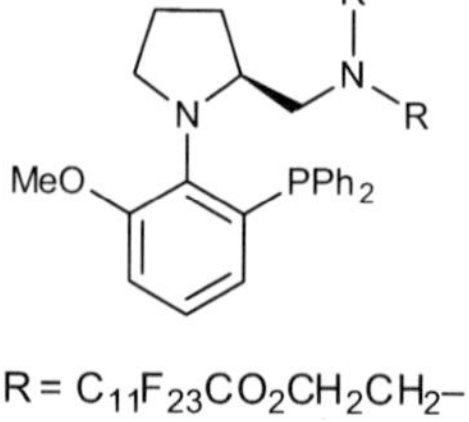

48 R^1 = H, R^2 = C$_7$F$_{15}$CH$_2$O-

50 R^1 = (C$_6$F$_{13}$CH$_2$CH$_2$)$_3$Si-, R^2 = H

49 R = C$_{11}$F$_{23}$CO$_2$CH$_2$CH$_2$–

Figure 6.8 Mono- and bidentate fluorous phosphines for Pd-catalyzed reactions.

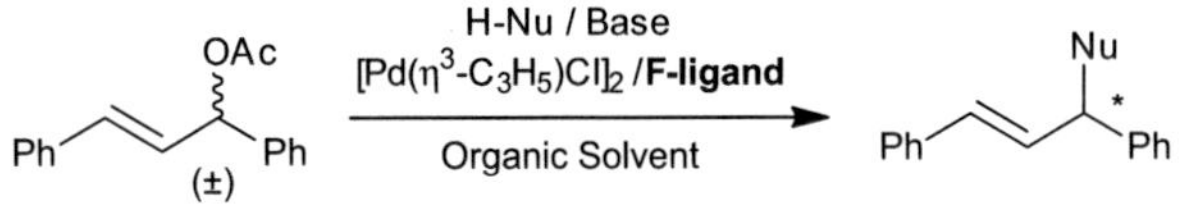

Scheme 6.10 Pd^0-catalyzed asymmetric allylic alkylation of 1,3-diphenyl-2-propenyl acetate.

in moderate to excellent yields (67–99%) with enantioselectivities up to 87% ee being attained using dimethyl malonate as a nucleophile in toluene as a solvent. Palladium catalysts based on ligands **46** and **47** (Figure 6.8) were also tested in the same C–C bond-forming reaction, showing reduced enantioselectivities and activities. A strong base such as NaH was required to activate dimethyl malonate and the allylated product was obtained in 37 and 24% ee, respectively.[79] Fluorous MOPs and their corresponding palladium complexes were removed from homogeneous reaction mixtures by liquid–liquid extraction with perfluorooctane. In the case of **45**, the recovered fluorous materials did not show any catalytic activity when tested in the second run probably due to the oxidation of the phosphine moiety to phosphine oxide during workup.[78]

The synthetic approach that allowed the introduction of three fluorous ponytails in the structure of MOP **45** was extended to the preparation of the fluorous 2,2'-bis(diphenylphosphino)-1,1'-binaphthyl (BINAP) analogue **48** (Figure 6.8).[80] Its behaviour was studied in several standard asymmetric reactions, including the palladium-catalyzed allylic alkylation of 1,3-diphenyl-2-propenyl acetate with dimethyl malonate where it was found to be inferior to the monodentate ligand **45**. Quantitative conversion of the substrate was obtained when the reaction was run in THF at 50 °C in the presence of NaH, but the enantioselectivity was rather low (ee = 32%). The fluorous catalyst recovered by liquid–liquid extraction of the reaction mixture with FC-72 gave the same enantioselectivity in the second reaction cycle, but a definitely lower conversion (20%).

A most successful fluorous catalytic system for the asymmetric allylic alkylation of 1,3-diphenyl-2-propenyl acetate with dialkyl malonates was developed by Mino and co-workers.[81] The complex prepared from [Pd(η^3-C$_3$H$_5$)Cl]$_2$ and diaminophosphine **49** (Figure 6.8) exhibited a clear thermomorphic behaviour in Et$_2$O and hexane, and reactions could be conveniently performed under homogeneous conditions in a temperature range 0–30 °C. Enantioselectivities up to 97% ee were achieved working at 0 °C in Et$_2$O in the presence of BSA/LiOAc as a base, diethylmalonate as a nucleophile and a ligand loading of 10 mol%. Palladium complexes of fluorous Box ligands **7–12** (see Figure 6.2) afforded comparable activities and enantioselectivities for this reaction,[44–46] but their separation and recovery was problematic, as formation of palladium black occurred readily throughout the process. Separation of the free Box ligands from the products was achieved by evaporation of the organic solvent followed by extraction of the residue with FC-72 or by F-SPE, but only **12** could be recovered in quantitative yield.[45] Fresh [Pd(η^3-C$_3$H$_5$)Cl]$_2$ had to be added to the ligands to regenerate the catalytic systems for reuse in a second run. In the case of **49**, recycling experiments were performed with a ligand loading of 10 mol% at 30 °C in order to speed up reactions, resulting in a slightly reduced initial enantioselection level (ee = 90%) with respect to the optimal reaction conditions described above.[81] The fluorous catalyst was easily recovered from the reaction mixture by cooling to 0 °C, followed by the removal of Et_2O and extraction of the organic products with cold hexane. The residual solid fluorous material (most likely a cationic palladium enolate complex with ligand **49** and diethylmalonate) was reused as such. Five consecutive reaction cycles were performed without deterioration of its activity and enantioselectivity.

Finally, the asymmetric Heck reaction between 2,3-dihydrofuran and 4-chlorophenyltriflate (Scheme 6.11) catalyzed by a fluorous palladium complex based on the BINAP analogue **50** (Figure 6.8) was reported by Nakamura and co-workers.[82] Catalytic tests were conducted in the presence of **50** and $Pd(OAc)_2$ under either homogeneous or fluorous biphasic conditions. The major product 2-(4-chlorophenyl)-2,3-dihydrofuran was obtained in 59% yield and 92% ee using benzene as a solvent (isomer ratio 72/28). The same reaction proceeded about three times faster when run in the presence of the parent BINAP in benzene, affording the major isomer in 71% yield and 91% ee. The fluorous ligand used in a fluorous biphasic system benzene/FC-72 afforded the major isomer with an excellent 93% ee, but in modest yield (39%) and regioselectivity (isomer ratio = 69/31). Attempts to reuse the fluorous phase in further reactions failed due to the oxidation of **50** to the corresponding phosphine oxide. The arylation of 2,3-dihydrofuran with various triflates was also investigated using a palladium complex of fluorous BINAP **48**,[80] under conditions otherwise equivalent to those reported by Nakamura's group. Best results were obtained with 4-chlorophenyl triflate as arylating agent in BTF as a solvent. In this case too, 2-(4-chlorophenyl)-2,3-dihydrofuran was obtained as the major product, with moderate enantioselectivity (68% ee), but very high regioselectivity (isomer ratio = 97/3). Ligand **48** did not contain enough fluorine to be used with success under fluorous biphasic conditions, but it could be quickly separated from the organic products by liquid–liquid extraction with FC-72.

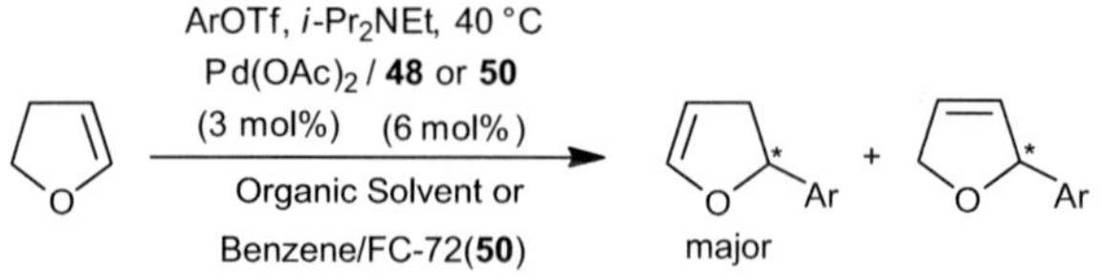

Scheme 6.11   Asymmetric Heck arylation of 2,3-dihydrofuran.

Catalytically active chiral complexes generated *in situ* from inexpensive copper precursors and fluorous chiral bidentate nitrogen ligands have been tested in various asymmetric C–C bond-forming reactions, starting with examples reported by Benaglia and co-workers who developed fluorous chiral

Box **51–54** featuring a different number of R_F substituents attached to the bridging carbon atom through oxygenated spacers (Figure 6.9).[47,48] Complexes generated from ligands **52** and **54** and an equimolar amount of CuOTf were evaluated as catalysts in the cyclopropanation of styrene with ethyl diazoacetate (Scheme 6.12).[47] The use of the C_2-symmetric ligand **51** (10 mol%) under fluorous biphasic conditions CH_2Cl_2/perfluorooctane at 20 °C led to a *trans/cis* mixture of cyclopropanes (73/27) in 55% overall yield, the major *trans* isomer being obtained with 60% ee. The complex based on the less fluorinated C_1-symmetric ligand **53** was found to be completely soluble in CH_2Cl_2 and when the cyclopropanation was run in such a solvent a 65/35 mixture of *trans/cis* cyclopropanes was obtained in 68% yield, with enantioselectivity up to 78% ee for the major *trans* isomer. Under the same conditions, a poly(ethyleneglycol)-supported Box structurally similar to **53** gave the major *trans* isomer with up to 91% ee, whereas up to 99% ee was reported in the literature for C_2-symmetric *gem*-dimethyl-substituted Box.[83] This trend was tentatively explained as a result of the reduced complexation ability of Box ligands bearing bulky fluorinated substituents on the bridging carbon atom, leading to the presence in

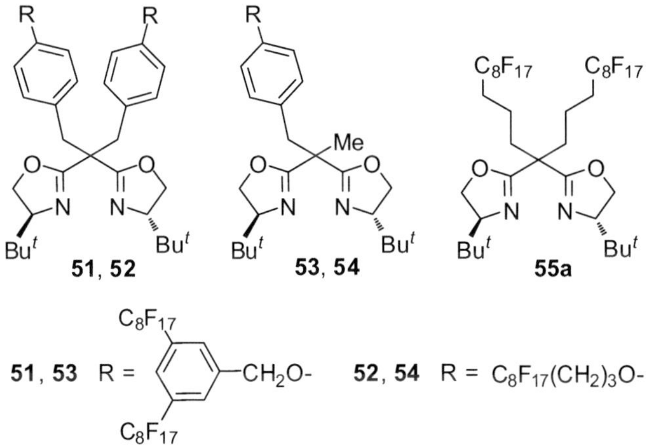

Figure 6.9 C_2- and C_1-symmetric fluorous bis(oxazolines) tested in Cu^I- and Cu^{II}-catalyzed reactions.

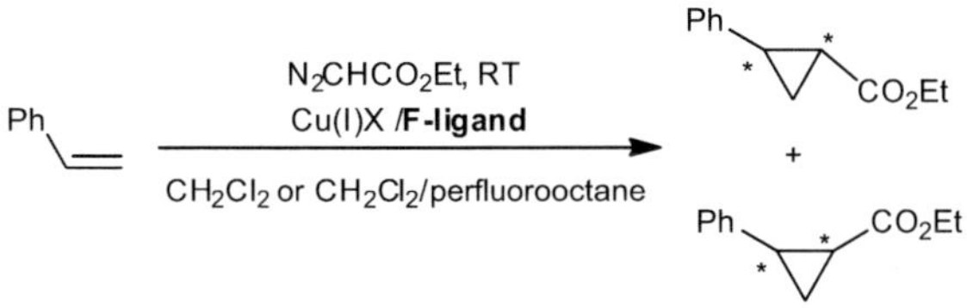

Scheme 6.12 Cu^I-catalyzed asymmetric cyclopropanation of styrene.

solution of Box-free catalytically active species that promote a poorly enantioselective transformation. Results obtained in the same reaction operated in
the presence of Box **55a** (Figure 6.9) bearing two uncomplicated fluorous
ponytails at the bridging carbon atom were in agreement with this assump-
tion.[49] The homogeneous catalytic system afforded a 68/32 mixture of *trans/cis*
cyclopropanes in 22% yield at 20 °C, with 84% ee and 81% ee for the major
and the minor isomer, respectively. The ligand loading was only 2 mol%, and
the isolated yield could be increased up to 63% with virtually identical dia-
stereo- and enantioselectivity by simply running the reaction at 40 °C. The
fluorous nature of the Cu-catalyst obtained from ligand **55a** allowed its efficient
recovery and reuse by means of precipitation induced by the addition of cold
hexane to the concentrated reaction mixture. The recovered solid could be
reused without further addition of Cu^IX or ligand and the diastereoselectivities
and enantioselectivities were maintained over five runs, although a slight
decrease in chemical yield was observed.

Other enantiopure fluorous bidentate nitrogen ligands have been investigated
as Cu^I ligands in the asymmetric cyclopropanation of styrene under both
homogeneous and FBS conditions.[84,85] Among them, relatively rigid diamines
derived from *trans*-1,2-diaminocyclohexane **55b**–**59** (Figure 6.10) afforded the
best results. As an example, the complex generated from $Cu(CH_3CN)_4PF_6$
(10 mol%) and diamine **55b** (20 mol%) catalyzed the reaction between styrene
and ethyl diazoacetate in a biphasic system CH_2Cl_2/perfluoroctane at 20 °C
affording a 67/33 mixture of *trans/cis* cyclopropane adducts in 77% overall
yield, the ee value of the major *trans* isomer being 62%.[84] The complex was
easily removed from the products by simply decanting the fluorous phase
that was reused as such in a second run maintaining the same yield and

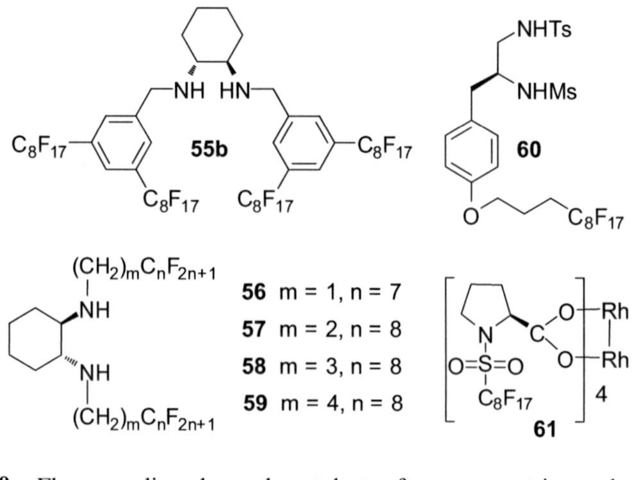

Figure 6.10   Fluorous ligands and catalysts for asymmetric cyclopropanation
reactions.

diastereoselectivity. However, the enantioselectivity fell to 46% ee due to partial decomposition of the ligand.[84]

Two additional examples of transition metal-based fluorous catalysts for cyclopropanation reactions not based on Cu[I] have been reported. The fluorous disulfonamide **60** (Figure 6.10) was synthesized by Imai and co-workers and successfully employed in the enantioselective Simmons–Smith reaction performed on allylic alcohols (Scheme 6.13).[86] *Trans*-oriented alcohols were converted to the corresponding cyclopropane derivatives in excellent yields (>90%) and good enantioselectivities (67–78% ee) and the ligand recovered by F-SPE retained the initial enantioselectivity in two further reaction cycles.

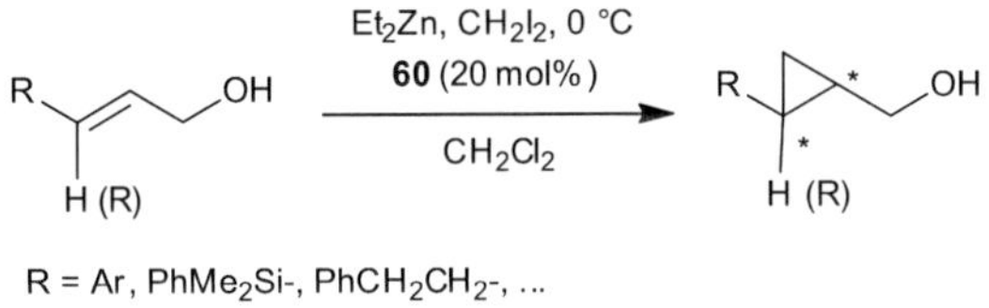

R = Ar, PhMe₂Si-, PhCH₂CH₂-, ...

Scheme 6.13 Simmons–Smith asymmetric cyclopropanation of allylic alcohols.

Biffis and co-workers prepared the fluorous tetrakis-dirhodium(II)-prolinate complex **61** (Figure 6.10) which was applied in the cyclopropanation of styrene with methyl phenyl diazoacetate (Scheme 6.14).[87] Different fluorous techniques for the recovery and recycling of this catalyst were compared, including its adsorption onto fluorous silica prior to the reaction stage that was thus performed in pentane under solid/liquid heterogeneous conditions: the reaction proceeded in 66% chemical yield with almost complete diastereoselectivity and moderate enantioselectivity (ee = 60%). In an alternative, a solution of **61** in PFMCH was added to an excess of styrene and the reaction was run under fluorous biphasic conditions. The products formed upon addition of methyl phenyl diazoacetate were soluble in styrene, whereas the catalyst remained confined in the fluorous phase that was removed from the organic phase by simple decantation at the end of the reaction. Under these conditions, the reaction yield attained 79% and the enantioselectivity 62% ee. The enantioselectivity rose to 72% ee in a successive run carried out with the recovered fluorous phase, with no substantial variation of chemical yield, while in the third run the yield fell to 65%.

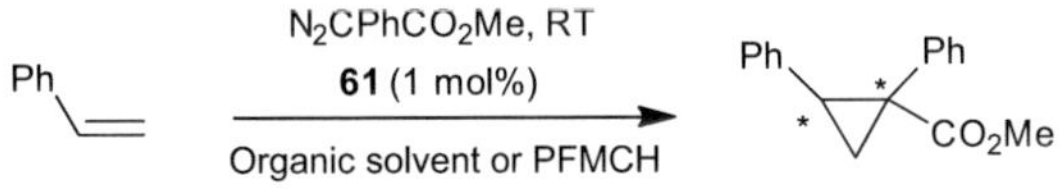

Scheme 6.14 Rh[II]-catalyzed asymmetric cyclopropanation of styrene.

Fluorous chiral Box ligands proved to be useful in Cu[II]-catalyzed asymmetric reactions such as the Mukaiyama aldol addition of a silylketene thioacetal to methyl pyruvate (Scheme 6.15)[48] and the glyoxylate ene reaction (Scheme 6.16).[47,48,50]

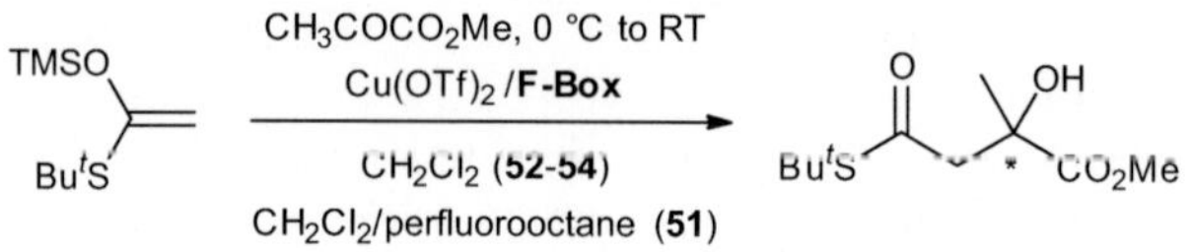

Scheme 6.15 Cu[II]-catalyzed enantioselective Mukaiyama aldol reaction.

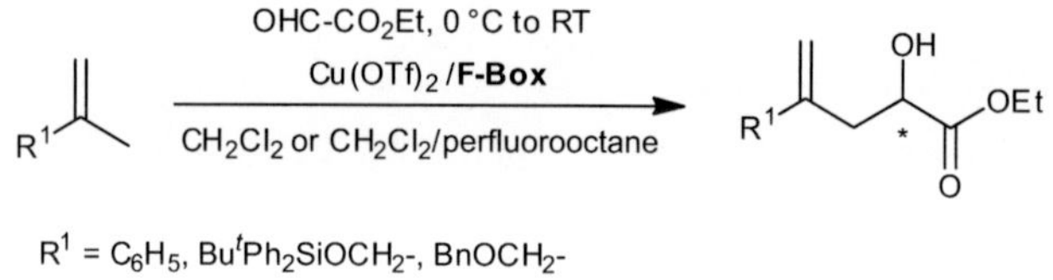

R^1 = C_6H_5, Bu[t]Ph_2SiOCH_2-, $BnOCH_2$-

Scheme 6.16 Cu[II]-catalyzed enantioselective glyoxylate ene reaction.

In both cases C_1-symmetric ligands **53** and **54** performed better than C_2-symmetric ligands **51** and **52** (Figure 6.9). The latter ones afforded an essentially racemic product (ee < 10%) in low chemical yields (<15%) in the aldol reaction, while catalytic systems based on **53** and **54** showed fair activities (55% and 65% yield, respectively) and good enantioselectivities (ee = 83% and 85%), close to those observed with the benchmark Evan's Box (71% yield and 92% ee). Copper-free ligand **54** could be recovered in 85% yield upon treatment of the reaction mixture with aqueous KCN followed by F-SPE.[48] The C_1-symmetric ligands gave equivalent results also in the ene reaction between α-methylstyrene and ethyl glyoxalate (Scheme 6.16). Indeed, the use of Box **53** and **54** in CH_2Cl_2 led to the α-hydroxy ester product in 65% yield, with ee values of 67% and 74%, respectively. The Cu[II]-catalyzed ene reaction proceeded smoothly also in the presence of C_2-symmetric Box, but in this case the nature of the fluorous substituents at the bridging carbon atom had a much greater influence on the enantioselectivities: Box **52** gave the product in 73% chemical yield, but only 7% ee under homogeneous reaction conditions, while the ee value reached 26% when the reaction was run under fluorous biphasic conditions in the presence of Box **51**.[48] Less encumbered C_2-symmetric fluorous Box **7** and **9** (Figure 6.2) investigated by Sinou and co-workers provided better results, affording enantioselectivities as high as 87% ee with α-methylstyrene as a substrate.[50] The ene reaction catalyzed by Cu[II] complexes of **7** and **9** afforded α-hydroxy esters in good yields and moderate to high ee values with other representative substrates as well. Moreover, these fluorous complexes could be efficiently recovered and reused using the same technique previously described in the case of styrene cyclopropanation. Up to five consecutive runs could be thus performed in the case of α-methylstyrene. Sinou and co-workers also explored the immobilization of the Cu[II]-Box complexes onto fluorous silica and their use as heterogeneous solid catalysts. This approach gave good results in terms of recovery and reuse, but reduced enantioselectivities were sometimes observed as a consequence of the adverse influence of the fluorous support.[50]

Cu^{II}-catalyzed enantioselective reactions have been recently used by Reiser and co-workers to study the effect of a combination of fluorous ponytails and triazole linking units on the performance of Cu^{II} complexes of chiral aza Box ligands **62** and **63** (Figure 6.11).[88] Indeed, supported chiral catalysts with triazole linkers often suffer from reduced enantioselectivities due to the ability of triazole to act as an additional coordination site. The concurrent presence of non-coordinating R_F substituents was expected to offset the interference of triazole, and this was actually observed for reactions that are better performed in CH_2Cl_2 or THF, as the enantioselective nitroaldol reaction between benzaldehyde and nitromethane where the Cu^{II} complexes generated *in situ* from $Cu(OAc)_2$ (5 mol%) and an equimolar amount of **62** or **63** in THF afforded (S)-2-nitro-1-phenylethanol with comparable enantioselectivities (90% ee and 86% ee, respectively) as observed with the standard aza-Box **64** (92% ee). Both fluorous catalysts could be recovered after completion of the reaction by elimination of the organic solvent followed by precipitation from a cold mixture of hexanes-CH_2Cl_2 1/1 and filtration. Some erosion of enantioselectivity was observed after three (ligand **62**) or two (ligand **63**) reaction cycles.

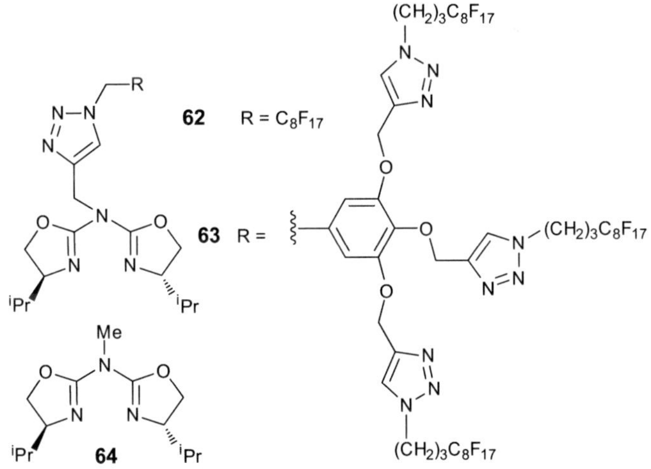

Figure 6.11 Fluorous aza bis(oxazolines) ligands.

Things turned out differently when a protic solvent was required, as in the case of the enantioselective alkylation of indole with benzylidenemalonate (Scheme 6.17). Here the fluorous ligands in combination with various Cu^{II} consistently gave inferior results with respect to **64**. Enantioselectivities up to 85% ee and 68% were attained with **62** and **63**, respectively, whereas the complex generated from $Cu(OTf)_2$ and **64** afforded the product in 95% ee. Moreover, the activity of the recovered fluorous catalysts was maintained in a

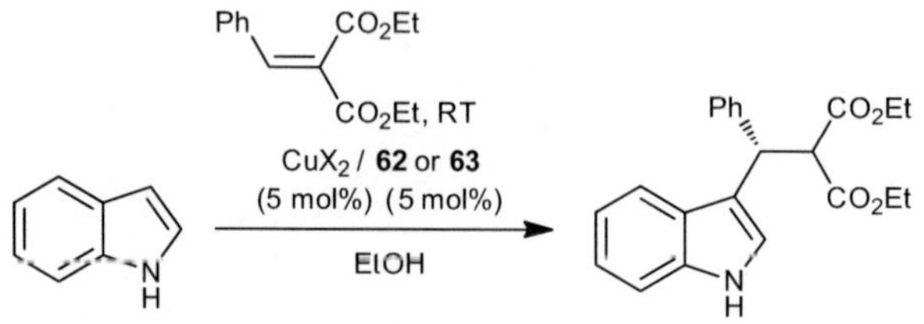

Scheme 6.17 Cu^{II}-catalyzed enantioselective alkylation of indole with benzylidenemalonate.

second reaction cycle, but enantioselectivities dropped, indicating the complexation of Cu^{II} outside the chiral azabis(oxazoline) environment.

6.2.3 Reduction Catalysts

An impressive number of homogeneous catalytic systems have been reported for the asymmetric catalytic hydrogenation of prochiral substrates containing unsaturated C=C, C=O and C=N bonds with hydrogen gas, and transition metal complexes of chiral phosphorus ligands have achieved great success in this area.[89] Early examples of their fluorous analogues were developed with the aim to efficiently perform hydrogenation reactions in scCO$_2$ (Figure 6.12). Leitner, Pfaltz and co-workers thus prepared the cationic Ir^{I} complexes **65a–c** featuring the same enantiopure fluorous phosphinodihydrooxazole ligand, but different fluorinated counter anions.[90] These complexes were used as catalysts for the hydrogenation of prochiral imines (Scheme 6.18) under both classical homogeneous conditions in CH$_2$Cl$_2$ or in scCO$_2$. The nature of the counter anion had little influence on the hydrogenation of the model substrate N-(1-phenylethylidene)aniline in CH$_2$Cl$_2$, which occurred with quantitative conversions and enantioselectivities ranging between 80% and 86% ee, as observed in reactions catalyzed by the corresponding non-fluorous complexes. On the other hand, the level of enantioselection strongly depended on the counter anion for reactions carried out in scCO$_2$, the highest enantioselectivity (80% ee) being obtained using catalyst **65a**.

The beneficial role of the tetrakis-[3,5-bis(trifluoromethyl)phenyl]borate anion (BARF) was confirmed in the asymmetric hydrogenation of dimethyl itaconate in scCO$_2$ (Scheme 6.19) catalyzed by cationic rhodium complexes of the bidentate ligand **66** (Figure 6.12).[91] The initial ee value of 13% was increased to 72% ee simply by exchanging the original BF$_4^-$ counter anion with BARF.

Ligand **14** (Figure 6.4), developed by Leitner's group, was analogously applied in the Rh-catalyzed hydrogenation of dimethyl itaconate and 2-acetamido methyl acrylate in compressed CO$_2$ (Scheme 6.19).[56] In contrast to the previous examples, the addition of **14** to [Rh(COD)$_2$]BF$_4$ formed an active hydrogenation catalyst that gave very high levels of enantiocontrol. Both substrates were reduced quantitatively in the presence of 0.01 mol% of catalyst,

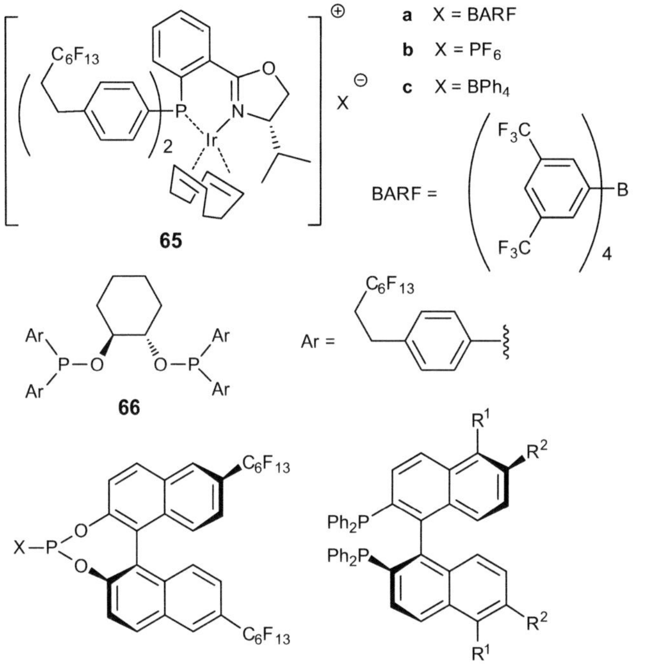

67 X = Ph
68 X = Me_2N-
69 X = PhO-
70 X = *p*-C_6F_{13}PhO-

71 $R^1 = C_6F_{13}$, R^2 = H
72 $R^1 = C_8F_{17}$, R^2 = H
73 R^1 = H, $R^2 = C_6F_{13}$
74 R^1 = H, $R^2 = C_6F_{13}(CH_2)_2$-
75 R^1 = H, $R^2 = [C_8F_{17}(CH_2)_2]_3Si$-

Figure 6.12 Fluorous catalysts and ligands for hydrogenation reactions.

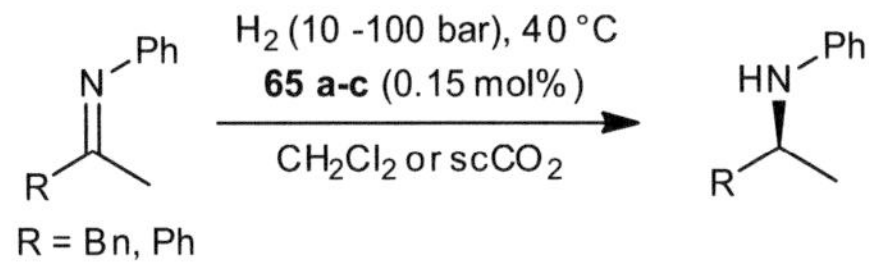

Scheme 6.18 Enantioselective hydrogenation of imines.

with enantioselectivities up to 97% ee. More recently, the asymmetric hydrogenation of water-soluble substrates in inverted $scCO_2$-water biphasic system have been reported by the same group.[92,93] In this case the fluorous complex generated from **14** and $[Rh(COD)_2]$BARF was immobilized in the $scCO_2$ phase whereas the polar organic compounds were contained in the aqueous phase. This approach allowed the straightforward separation of the products from the catalyst upon completion of the reaction. Full conversion and an ee value of

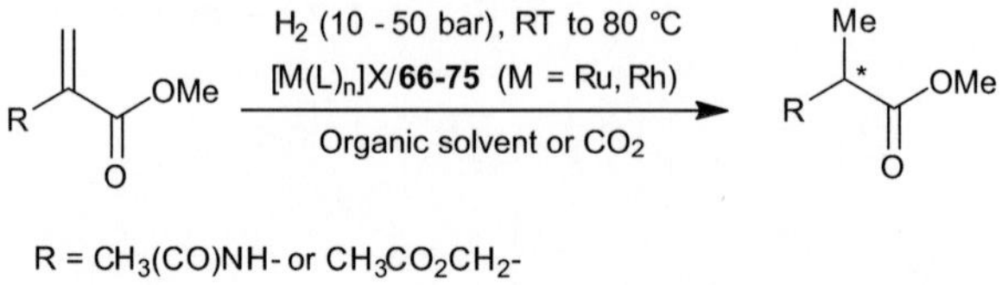

R = CH$_3$(CO)NH- or CH$_3$CO$_2$CH$_2$-

Scheme 6.19   Enantioselective hydrogenation of dimethyl itaconate and 2-acetamido methyl acrylate.

97% were achieved for the hydrogenation of 2-acetamido methyl acrylate using a catalyst loading of 0.5 mol% catalyst.[92] The product was isolated in quantitative yields directly from the water layer whereas the catalyst dissolved in scCO$_2$ remained in the reactor and could be recycled five times with an average ee value of 98.4% and without any significant loss of catalytic efficiency. Asymmetric hydrogenation of itaconic acid was also carried using the same fluorous complex.[93] The catalytic system was stable over seven cycles yielding product with up to 94% ee, after which its activity, but not its enantioselectivity, started to decrease.

Despite these promising results, the viability of asymmetric hydrogenations catalyzed by fluorous chiral complexes in scCO$_2$ is yet to be fully demonstrated. On the contrary, systematic investigations performed with ruthenium and rhodium complexes of fluorous chiral phosphorus ligands possessing a binaphthyl backbone revealed that the scCO$_2$ environment can have a negative effect on their catalytic behaviour.[94–97]

Hope and co-workers have studied the application of monodentate ligands **67–70** (Figure 6.12) deriving from enantiopure binaphthol in the rhodium-catalyzed hydrogenation of dimethyl itaconate in CH$_2$Cl$_2$ and scCO$_2$.[94] In the conventional halogenated solvent the fluorous complexes deriving from **67–69** displayed improved enantioselectivities and almost equivalent catalytic activities to those of their non-fluorous analogues, affording ee values up to 99% and complete substrate conversion. This remarkable catalytic behaviour was not retained in scCO$_2$ where ee values and conversions lower than 10% were attained. The addition of NaBARF improved these results to a little extent, although in the case of ligand **70** the ee value reached 65% with 28% conversion.

Fluorous BINAP analogues **71** and **72** (Figure 6.12), developed by Lemaire and co-workers, gave excellent results in the ruthenium-catalyzed asymmetric hydrogenation of β-keto esters in MeOH or EtOH, affording quantitative conversions and ee values ranging from 96% to 99%. In the presence of the same ligands the ruthenium-catalyzed asymmetric hydrogenation of 2-acetamido methyl acrylate conducted in scCO$_2$ did not proceed at all.[95] The catalytic activity was recovered by addition of a co-solvent (hexafluoroisopropanol or BTF) to the scCO$_2$ system. Complete conversions were thus obtained, with moderate enantioselectivities (*e.g.* 74% ee with both **71** and **72** in the case of addition of BTF). Such a behaviour was tentatively ascribed to the improved solubility of the fluorous catalysts in the mixed $scCO_2$/fluorinated solvent system, but this hypothesis was ruled out by a concurrent study due to the groups of Hope, Stuart and Xiao who investigated the activity of ruthenium complexes of the fluorous BINAP **73** and **74** in the asymmetric hydrogenation of dimethyl itaconate.[96,97] They found that the level of enantioselectivity achieved with these complexes ($>95\%$ ee) was similar to that obtained with BINAP for reactions performed in MeOH, in agreement with the results of Lemaire.[95] They also noted that the inclusion of ethylene spacers between the R_F substituents and the binaphthyl framework allowed to attain a reaction rate similar or faster compared to that obtained with BINAP (complete conversions in about 15 min).[96] In $scCO_2$ the fluorous complexes showed reduced enantioselectivities (up to 76% ee with ligand **74**), and longer reaction times (24 h) were required to achieve complete conversions. However, catalyst solubility could not be the cause of this behaviour, since best results were obtained under operating conditions that induced the presence of two distinct phases in the system, and not when a single homogeneous phase was observed. Further experiments performed in MeOH under various CO_2 pressures supported the alternative view that the poor reactivities and enantioselectivities observed in $scCO_2$ were actually due to the low polarity of this medium.[97]

Having shown that rhodium and ruthenium complexes of fluorous BINAP analogues were more conveniently employed in standard organic solvents, Hope and co-workers next studied the recovery of these catalytic species using F-SPE techniques. The activities and enantioselectivities of ruthenium complexes of ligands **73** and **74** were thus assessed in the catalyzed asymmetric hydrogenation of methyl acetoacetate (Scheme 6.20) run in dichloromethane. In the presence of 0.1 mol% of catalyst the substrate was quantitatively consumed in 1 hour after which the product was readily isolated using F-SPE with ee values ranging between 76% (ligand = **73**) and 80% (ligand = **74**), to be compared with 78% ee achieved with (*R*)-BINAP. Recovery of the chiral ruthenium catalyst species using F-SPE turned out to be unfeasible, because the metal bound irreversibly to the surface of the fluorous silica. However, the fluorous BINAP ligands contaminated with small amounts of the corresponding phosphine oxides could be recovered by elution with degassed CH_2Cl_2, and reused in a second run after addition of fresh aliquots of $[RuCl_2(C_6H_6)]_2$ with no apparent effect on the activities and enantioselectivities.[22]

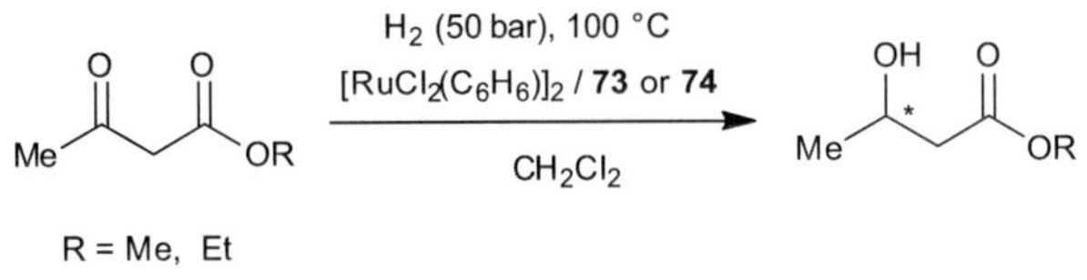

Scheme 6.20 Enantioselective hydrogenation of methyl acetoacetate.

Attempts to recycle transition metal catalysts containing fluorous chiral phosphines are often foiled by the oxidation of the phosphine moiety to phosphine oxide.[78–80,82] For asymmetric hydrogenations this side reaction occurs mostly during work-up operations. Horn and Bannwarth have shown that non-covalent immobilization of a ruthenium complex of fluorous BINAP **75** (Figure 6.12) onto fluorous silica prior to the reaction stage facilitates its recovery and recycling.[98] The complex was tested as a heterogeneous catalyst in the asymmetric hydrogenation of the two model substrates dimethyl itaconate and 2-acetamido methyl acrylate run in MeOH/BTF mixtures, affording the corresponding products in almost quantitative conversions and high enantioselectivities ($>90\%$ ee). In comparison, homogeneous reactions performed in MeOH in the presence of (*S*)-BINAP as a ligand required shorter reaction times, but gave lower ee values (*e.g.* 76% ee in the hydrogenation of 2-acetamido methyl acrylate). Work-up of the heterogeneous reaction mixtures, carried out under an inert atmosphere, consisted simply in the removal of the solvent under reduced pressure followed by extraction of the organic products from the solid catalyst with aqueous MeOH. The fluorous complex was retained on the support and could be reused as such for the hydrogenation of fresh substrate. In these recycling experiments metal leaching into the products was very low, ranging between 1.6 to 4.5 ppm. Nevertheless, the observed activities declined after three subsequent reaction cycles due to unavoidable catalyst degradation.

The asymmetric reduction of ketones to the corresponding alcohols can be conveniently performed by catalytic transfer hydrogenation from a hydrogen donor molecule, usually propan-2-ol or HCO_2H. Enantioselective versions of this useful functional group transformation have been developed, with outstanding results obtained in the case of catalytic systems based on combinations of platinum group metals and chiral nitrogen-containing ligands.[99] Sinou, Pozzi and co-workers have extensively studied the potential of C_2-symmetric fluorous diamines, diimines and salen ligands in the fluorous biphasic hydrogen transfer reduction of aryl alkyl ketones (Scheme 6.21).[85,100–103] They first demonstrated that fluorous chiral salen **3**, **4** and **7** (Figure 6.1, 10 mol%) in combination with $[Ir(COD)Cl]_2$ (5 mol%) were active as catalysts in the reduction of acetophenone performed in a fluorous biphasic system where the hydride source propan-2-ol acted also as the organic solvent, in the presence of KOH as a promoter.[100] Enantioselectivities of up to 60% ee were achieved using salen **3** in the reduction of ethyl phenyl ketone, which were slightly higher than those obtained using comparable non-fluorous chiral aldimines under

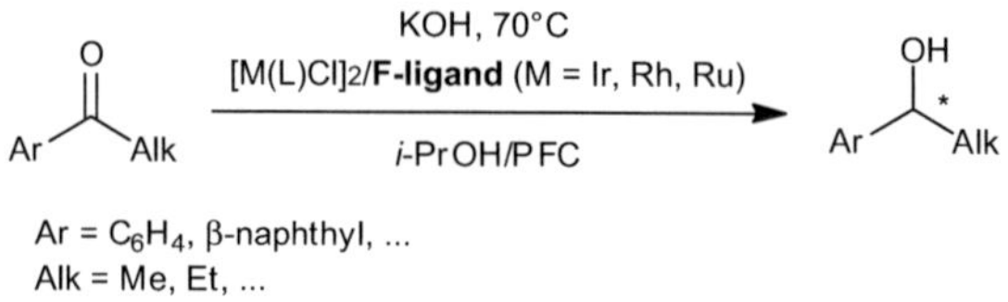

Ar = C_6H_4, β-naphthyl, ...
Alk = Me, Et, ...

Scheme 6.21 Hydrogen transfer reduction of ketones under under fluorous biphasic conditions.

homogeneous conditions. However, recycling of the fluorous layer gave lower poor results, as partial decomposition of the C=N imine bonds of the ligand occurred under basic reaction conditions and the metal was massively lost into the organic phase. In order to circumvent this problem, fluorous diamines derived from *trans*-1,2-diaminocyclohexane **55–59** (Figure 6.10) were used as ligands of iridium, rhodium and ruthenium.[85,101] Best results were obtained with a catalyst generated from diamine **55** and [Ir(COD)Cl]₂, which displayed enantioselectivities up to 79% ee in the hydrogen transfer reduction of acetophenone.[101] The catalytically active fluorous layer could be recycled up to four times with moderate loss of Ir (4% in the first run, then ≤1%).

Hydrogen transfer reduction of acetophenone under fluorous biphasic conditions was also readily achieved in the presence of less rigid fluorous chiral diamines, diimines and β-amino alcohols (for instance **76–79**, Figure 6.13) associated with [Ir(COD)Cl]₂, [Rh(C₆H₁₀)Cl]₂ or [Ru(*p*-cymene)Cl₂]₂.[102,103] Although excellent reduction yields were observed with all the ligands used and whatever the metal source, much lower enantioselectivities (up to 35% ee in the case of **77**/[[Rh(C₆H₁₀)Cl]₂) were obtained. Recyclability of the fluorous layer was possible in the case of diamine **77** and its efficiency was found to depend on the nature of the metal source, with [Ru(*p*-cymene)Cl₂]₂ affording the most promising results.[103]

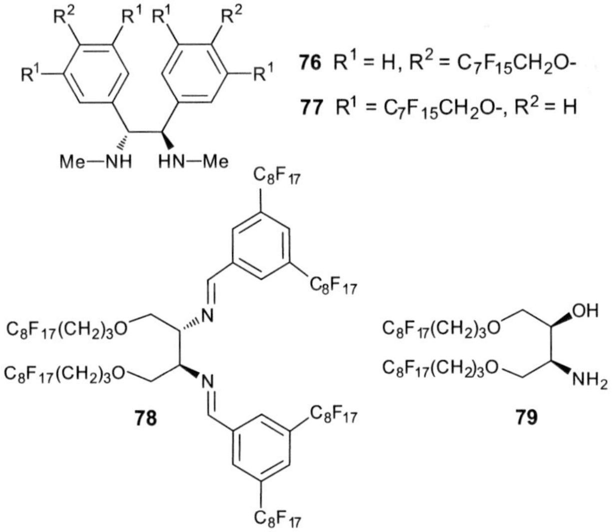

Figure 6.13 Fluorous nitrogen ligands for hydrogen transfer reduction of ketones.

6.3 Metal-free Catalytic Systems

Catalytic systems based on transition metal-free organic molecules (organic catalysts) have received great attention in the last decade and their application in stereoselective synthesis has become a very rapidly growing field of research.[104] Starting from early examples of structurally simple catalysts, a number of enantiomerically pure, bi- or even multifunctional chiral organic molecules have been thus designed to cover a broad range of useful asymmetric catalytic reactions. At the same time, the increased structural complexity and synthetic cost of organic catalysts, often effective only at high loading, call for the development of suitable separation and recovery methods.[105] In this context the fluorous approach remains still relatively undeveloped in comparison with more traditional immobilization techniques, but its viability is witnessed by some interesting examples that have appeared in the literature.

6.3.1   Proline Derivatives

(S)-Proline and (S)-proline derivatives (*e.g.* substituted prolinamides or pyrrolidines) have been widely investigated as chiral organic catalysts in several enantioselective reactions.[106] The immobilization of such compounds onto various supports have been concurrently pursued in the attempt to simplify their separation and recycling, but also to enable the exploration of new solubility profiles and the fine tuning of the catalytic properties.[107] To this end, phase tagging strategies, among which functionalization of (S)-proline derivatives with fluorous ponytails, have been explored as well.

The first report of a fluorous proline derivative was due to Fache and Piva who tested proline **13** (see Figure 6.3) not only as a copper ligand (see Section 6.2.1), but also as an organocatalyst in the intermolecular aldol reaction between a large excess of acetone and *p*-nitrobenzaldehyde in BTF as a solvent.[52] The product, 4-hydroxy-4-(4-nitrophenyl)butan-2-one, was isolated in good yield (72%) and enantioselectivity (73% ee) when the reaction was performed in the presence of **13** (25 mol%). These results were fully comparable to those obtained with proline in DMSO, with the advantage of simpler work-up of the reaction mixture. However, recovery and reuse of the fluorous catalyst were not attempted.

Soós and co-workers described the synthesis of the fluorous α,α-diarylprolinol **80** (Figure 6.14) and demonstrated its use as a recoverable precatalyst in the enantioselective Corey–Bakshi–Shibata (CBS) borane reduction of ketones.[108] The fluorous prolinol **80** (10 mol%) treated *in situ* with boron sources such as $B(OMe)_3$ or $BH_3 \cdot THF$ smoothly generated the corresponding oxazaborolidines which efficiently promoted the reduction of prochiral aryl alkyl, biaryl and cycloalkyl ketones (Scheme 6.22). These reactions were conducted in THF, and the precatalyst was recovered by a work-up process that consisted of quenching with MeOH and H_2O, followed by F-SPE of the resulting mixture. Secondary alcohols were quickly removed from the fluorous support by washing with aqueous CH_3CN and isolated in excellent yields (up to 93%) and enantioselectivities (up to 95% ee). Hydrolysis of the fluorous chiral oxaborolidine catalysts occurred under work-up conditions: the fluorous precatalyst **80** was thus recovered quantitatively in a second-pass elution using

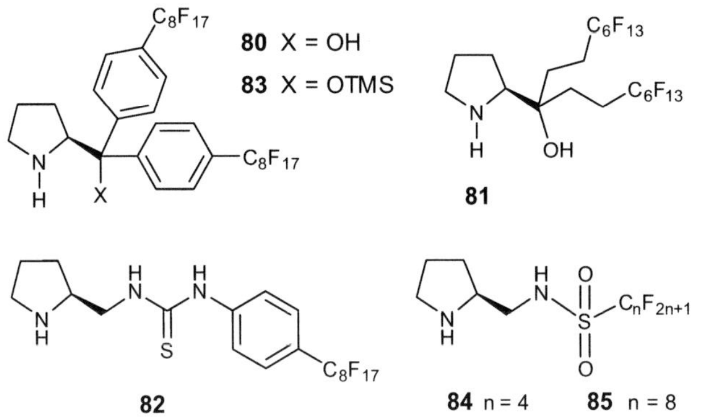

Figure 6.14 Fluorous proline derivatives.

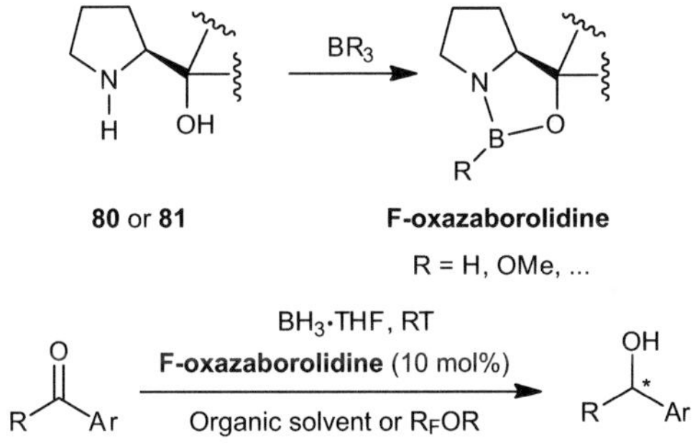

Scheme 6.22 Fluorous CBS reduction of ketones.

THF and recycled two more times without loss of catalytic activity and enantioselectivity.

Two alternative recovery methods for fluorous α,α-disubstituted prolinol CBS precatalysts were later reported. Funabiki and co-workers took advantage of the temperature-dependent solubility profile of dialkyl prolinol **81** (Figure 6.14) in toluene in order to achieve its separation by simple filtration without the use of any fluorous solvent or support.[109] Reduction of various ketones performed under homogeneous conditions at room temperature gave the highest enantioselectivities (up to 93% ee in the case of acetophenone) when the CBS catalyst was made with B(OMe)$_3$. However, precipitation of **81** upon cooling of the quenched reaction mixture was more efficient when BH$_3$·THF was used both to generate the catalyst and as reducing agent. The fluorous α,α-disubstituted prolinol was thus recovered in 32–84% yields depending on the substrate and on the boron source, but for the model reduction of acetophenone the enantioselectivity fell from 93% ee to 37% ee when **81** was reused in a second reaction run.

The immobilization of the oxazaborolidine obtained from **80** and BH_3 in the hydrofluoroether $(i\text{-}C_3F_7)(n\text{-}C_3F_7)CFOC_2H_5$ (HFE-7500) was disclosed by a research team of Fluorous Technologies Inc.[110] They studied the CBS reduction of acetophenone in this attractive fluorinated fluid which display most of the advantages of PFC associated with a much reduced atmospheric lifetime (about 2.2 years) and global warming potential. The reaction proceeded rapidly, with quantitative conversion of the starting ketone to 2-phenylethanol and similar high enantioselectivities (92–94% ee) both in the presence and in the absence of THF coming from the solution of borane adduct. Separation of 2-phenylethanol from the reaction mixture, previously quenched with MeOH, was achieved by extraction with a polar organic solvent, after which the residual HFE-7500 phase containing **80** was reused in a subsequent reaction cycle after addition of fresh borane solution. The optimized recycling procedure involved the use of 10 M BH_3·dimethylsulfide as a source of boron and extraction with DMSO. Nine consecutive runs were thus performed starting with 10 mol% of fresh **80**, with ee values remaining constants through the first five runs, then slowly dropping down to 88% by the eighth run and finally to 74% in the last run. The fluorous prolinol was then recovered in 54% yield and 90% purity after evaporation of the HFE-7500 phase.

Zhao, Zhu and co-workers have recently evaluated the behaviour of **80** and **81** as catalysts in the enantioselective epoxidation of α,β-enones with *t*-butyl-hydroperoxide (TBHP) (Scheme 6.23).[111] Both fluorous prolinols showed a moderate activity in the epoxidation of 1,3-diphenylpropenone in various organic solvents, but only the α,α-diarylprolinol **80** afforded satisfactory levels of enantioselectivity (up to 85% ee in CCl_4) comparable to those previously observed with α,α-diphenyl-L-prolinol.[112] The epoxidation of related calchones under optimized reaction conditions gave the corresponding epoxides in 31–67% yields and ee values from 65 to 83%. About 75% of the initial amount of catalyst **80** employed in the epoxidation of 1,3-diphenylpropenone could be recovered by evaporation of the reaction solvent, followed by complete dissolution of the residue in hot MeOH and selective precipitation of upon cooling to –5 °C. Despite this significant loss of material, the authors were able to perform three additional consecutive runs with no significant erosion of epoxide yields and enantioselectivities.

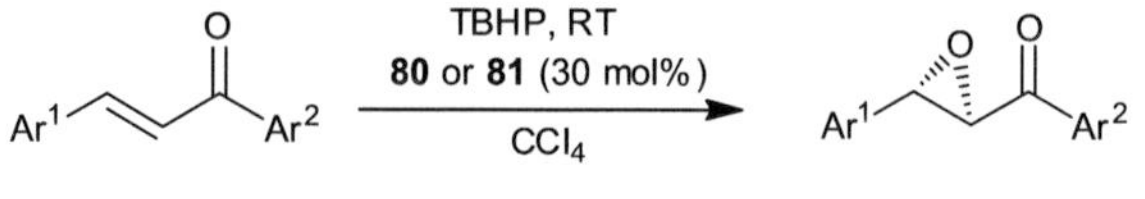

Scheme 6.23   Enantioselective epoxidation of α,β-enones.

In the last example, prolinol acted as a bifunctional catalyst performing the simultaneous activation of the enone substrate and the oxidant by the hydroxy and amino groups, respectively.[112] A different fluorous bifunctional catalyst containing (*S*)-pyrrolidine and thiourea (**82**, Figure 6.14) has been specifically designed by Cai, Zhang and co-workers for the enantioselective α-chlorination of aldehydes with *N*-chlorosuccinimide (NCS) as the chlorine source (Scheme 6.24).[113] Indeed, hydrogen bonding interactions between NCS and the thiourea moiety of **82**, enhanced by the electron-withdrawing effect of the R_F substituent, facilitate the chlorine transfer to the substrate bound to the pyrrolidine nitrogen atom. This dual action was confirmed by comparison of the results obtained in the α-chlorination of hydrocinnamaldehyde performed in the presence of **82**, (*S*)-proline, (*S*)-prolinamide or thiourea. Among all these catalysts, **82** afforded the highest activity (99% yield) and enantioselectivity (85% ee). The fluorous catalyst could be recovered in 87% yield by F-SPE of the reaction mixture, and the second-round reaction using the recovered catalyst gave similar product yield and ee value. The α-chlorination of other aldehydes also proceeded efficiently to furnish the products in excellent yields (91–99%) and good enantioselectivities (85–95% ee).

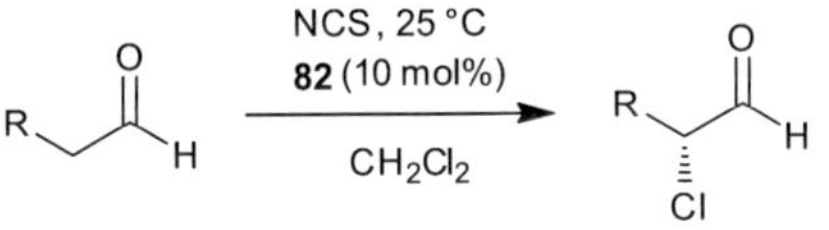

Scheme 6.24   Enantioselective α-chlorination of aldehydes.

Wang and co-workers reported the use of fluorous proline derivatives as chiral catalysts for the enantioselective Michael addition reaction of carbonyl compounds with nitroolefins (Scheme 6.25).[114,115] α,α-Diarylprolinol trimethylsilyl ether **83** (Figure 6.14) was found to promote Michael addition reactions between aliphatic aldehydes and *trans*-β-nitrostyrene derivatives at room temperature in BTF as a solvent, affording the corresponding γ-nitro carbonyl compounds in 81–91% yields, with enantioselectivities generally higher than 99% ee and high degree of diastereoselectivity (up to 29/1 *syn*/*anti* ratio).[114] The fluorous catalyst could be recovered in about 90% yield using F-SPE and reused. The loss of material was mirrored by a steady decrease of catalytic activity. Nevertheless, stereoselectivities were not affected even after six consecutive runs.

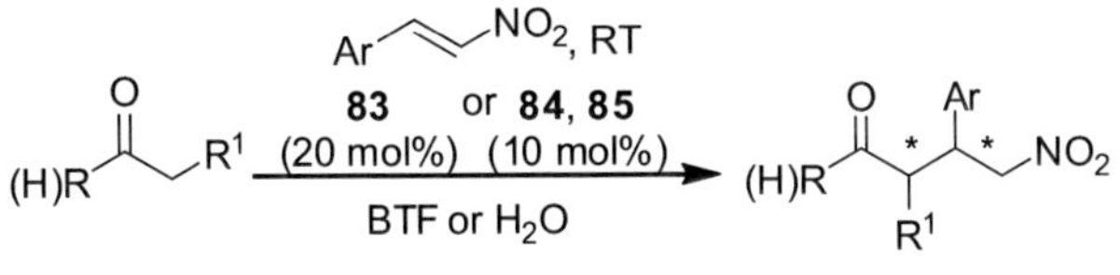

Scheme 6.25   Michael addition reactions of carbonyl compounds with nitroolefins.

Pyrrolidine sulfonamides **84** and **85** (Figure 6.14) were tested for analogous Michael addition reactions in an aqueous environment (Scheme 6.25), in the working hypothesis that the hydrophobic surroundings created by the presence of fluorous ponytails could enhance the interactions between the catalyst and the substrate.[115] Actually, both fluorous sulfonamides promoted the Michael addition of cyclohexanone to *trans*-β-nitrostyrene affording the product in high yields (up to 95%), enantioselectivities (up to 90% ee) and diastereoselectivities (up to 27/1 *syn/anti* ratio). However, reaction rates were very slow in the case of **85**, possibly as a result of the steric bulk of the C_8F_{17} substituent. Other cyclic ketones, aldehydes and nitrolefins underwent the catalytic process efficiently in the presence of **84** to give products with good to excellent enantioselectivities (68–95% ee) and up to 50/1 diastereomeric ratios.

Pyrrolidine sulfonamide **84** was next applied in the asymmetric aldol reaction of cyclic ketones with aromatic aldehydes, again in aqueous environment (Scheme 6.26).[116] Results obtained in the model aldol reaction of cyclohexanone with 4-nitrobenzaldehyde highlighted the importance of water as a component of the reaction system, in agreement with a proposed transition state model involving multiple hydrogen bonding interactions favoured, among others, by the R_F substituent present in the catalyst structure. Under optimized reaction conditions, cyclohexanone reacted with a variety of aromatic aldehydes affording the corresponding aldol adducts in very good yields (73–92%), enantioselectivities (86–94% ee) and diastereoselectivities (5/1 to 20/1 *anti/syn* ratio). Extension of the aldol reaction to other donors such as cyclopentanone and propanal was also demonstrated with excellent results.

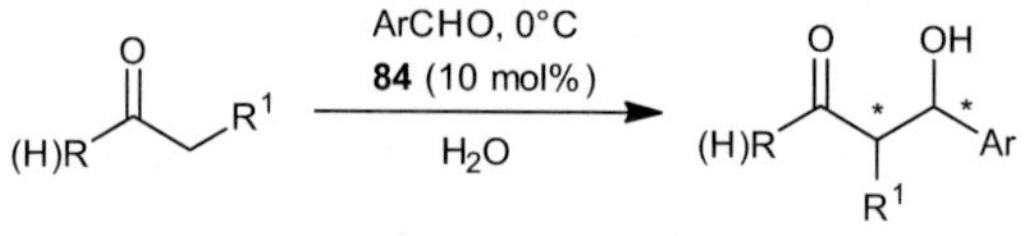

Scheme 6.26   Enantioselective aldol reaction on water.

Recovery of pyrrolidine sulfonamide **84** was achieved using F-SPE, with comparable efficiencies (about 90%) for Michael and aldol reactions. The catalyst was thus subsequently reused five and six more times, respectively, with substantial retention of the enantioselection levels and progressive reduction of reaction rates. Eroded diastereoselectivites were observed only after four Michael reaction cycles. Although promising, these results should be considered with some caution, since recycling experiments were performed with an exceedingly high initial catalyst loading (20 mol%) with respect to the optimal reaction conditions.

6.3.2   Miscellaneous Catalysts

Besides (*S*)-proline derivatives, a restricted number of chiral fluorous molecules (Figure 6.15) have been applied as organic catalysts in key asymmetric carbon–carbon bond forming reactions and, in one case, in the enantioselective reduction of imines with Cl_3SiH.

In a pioneering work due to Fache and Piva, a family of fluorous cinchona derivatives were synthesized and tested in the base-catalyzed Diels–Alder reaction between anthrone and *N*-methyl maleidimide (Scheme 6.27) in various

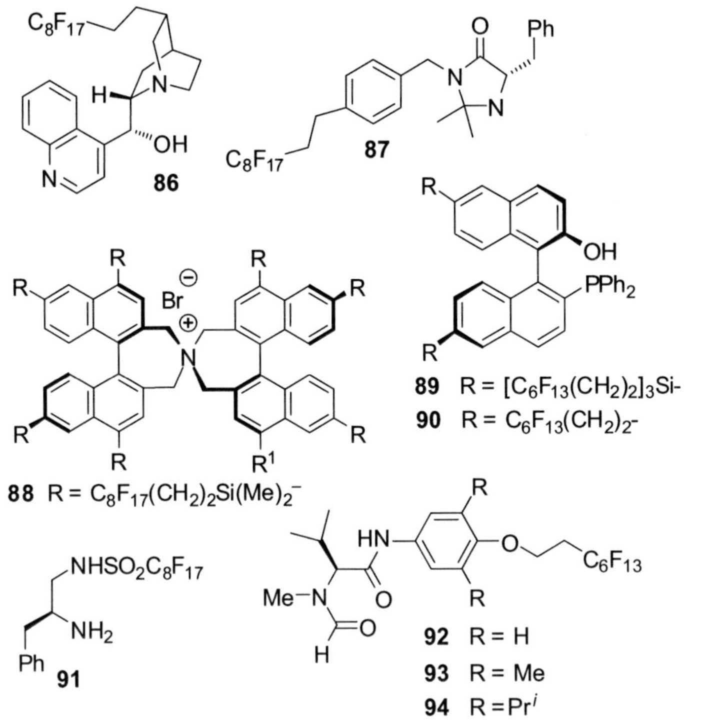

Figure 6.15 Miscellaneous fluorous organocatalysts.

solvent systems.[117] Enantioselectivities up to 40% ee and quantitative conversions were achieved using cinchonidine **86** (Figure 6.15) in BTF at room temperature. Both the catalyst and the reagents were freely soluble in such a solvent, but not the Diels–Alder adduct, about 75% percent of which precipitated out and could be recovered by simple filtration. The liquid phase maintained its catalytic activity in a second run, but the enantioselectivity diminished to less than 20% ee.

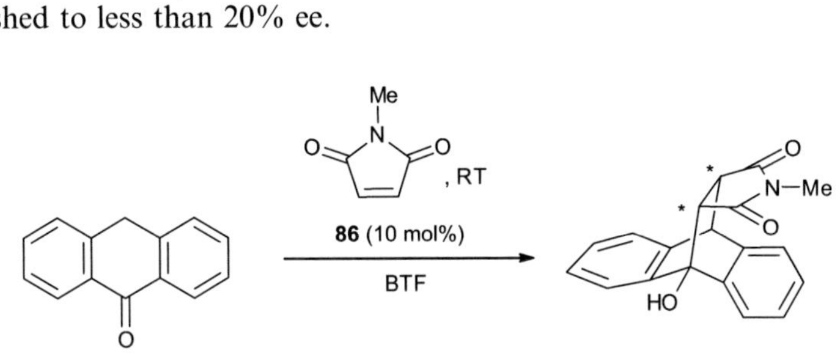

Scheme 6.27 Diels–Alder reaction between anthrone and *N*-methyl maleidimide.

Zhang and co-workers later developed the chiral imidazolidinone **87** (Figure 6.15), an analogue of the well-known MacMillan catalyst, which proved to be a more efficient catalyst than **86** for Diels–Alder reactions.[118] Compared to MacMillan catalyst, **87** afforded slightly higher *endo/exo* ratio (93/7 versus 90/10) and enantioselectivity (93% ee versus 88% ee for the *endo* isomer) in the model reaction between acrolein and cyclohexadiene (Scheme 6.28). Other α,β-unsaturated aldehydes and dienes Diels–Alder adducts were also obtained with consistently high stereoselectivities thus showing the potential of this fluorous catalyst. In addition, **87** could be readily separated from the reaction products by F-SPE and recovered in 80–84% yields with excellent purity.

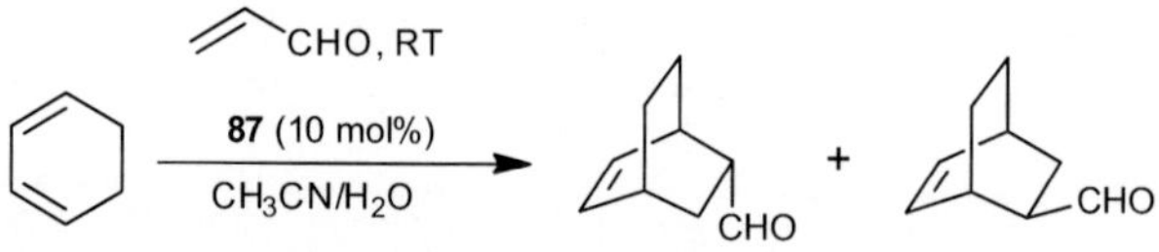

Scheme 6.28   Diels–Alder reaction between acrolein and cyclohexadiene.

Fluorous quaternary ammonium and phosphonium salts, as well as fluorous crown and aza-crown ethers, have been gradually emerging as viable alternative to classic phase transfer catalysts.[119] This concept was first demonstrated by Maruoka and co-workers, who applied the C_2-symmetric fluorous ammonium bromide **88** (Figure 6.15) as a phase-transfer catalyst in the asymmetric synthesis of both natural and unnatural α-amino acids.[120] The enantioselective alkylation of a protected glycine derivative with various benzyl- and alkyl bromides (Scheme 6.29) was thus conducted in a toluene/water system in which **88** was poorly soluble, affording the alkylated products in good yields (from 81 to 93%) with enantioselectivities ranging from 87 to 93% ee. Closely related C_2-symmetric non-fluorous ammonium salts showed only slightly better enantioselectivities but, due to their solubility in toluene, provided much faster reaction rates. On the other hand, compound **88** was easily separated from the organic products by extraction of the heterogeneous reaction mixture with FC-72. The fluorous salt recovered after evaporation of FC-72 could be reused without any loss of activity and selectivity for at least two additional runs.[121]

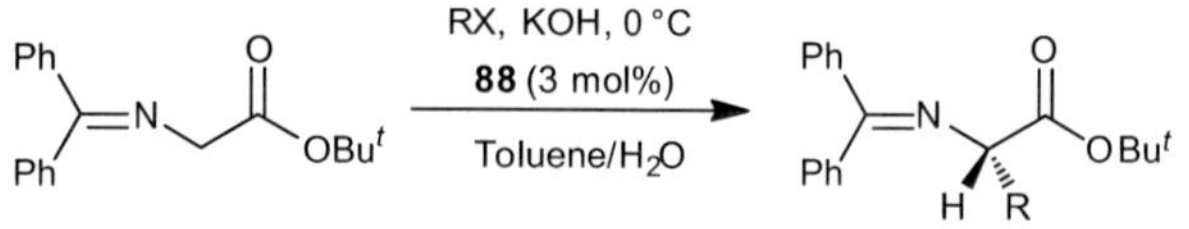

Scheme 6.29   Enantioselective alkylation of a protected glycine derivative.

In their continuous effort to develop efficient chiral Lewis base catalysts for the challenging asymmetric aza-Morita-Baylis-Hillman reaction of *N*-tosylated imines with methyl vinyl ketone (Scheme 6.30), Shi and co-workers synthesized the fluorous phosphines **89** and **90** (Figure 6.15) bearing a phenolic hydroxy group on the 2-position of the binaphthyl framework.[122] Both phosphines exhibited a temperature-dependent catalytic behaviour, best results being

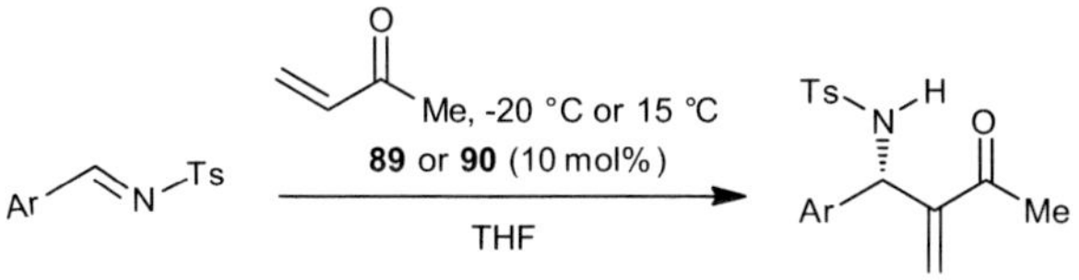

Scheme 6.30   Asymmetric aza-Morita-Baylis-Hillman reaction.

obtained at $-20\,^\circ$C in the case of **89** and $15\,^\circ$C in the case of **90**. Under optimized conditions, a variety of N-tosylated imines reacted smoothly with methyl
vinyl ketone in the presence of **89**, affording the corresponding aza-Morita-
Baylis-Hillman (S)-adducts in good to excellent yields (70–98%) and enantioselectivities (71–95% ee). These results compared favourably not only to
those obtained using **90**, but also to the ones achieved with a similar chiral
phosphine without fluorous ponytails. Unfortunately, complete oxidation of **89**
and **90** occurred in the reaction system and the corresponding phosphine oxides
were recovered in only 60–70% yield by flash column chromatography.

Miura and co-workers have recently reported the asymmetric aldol reaction
of cyclohexanone with aromatic aldehydes in the presence of a catalytic system
based on the fluorous sulfonamide **91** (Figure 6.15) derived from (S)-pheny-
lalanine and trifluoroacetic acid (TFA).[123] The aldol reactions were performed
in an aqueous environment (Scheme 6.31) and, as in the case of pyrrolidine
sulfonamide **84** (Figure 6.14),[116] the enhanced acidity of the sulfonamidic
N–H group due to the electron-withdrawing effect of the R_F substituent was
proposed to play a stabilizing role in the transition state. Also the aromatic
aldehydes and ketone chosen to evaluate the scope and limitation of the **91**/
TFA and **84** were similar, hence an indirect comparison between these two
catalytic systems is quite informative. First, the catalytic activity of **91** in the
absence of TFA was very poor with respect to the one displayed by **84**. Even
upon addition of the TFA co-catalyst, reactions promoted by **91** had to be
conducted at room temperature in order to achieve the best compromise
between activity and stereoselectivity. Reaction yields (41–100%) and enantioselectivities (74–96%) were similar to those obtained using **84**, but in many
instances the latter showed improved diastereoselectivities (*e.g. anti/syn* ratio
20/1 with respect to 4/1, in the case of the reaction between 4-bromo-
benzaldehyde and cyclohexanone). Recovery of **91** by F-SPE was demonstrated
upon six consecutive reaction runs, with an average 93% efficiency in each run.

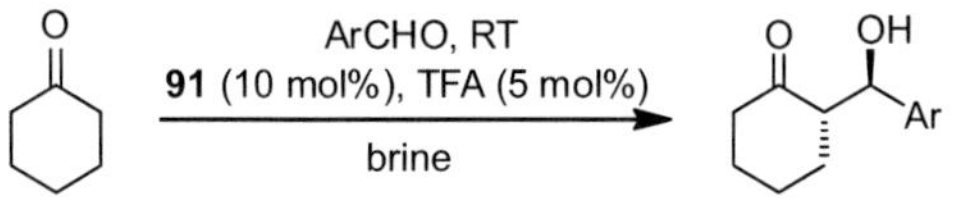

Scheme 6.31   Enantioselective aldol reaction of cyclohexanone with aromatic
aldehydes.

Stereoselection levels were maintained in the recycling experiments, with a progressive erosion of the catalytic activity.

Fluorous (*S*)-valine-derived *N*-methylformamides **92–94** (Figure 6.15) have been introduced by Malkov, Kočovský and co-workers as catalysts for the enantioselective reduction of imines derived from acetophenone and its congeners with HSiCl$_3$ (Scheme 6.32).[124] In this process the difference between the efficiency and selectivity of the fluorous catalysts **92–94** and those of their non-fluorous analogues bearing H or OR groups in the 4-position of the aryl ring were negligible. At the same time, the nature of the aryl substituents on the 3,5-positions had significant effects on the catalytic performance of **92–94** and also on their recoverability. As an example, the *p*-methoxyanilide of acetophenone was reduced to the corresponding amine in 80% yield and 84% ee in the presence of **92**, whereas **93** and **94** afforded the same amine in 90% and 98% yield, with 91% ee and 89% ee, respectively. All these fluorous catalysts could be isolated from reaction mixtures using F-SPE and reused as such for subsequent reduction cycles of the model substrate. However, **94** was recovered with 99% to 88% efficiency in each single F-SPE step and the loss of catalyst had only a marginal effect on the enantioselectivies observed in five consecutive runs. In the case of **92** the recovery efficiencies varied between 99% to 70% and the observed ee values dropped from 84% to 74% in four consecutive runs.

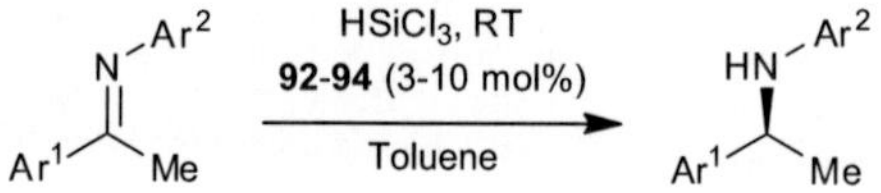

Scheme 6.32   Enantioselective reduction of imines.

6.4   Conclusions

The development of recoverable and reusable fluorous chiral catalysts has been the subject of intense study over the past decade, and the viability of this approach has been investigated with variable success in many typical asymmetric organic reactions. While early studies were focused on the application of PFC-soluble transition metal complexes suitable for fluorous biphasic reaction conditions, the attention has progressively turned to light fluorous chiral catalysts that operate under the established conditions for their non-fluorous analogues. The recent explosion of organocatalysis has favoured this shift, and significant examples of light fluorous, transition metal-free chiral catalysts have recently appeared.

This wave of research has provided useful insights on the effects of the incorporation of medium-sized perfluoroalkyl groups on the phase properties and catalytic behaviour of molecules belonging to many major classes of chiral ligands and catalysts. Accordingly, the design of efficient fluorous chiral catalysts must include a careful evaluation of the location and number of perfluoroalkyl substituents and, for cationic fluorous transition metal complexes, of the characteristics of the counter anion. The efficient shielding of the catalytically active site from the electron-withdrawing effect of the R_F substituents need also to be ensured in order to attain activities and stereoselectivities comparable or even superior to those attained in presence of the parent non-fluorous catalyst. Notable exceptions to this broad criterion exist, especially in the field of organocatalysis where the hydrophobic characteristics and/or enhanced acidity provided by the presence of R_F substituents can be highly beneficial.

Finally, there are several reliable approaches available to isolate fluorous chiral catalysts in post-reaction operations, which greatly simplify the purification of products. The biggest challenge remains the efficient recovery and repeated use of these catalysts without deterioration of their original activity and selectivity. This is claimed to be achieved in a significant number of literature examples but, as for many other homogeneous catalysts designed for recovery and recycling, the evidence used to support such claims does not always hold up to close scrutiny. Further progress in this field will be made possible by rigorous and detailed studies of reaction kinetics, catalyst stability and deactivation pathways.

References

1. D. O'Hagan, *Chem. Soc. Rev.*, 2008, **37**, 308.
2. P. Kirsch, *Modern Fluoroorganic Chemistry: Synthesis, Reactivity, Applications*, Wiley-VCH, Weinheim, 2004.
3. D. M. Lemal, *J. Org. Chem.*, 2004, **69**, 1.
4. G. P. Moss, P. A. S. Smith and D. Tavernier, *Pure Appl. Chem.*, 1995, **67**, 1307.
5. R. E. Banks and J. C. Tatlow, in *Organofluorine Chemistry: Principles and Commercial Applications*, ed. R. E. Banks, B. E. Smart and J. C. Tatlow, Plenum Press, New York, 1994, p. 4.
6. D.-W. Zhu, *Macromolecules*, 1996, **29**, 2813.
7. A. R. Ravishankara, S. Solomon, A. A. Turnipseed and R. F. Warren, *Science*, 1993, **259**, 194.
8. L. C. Clark and F. Gollan, *Science*, 1966, **152**, 1755.
9. J. H. Hildebrand and D. R. F. Cochran, *J. Am. Chem. Soc.*, 1949, **71**, 22.
10. D.-W. Zhu, *Synthesis*, 1993, 953.
11. I. T. Horváth and J. Rábai, *Science*, 1994, **266**, 72.
12. *Handbook of Fluorous Chemistry*, ed. J. A. Gladysz, D. P. Curran and I. T. Horváth, Wiley-VCH, Weinheim, 2004.
13. J. A. Gladysz and D. P. Curran, *Tetrahedron*, 2002, **58**, 3823.
14. D. P. Curran, in *Handbook of Fluorous Chemistry*, ed. J. A. Gladysz, D. P. Curran and I. T. Horváth, Wiley-VCH, Weinheim, 2004, p. 128.
15. D. P. Curran, *Aldrichim. Acta*, 2006, **39**, 3.
16. W. Zhang, *Green. Chem.*, 2009, **11**, 911.
17. G. Pozzi, F. Montanari and S. Quici, *Chem. Commun.*, 1997, 69.

18. G. Pozzi, F. Cinato, F. Montanari and S. Quici, *Chem. Commun.*, 1998, 877.
19. D. P. Curran, S. Hadida and M. He, *J. Org. Chem.*, 1997, **62**, 6714.
20. D. P. Curran and Z. Luo, *J. Am. Chem. Soc.*, 1999, **121**, 9069.
21. Y. Nakamura, S. Takeuchi, Y. Ohgo and D. P. Curran, *Tetrahedron Lett.*, 2000, **41**, 57.
22. E. G. Hope, A. M. Stuart and A. J. West, *Green Chem.*, 2004, **6**, 345.
23. W. Zhang, Y. Lu and T. Nagashima, *J. Comb. Chem.*, 2005, **7**, 893.
24. C. C. Tzschucke, C. Markert, H. Glatz and W. Bannwarth, *Angew. Chem., Int. Ed. Engl.*, 2002, **41**, 4501.
25. A. Biffis, M. Braga and M. Basato, *Adv. Synth. Catal.*, 2004, **346**, 451.
26. M. Wende, R. Meier and J. A. Gladysz, *J. Am. Chem. Soc.*, 2001, **123**, 11490.
27. K. Ishihara, S. Kondo and H. Yamamoto, *Synlett*, 2001, 1371.
28. J. A. Gladysz and V. Tesevic, *Top. Organomet. Chem.*, 2008, **23**, 67.
29. M. Wende and J. A. Gladysz, *J. Am. Chem. Soc.*, 2003, **125**, 5861.
30. L. V. Dinh and J. A. Gladysz, *Angew. Chem., Int. Ed. Engl.*, 2005, **44**, 4095.
31. F. O. Seidel and J. A. Gladysz, *Adv. Synth. Catal.*, 2008, **350**, 2443.
32. F. Montanari, G. Pozzi and S. Quici, in *Green Chemistry: Challenging Perspectives*, ed. P. Tundo and P. Anastas, Oxford University Press, Oxford, 2000, p. 145.
33. G. Pozzi, M. Cavazzini, F. Cinato, F. Montanari and S. Quici, *Eur. J. Org. Chem.*, 1999, 1947.
34. M. Cavazzini, A. Manfredi, F. Montanari, S. Quici and G. Pozzi, *Chem. Commun.*, 2000, 2171.
35. M. Cavazzini, A. Manfredi, F. Montanari, S. Quici and G. Pozzi, *Eur. J. Org. Chem.*, 2001, 4639.
36. T. Katsuki, *J. Mol. Catal. A: Chem.*, 1996, **113**, 87.
37. L. Canali and D. C. Sherrington, *Chem. Soc. Rev.*, 1999, **28**, 85.
38. M. Cavazzini, S. Quici and G. Pozzi, *Tetrahedron*, 2002, **58**, 3943.
39. I. Shepperson, M. Cavazzini, G. Pozzi and S. Quici, *J. Fluorine Chem.*, 2004, **125**, 175.
40. E. N. Jacobsen, *Acc. Chem. Res.*, 2000, **33**, 421.
41. P. G. Cozzi, *Chem. Soc. Rev.*, 2004, **33**, 410.
42. M. Cavazzini, G. Pozzi, S. Quici and I. Shepperson, *J. Mol. Catal. A: Chem.*, 2003, **204-205**, 433.
43. R. Rasappan, D. Laventine and O. Reiser, *Coord. Chem. Rev.*, 2008, **252**, 702.
44. J. Bayardon and D. Sinou, *Tetrahedron Lett.*, 2003, **44**, 1449.
45. J. Bayardon and D. Sinou, *J. Org. Chem.*, 2004, **69**, 3121.
46. J. Bayardon and D. Sinou, *Tetrahedron: Asymmetry*, 2005, **16**, 2965.
47. R. Annunziata, M. Benaglia, M. Cinquini, F. Cozzi and G. Pozzi, *Eur. J. Org. Chem.*, 2003, 1191.
48. B. Simonelli, S. Orlandi, M. Benaglia and G. Pozzi, *Eur. J. Org. Chem.*, 2004, 2669.

49. J. Bayardon, O. Holczknecht, G. Pozzi and D. Sinou, *Tetrahedron: Asymmetry*, 2006, **17**, 1568.
50. R. Kolodziuk, C. Goux-Henry and D. Sinou, *Tetrahedron: Asymmetry*, 2007, **18**, 2782.
51. M. B. Andrus and J. C. Lashley, *Tetrahedron*, 2002, **58**, 845.
52. F. Fache and O. Piva, *Tetrahedron: Asymmetry*, 2003, **14**, 139.
53. P. G. Jessop, *J. Supercritic. Fluids*, 2006, **38**, 21.
54. D. Koch and W. Leitner, *J. Am. Chem. Soc.*, 1998, **120**, 13398.
55. G. Franciò and W. Leitner, *Chem. Commun.*, 1999, 1663.
56. G. Franciò, K. Wittmann and W. Leitner, *J. Organomet. Chem.*, 2001, **621**, 130.
57. D. Bonafoux, Z. Hua, B. Wang and I. Ojima, *J. Fluorine Chem.*, 2001, **112**, 101.
58. R. Noyori and S. Hashiguchi, *Acc. Chem. Res.*, 1997, **30**, 40.30.
59. H. Kleijn, E. Rijnberg, J. T. B. H. Jastrzebski and G. van Koten, *Org. Lett.*, 1999, **1**, 853.
60. H. Kleijn, J. T. B. H. Jastrzebski, B.-J. Deelman and G. van Koten, *J. Mol. Catal. A: Chem.*, 2008, **287**, 65.
61. L. Pu and H.-B. Yu, *Chem. Rev.*, 2001, **101**, 757.
62. Y. Nakamura, S. Takeuchi, K. Okumura and Y. Ohgo, *Tetrahedron*, 2001, **57**, 5565.
63. J. K. Park, H. G. Lee, C. Bolm and B. M. Kim, *Chem. Eur. J.*, 2005, **11**, 945.
64. Y. S. Sokeirik, M. Omote, K. Sato, I. Kumadaki and A. Ando, *Tetrahedron: Asymmetry*, 2006, **17**, 2654.
65. M. Omote, Y. Eto, A. Tarui, K. Sato and A. Ando, *Tetrahedron: Asymmetry*, 2007, **18**, 2768.
66. M. Omote, Y. Eto, A. Tarui, K. Sato and A. Ando, *Tetrahedron: Asymmetry*, 2009, **20**, 602.
67. Y. Tian and K. S. Chan, *Tetrahedron Lett.*, 2000, **41**, 8813.
68. Y. Tian, Q. C. Yang, T. C. W. Mak and K. S. Chan, *Tetrahedron*, 2002, **58**, 3951.
69. Y. Nakamura, S. Takeuchi, K. Okumura, Y. Ohgo and D. P. Curran, *Tetrahedron*, 2002, **58**, 3963.
70. J. Fawcett, E. G. Hope, A. M. Stuart and A. J. West, *Green Chem.*, 2005, **7**, 316.
71. Y.-Y. Yin, G. Zhao, Z.-S. Qian and W.-X. Yin, *J. Fluorine Chem.*, 2003, **120**, 117.
72. M. Omote, Y. Nishimura, K. Sato, A. Ando and I. Kumadaki, *Tetrahedron Lett.*, 2005, **46**, 319.
73. M. Omote, Y. Nishimura, K. Sato, A. Ando and I. Kumadaki, *Tetrahedron*, 2006, **62**, 1886.
74. M. Omote, N. Tanaka, A. Tarui, K. Sato, I. Kumadaki and A. Ando, *Tetrahedron Lett.*, 2007, **48**, 2989.
75. Y. S. Sokeirik, H. Mori, M. Omote, K. Sato, A. Tarui, I. Kumadaki and A. Ando, *Org. Lett.*, 2007, **9**, 1927.

76. Y. S. Sokeirik, A. Oshina, M. Omote, K. Sato, A. Tarui, I. Kumadaki and A. Ando, *Chem. Asian J.*, 2008, **3**, 1850.
77. E. G. Hope and A. M. Stuart, in *Handbook of Fluorous Chemistry*, ed. J. A. Gladysz, D. P. Curran and I. T. Horváth, Wiley-VCH, Weinheim, 2004, p. 247.
78. M. Cavazzini, S. Quici, G. Pozzi, D. Maillard and D. Sinou, *Chem. Commun.*, 2001, 1220.
79. D. Maillard, J. Bayardon, J. D. Kurichiparambil, C. Nguefack-Fournier and D. Sinou, *Tetrahedron: Asymmetry*, 2002, **13**, 1449.
80. J. Bayardon, M. Cavazzini, D. Maillard, G. Pozzi, S. Quici and D. Sinou, *Tetrahedron: Asymmetry*, 2003, **14**, 2215.
81. T. Mino, Y. Sato, A. Saito, Y. Tanaka, H. Saotome, M. Sakamoto and T. Fujita, *J. Org. Chem.*, 2005, **70**, 7979.
82. Y. Nakamura, S. Takeuchi, S. Zhang, K. Okumura and Y. Ohgo, *Tetrahedron Lett.*, 2002, **43**, 3053.
83. D. A. Evans, K. A. Woerpel, M. M. Hinman and M. M. Faul, *J. Am. Chem. Soc.*, 1991, **113**, 726.
84. I. Shepperson, S. Quici, G. Pozzi, M. Nicoletti and D. O'Hagan, *Eur. J. Org. Chem.*, 2004, 4545.
85. J. Bayardon, D. Sinou, O. Holczknecht, L. Mercs and G. Pozzi, *Tetrahedron: Asymmetry*, 2005, **16**, 2319.
86. T. Miura, K. Itoh, Y. Yasaku, N. Koyata, Y. Murakami and N. Imai, *Tetrahedron Lett.*, 2008, **49**, 5813.
87. A. Biffis, M. Braga, S. Cadamuro, C. Tubaro and M. Basato, *Org. Lett.*, 2005, **7**, 1841.
88. R. Rasappan, T. Olbrich and O. Reiser, *Adv. Synth. Catal.*, 2009, **351**, 1961.
89. W. S. Knowles and R. Noyori, *Acc. Chem. Res.*, 2007, **40**, 1238.
90. S. Kainz, A. Brinkmann, W. Leitner and A. Pfaltz, *J. Am. Chem. Soc.*, 1999, **121**, 6421.
91. S. Lange, A. Brinkmann, P. Trautner, K. Woelk, J. Bargon and W. Leitner, *Chirality*, 2000, **12**, 450.
92. K. Burgemeister, G. Franciò, H. Hugl and W. Leitner, *Chem. Commun.*, 2005, 6026.
93. K. Burgemeister, G. Franciò, W. H. Gego, L. Greinen, H. Hugl and W. Leitner, *Chem. Eur. J.*, 2007, **13**, 6026.
94. D. J. Adams, W. Chen, E. G. Hope, S. Lange, A. M. Stuart, A. West and J. Xiao, *Green Chem.*, 2003, **5**, 118.
95. M. Berthod, G. Mignani and M. Lemaire, *Tetrahedron: Asymmetry*, 2004, **15**, 1121.
96. D. J. Birdsall, E. G. Hope, A. M. Stuart, W. Chen, Y. Hu and J. Xiao, *Tetrahedron Lett.*, 2001, **42**, 8551.
97. Y. Hu, D. J. Birdsall, A. M. Stuart, E. G. Hope, and J. Xiao, *J. Mol. Catal. A: Chem.*, 2004, **219**, 57.
98. J. Horn and W. Bannwarth, *Eur. J. Org. Chem.*, 2007, 2058.
99. S. Gladiali and E. Alberico, *Chem. Soc. Rev.*, 2006, **35**, 226.

100. D. Maillard, C. Nguefack, G. Pozzi, S. Quici, B. Valadé and D. Sinou, *Tetrahedron: Asymmetry*, 2000, **11**, 2881.
101. D. Maillard, G. Pozzi, S. Quici and D. Sinou, *Tetrahedron*, 2002, **58**, 3971.
102. J. Bayardon, D. Maillard, G. Pozzi and D. Sinou, *Tetrahedron: Asymmetry*, 2004, **15**, 2633.
103. J. Bayardon and D. Sinou, *ARKIVOC*, 2008 (vii), 26.
104. P. J. Dalko, *Enantioselective Organocatalysis*, Wiley-VCH, Wenheim, 2007.
105. M. Benaglia, in *Recoverable and Recyclable Catalysts*, ed. M. Benaglia, John Wiley & Sons, Chichester, 2009, p. 301.
106. G. Guillena, C. Najera and D. J. Ramon, *Tetrahedron: Asymmetry*, 2007, **18**, 2249.
107. M. Gruttadauria, F. Giacalone and R. Noto, *Chem. Soc. Rev.*, 2008, **37**, 1666.
108. Z. Dalicsek, F. Pollreisz, A. Gömöry and T. Soós, *Org. Lett.*, 2005, **7**, 3243.
109. S. Goushi, K. Funabiki, M. Ohta, K. Hatano and M. Matsui, *Tetrahedron*, 2007, **63**, 4061.
110. Q. Chu, M. S. Yu and D. P. Curran, *Org. Lett.*, 2008, **10**, 749.
111. F H. Cui, Y. Li, C. Zheng, G. Zhao and S. Zhu, *J. Fluorine Chem.*, 2008, **129**, 450.
112. A. Lattanzi, *Org. Lett.*, 2005, **7**, 2579.
113. L. Wang, C. Cai, D. P. Curran and W. Zhang, *Synlett*, 2010, 433.
114. L. Zu, H. Li, J. Wang, X. Yu and W. Wang, *Tetrahedron Lett.*, 2006, **47**, 5131.
115. L. Zu, J. Wang, H. Li and W. Wang, *Org. Lett.*, 2006, **8**, 3077.
116. L. Zu, H. Xie, H. Li, J. Wang and W. Wang, *Org. Lett.*, 2008, **10**, 1211.
117. F. Fache and O. Piva, *Tetrahedron Lett.*, 2001, **42**, 5655.
118. Q. Chu, W. Zhang and D. P. Curran, *Tetrahedron Lett.*, 2006, **47**, 9287.
119. G. Pozzi, S. Quici and R. H. Fish, *J. Fluorine Chem.*, 2008, **129**, 920.
120. S. Shirakawa, Y. Tanaka and K. Maruoka, *Org. Lett.*, 2004, **6**, 1429.
121. S. Shirakawa, M. Ueda, Y. Tanaka, T. Hashimoto and K. Maruoka, *Chem. Asian J.*, 2007, **2**, 1276.
122. M. Shi, L.-H.Chen and W.-D. Teng, *Adv. Synth. Catal.*, 2005, **347**, 1781.
123. T. Miura, K. Imai, M. Ina, N. Tada, N. Imai and A. Itoh, *Org. Lett.*, 2010, **12**, 1620.
124. A. V. Malkov, M. Figlus, S. Stončius and P. Kočovský, *J. Org. Chem.*, 2007, **72**, 1315.

CHAPTER 7

Aqueous Phase Asymmetric Catalysis

SZYMON BUDA,[a] MONIKA PASTERNAK[a] AND JACEK MLYNARSKI[a,b]

[a] Jagiellonian University, Faculty of Chemistry, Ingardena 3, 30-060, Krakow, Poland; [b] Institute of Organic Chemistry, Polish Academy of Sciences, Kasprzaka 44/52, 01-224, Warsaw, Poland

7.1 Introduction

Water is the solvent used by nature for life reactions. Indeed, 'water is life's mater and matrix, mother and medium. There is no life without water.'[1] Many enzymes catalyze reactions in a water environment under mild conditions with high efficiency and often with impressive stereoselectivity. This highly effective and environmentally benign methodology is also regarded as a goal in modern organic chemistry.[2] Following nature's lead, we also start to regard water as a versatile solvent for asymmetric transformations.[3] Thus, the development of enantioselective reactions using water as a solvent is now intensively investigated, although enzyme-catalyzed reactions were long thought to be the only possibility. It is highly desirable to develop a chemical system that, like an enzyme, can effect organic reactions in water with excellent efficiency and stereoselectivity.[4] In this respect, the development of water-tolerant catalysts and water-soluble ligands has rapidly become an area of intense research.[5] Only recently, catalytic asymmetric reactions promoted by water-compatible Lewis acids with chiral ligands have been developed.[6] As a result, asymmetric organometallic catalysis in aqueous media is now a well-used methodology in

RSC Green Chemistry No. 15
Enantioselective Homogeneous Supported Catalysis
Edited by Radovan Šebesta
© Royal Society of Chemistry 2012
Published by the Royal Society of Chemistry, www.rsc.org

organic synthesis.[7] More recently, efforts have also been devoted to asymmetric organocatalytic reactions performed using water as a reaction medium or as a reaction additive.[8]

Recent progress in the area initiated constructive discussion on the role and practical merits of water as a solvent. Thus water and water-based reaction media were debated with regards to terminology – that is, whether a reaction is carried out 'in water', 'in the presence of water' or 'in the presence of large excess of water'.[9] Janda and Hayashi have initiated discussion on the use of 'in water' or 'in the presence of water' terminology.[10] Hayashi proposed to use 'in water' when the reactants participating in the reactions are homogeneously dissolved whereas 'in the presence of water' should be used for a reaction that proceeds in a concentrated organic phase with water being presented as a second phase that influences the reaction in the organic phase. However, both reactions in/on water can come very close to the ideal green conditions. While the observed effect for 'on water' reactions are rate accelerations, the observed effect for 'in the presence of water' reaction is usually an increased enantios-electivity. For clear solutions of soluble organic reactants in water, the effects operating are: (i) the hydrophobic effect, which speeds reactions; (ii) hydrogen bonding effects on reactants and transition states, which may add to or oppose to hydrophobic effect; and (iii) water polarity effects, which may again increase or decrease reaction rates.[9]

From a practical viewpoint, the development of novel chiral catalysts is the most important aspect of this area of green chemistry. For both metal-assisted and metal-free strategies new water-soluble or water-phase-compatible chiral units are obviously of utmost importance. While the *de novo* design of water-tolerant chiral ligands and catalysts is an attractive but largely unexplored field, naturally occurring chiral units seem to be interesting and 'natural' sources of chirality.[11] In general, the reactions discussed here can be performed in pure water or in water and organic co-solvent mixtures. Good media should readily dissolve all or most of the participating reactants. On the basis of our knowledge of the chemical properties of the organic reactants, we can assume limited use of water as a solvent for homogeneous organic reactions. Nonetheless, the fact that many of the most desirable targets molecules (amino acids, carbohydrates, peptides, etc.) are readily soluble in water is an encouragement for developing new water-compatible strategies for their synthesis by using chiral catalysts or ligands. Furthermore, water solubility of lipophilic catalysts and ligands can be generally achieved by the modification of their structures, and particularly by the introduction of polar groups. Additionally, one of the more efficient and versatile methods of increasing solubility, and one that does not require modification of the solute, is to use an organic co-solvent. Another intriguing means of achieving aqueous solubility, or rather dispersion, is by using surfactants.

On the other hand, insoluble enzymes form a hydrophobic environment around their active site that diminishes contact between bulk water and the reaction transition states. Thus *in vitro* strategies can also include 'on water' methodology. Such a protocol can be regarded as more practical, making separation of substrate/products and water more efficient. Organic liquids that remained separate from water in a clear organic phase were ideal reactants, but solids could also be used, although being conceptually more demanding.[9] We can regard the reaction of two practically insoluble reactants 'on water' at ambient temperature as a model system close to the requirements of ideal green chemistry, and it is noteworthy that such reactions are possible, although rare.

Our intention was to collect the most versatile and interesting developments in asymmetric synthesis performed in aqueous media with and without addition of organic solvents, mostly in homogeneous solutions. A few example of biphasic 'on water' reactions are also included so as to demonstrate versatility of elaborated methodology. The examples highlighted in this chapter are a selection of the recent significant contributions, which in our opinion have had major impact in the area. Further, this review demonstrates the ideas and challenges that are essential for the progress in the field, leading to practical application of aqua asymmetric transformations.

7.2 Metal-assisted Asymmetric Reactions

Organometallic catalysis is still the most broadly used methodology in organic synthesis, and especially in the synthesis of drugs and fine chemicals. The activity of the catalyst depends on the sort of Lewis acid, while the asymmetric induction is related to the attachment of chiral organic ligands to the central metal atom. Various kinds of Lewis acids have been developed, and many have been applied in synthesis and in industry. However, asymmetric catalysis in water or aqueous solvents is difficult because many Lewis acid-type catalysts are not stable in the presence of water.[6] Advances in this field have required the development of novel, mostly transition metal, catalysts that exhibit at least kinetic stability towards water. The second challenge is designing a chiral ligand for reactions in aqueous media with appropriate binding properties to the central metal cation and as well as with satisfactory solubility in water. To address this issue some catalytic asymmetric reactions with water-compatible Lewis acids bearing chiral ligands have been developed.[7]

7.2.1 Hydrogenation

Hydrogenation of prochiral substrates in water or in a two-phase system is still one of the most studied reactions in the field of aqua asymmetric chemistry. One of the most popular catalysts of hydrogenation are rhodium complexes of water soluble ligands such as 1,2-diphosphines[12] and amino quaternized ammonium diphosphines (Figure 7.1).[13,14]

Those complexes are very effective at reducing the unsaturated α-amino acid precursors in water. Reduction of (Z)-α-acetamidocinnamic acid in the presence of ligand **2a** gave desired products with high enantioselectivity (up to 94% ee) in pure water (Scheme 7.1).

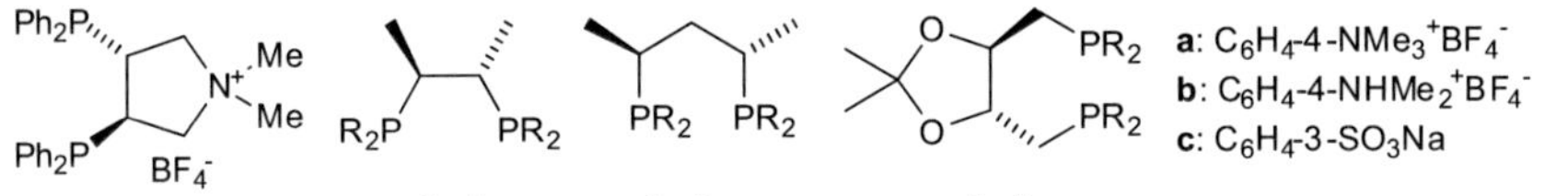

Figure 7.1 Water-soluble diphosphine ligands.

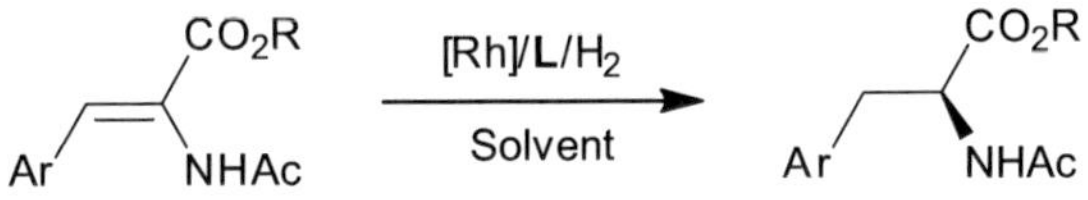

Ar	R	Ligand	Solvent	P_{H2} [atm]	ee [%]
C_6H_5	H	1	H_2O	1	90[a]
		2a	H_2O	14	94
		2b	H_2O	14	90
		3a	H_2O	14	67
		3b	H_2O	14	71
		3b	H_2O/MeOH (2/1)	14	70
		4b	H_2O	14	34
C_6H_5	Me	2a	H_2O	14	68
		3a	H_2O	14	40
		4a	H_2O	14	8
3-MeO-4-AcO-C_6H_3	H	2a	H_2O	14	93
		2b	H_2O	14	98
		3a	H_2O	14	76
		3b	H_2O	14	79
		4a	H_2O	14	42
		4b	H_2O	14	67

[a] reduced as the sodium salt.

Scheme 7.1 Asymmetric reduction of some enamides in the presence of rhodium complexes containing water-soluble diphosphines.

RajanBabu and co-workers[15] used a series of water-soluble, chelating bis-phosphinite ligands **5** (Figure 7.2) derived from α-salicin and bearing a qua-ternized ammonium function as ligands in the hydrogenation of some acet-amidoacrylic acid derivatives in aqueous media. The enantioselectivites obtained in neat water or a homogenous mixture of water and organic solvents was lower than at neat organic phase. The best results were observed for a mixture of water/ methanol (1/1), reaching up to 90% ee with quantitatively yield.

Another useful type of ligand used for rhodium complexes in the reduction of unsaturated amino acid precursor in water are non-ionic ligands such as polyoxadiphosphines or carbohydrate-based ligands. Selke *et al.* elaborated various D-hexopyranoside-2,3-*O*-bis(diphenylphosphinites) (**6**) in the reduction

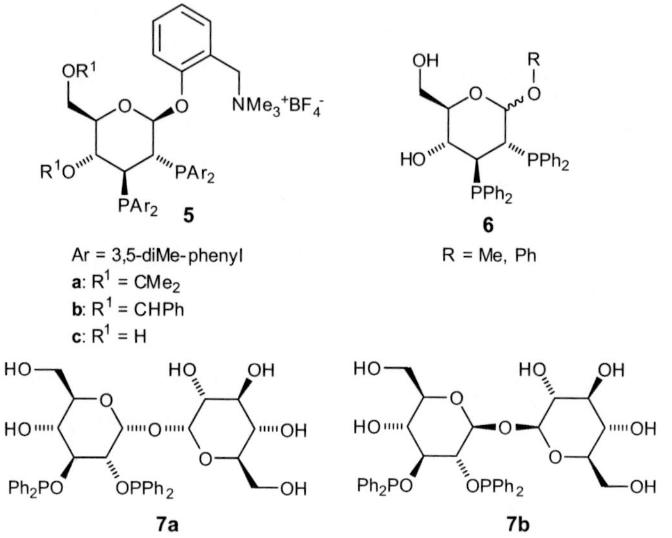

Figure 7.2 Carbohydrate-based ligands used for hydrogenations in water.

of (Z)-α-acetamidocinnamic acid and methyl ester in water.[16] The ligands bearing anomeric β-configuration delivered the highest enantioselectivity for the reaction carried out in water.

Ligands **7a** and **7b**, prepared from the corresponding α,α or β,β-trehaloses in the combination with rhodium, have been demonstrated as efficient catalyst for the asymmetric hydrogenation of dehydroamino acid and corresponding esters in water with very good yield and stereoselectivity up to 88% ee.[17,18]

The rhodium complex [Rh(COD)**8**]BF$_4$ reduced 2-acetamidoacrylic acid and its methyl ester in water with impressively high rate (99.6%) and enantioselectivity (93.6% ee) (Scheme 7.2).[19]

Asymmetric hydrogenation in a wet environment can also be efficiently carried out in the presence of surfactants. This type of reaction can be done with rhodium complexes, even with water-insoluble chiral ligands. In a series of his work, Oehme showed successful application of catalyst [Rh(COD)$_2$]BF$_4$/**9** for the reduction of some α-amino acid precursors in water. Both activity and enantioselectivity were enhanced significantly by adding the surfactants.[20,21] The level of enantioselectivity reached those observed in methanol, ranging up to 96% ee with shortened reaction time. Similar results, were obtained for phosphonate derivatives of α-amino acids.[22,23]

Asymmetric hydroxydiphosphines have been also applied for catalytic hydrogenation in water. While [Rh(COD)**4**]BF$_4$ showed a very low enantiomeric enhancement in the reduction of unsaturated amino acid precursors in water, the application of rhodium/**10** chelate as catalyst was far more successful, especially when applied along with surfactants.[24] Observed enantioselectivity jumped from 15 to 77% ee with SDS additive. It was even higher than

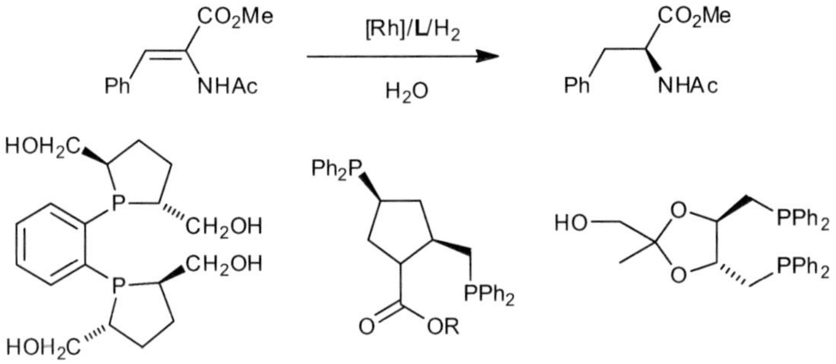

a: R = *t*-Bu
b: R = -$(CH_2CH_2O)_nC_{12}H_{25}$ (n = 10, 23)

Ligand	SDS	ee [%]
9a	−	78
	+	94
10	−	15
	+	77
7a	−	55
	+	90
7b	−	88
	+	>99
4	−	10
	+	45

Scheme 7.2 Asymmetric reduction of methyl α-acetamidocinnamate using rhodium complexes in the presence of surfactants.

the value obtained in pure methanol (50% ee). Uemura and co-workers[25] noticed also a huge increase in enantioselectivity in the reduction of methyl α-acetamidocinnamate in water in the presence of SDS, using water-soluble ligands **7a** and **7b** (Scheme 7.2).

Important unsaturated substrate precursors such as itaconic acid and dimethyl ester have been reduced using rhodium or ruthenium complexes associated with various water-soluble ligands. Reductions have been conducted in neat water[26] or in the homogenous mixture that contains water.[27]

The hydrogenation of various β-keto esters proceeded also in water in the presence of ruthenium complexes associated with water-soluble ligands such as BIFAPS **11**,[28] and BINAP derivative **12a,b** (Figure 7.3).[29,30]

Important substrates, aromatic prochiral ketones, can be reduced under hydrogen transfer conditions in the presence of [CpIrCl₂] or [CpRhCl₂] associated with water-soluble ligands **14** or **15**. Thorpe's team examined the influence of water content on the enantioselectivity of the above-mentioned reductions.[31,32] In general, the high ee values decrease slightly if the water content increases. They used water-soluble chiral amine sulfonamide ligands **15**

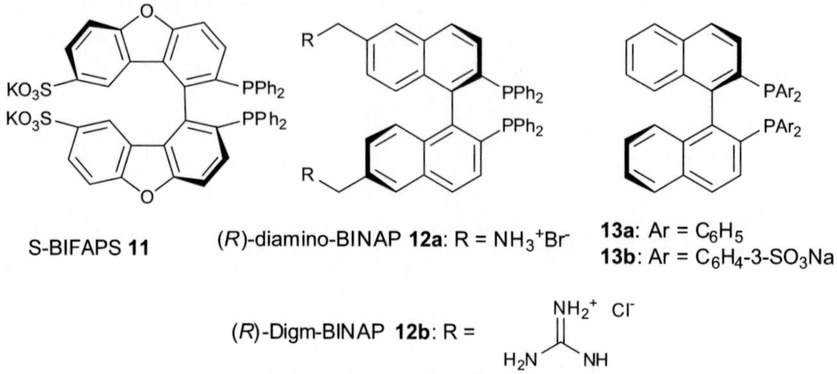
KO3S
KO3S
PPh2
PPh2
O
O
S-BIFAPS 11
R
R
PPh2
PPh2
(R)-diamino-BINAP 12a: R = NH3+Br-
PAr2
PAr2
13a: Ar = C6H5
13b: Ar = C6H4-3-SO3Na
(R)-Digm-BINAP 12b: R =
NH2+ Cl-
H2N
NH

Figure 7.3 BINAP derivatives.

in combination with rhodium and iridium salts as hydrogenation catalysts. Reaction carried out in a mixture of 2-propanol and water delivered a very good yield of up to 92% and ee up to 97% (Figure 7.4).

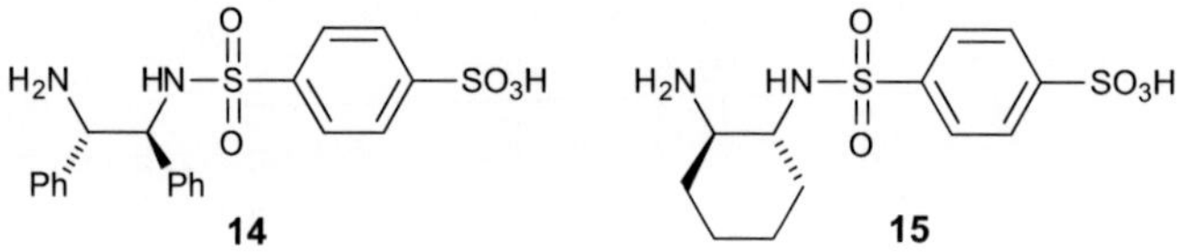
H2N
HN
O
S
O
SO3H
Ph
Ph
14
H2N
HN
O
S
O
SO3H
15

Figure 7.4 Chiral amine-sulfonamide ligand.

Rhodium in combination with nitrogen donors such as **15** and ruthenium with BINAP **17** complexes are particularly active hydrogenation catalysts for carbonyl compounds (Figure 7.5). Recently reported ee values for ruthenium

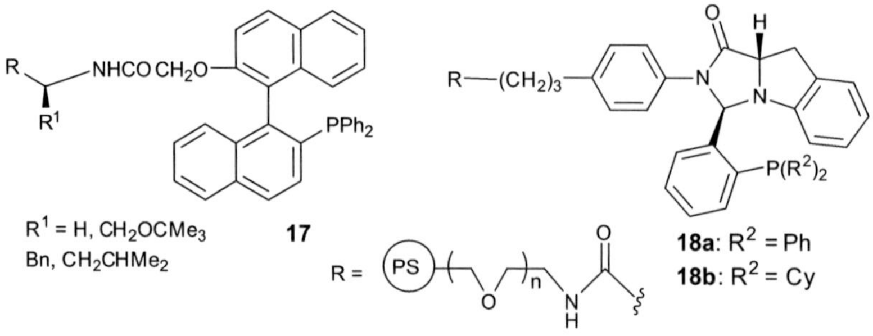
R
NHCO CH2O
R1
PPh2
R1 = H, CH2OCMe3
Bn, CH2CHMe2
17
R—(CH2)3
N
N
O
H
P(R2)2
18a: R2 = Ph
18b: R2 = Cy
R =
PS
O
n
N
H
O

Figure 7.5 Resin-supported catalysts.

catalysts were 94%,[29] whereas 50% ee was observed with a rhodium catalyst in aqueous alcohols.[33]

Asymmetric transfer hydrogenation using 2-propanol and formates are often tested in aqueous media as a result of high hydrophilicity of both reactants. Examples are reported with ruthenium and rhodium as active metals and unsaturated carboxylic acids and carbonyl compounds as substrates. Apart from achiral hydrogenations of model substrates like α-acetamidocinnamic acid and itaconic acid and their derivatives,[34] the asymmetric transfer hydrogenation of olefinic substrates have been also successful in aqueous systems. Rocha Gonsalves *et al.*[35] showed very good results with formic acid-sodium formate and rhodium with chiral ligands like SKEWPHOS, DEGUPHOS and CHIRAPHOS. Authors published up to 92% ee for the hydrogenation of (*Z*)-α-acetamidocinnamic acid and 57% ee for itaconic acid.

The synthesis of enantiomerically enriched alcohols by transfer hydrogenation with 2-propanol were reported by Thorpe *et al.*[31] Enantioselectivities up to 95% were reached for the hydrogenation of numerous methyl aryl ketones with ruthenium and water-soluble aminosulfonamide ligands **15**. Amino acids as chiral ligands in the transfer hydrogenation of acetophenone were used by Carmona *et al.*[36] A series of ruthenium, rhodium and iridium catalysts with (*S*)-proline as chiral modifier gave the best results with up to 71% ee.

Xiao and co-workers explored the asymmetric transfer hydrogenation of ketones using PEG-supported Noyori-Ikariya catalyst **16** using water as the reaction solvent.[37] The substrate scope consisted of aromatic ketones (Scheme 7.3) and cyclic ketones indanone and tetralone. For most examples the reaction gave excellent yields (up to 99%) and enantioselectivities (up to 94% ee). Further, Xiao applied the aqueous asymmetric transfer hydrogenation to the reduction of quinolines to tetrahydroquinolines.[38] This methodology seems to be flexible enough for the application to a range of heterocycles.

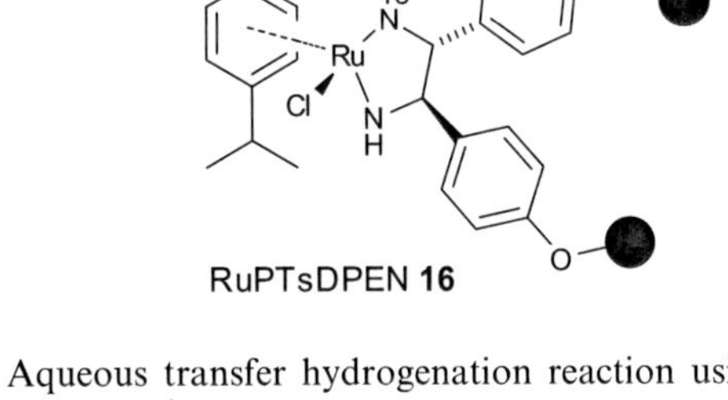

Scheme 7.3 Aqueous transfer hydrogenation reaction using PEG-supported ruthenium catalyst.

7.2.2 Alkylation Reactions

Palladium-catalyzed alkylation is now a common tool in organic synthesis. High enantioselectivities have been obtained for carbon–carbon as well as for carbon–heteroatom bond formation. Uozumi and co-workers prepared the amphiphilic resin-supported MOP ligand PEP-MOP 17[39] and *P,N*-chelating ligand PS-PEG 18a (Figure 7.5).[40] The palladium complexes of these ligands were found to be effective as catalysts for the asymmetric π-allyl substitution of both cyclic and acyclic substrates in water with up to 98% ee.

Sinou and co-workers studied palladium-catalyzed alkylation of 1,3-diphenyl-2-propenyl acetate with dimethyl malonate. Reactions proceeded in water using K_2CO_3 as the base and chiral non-water-soluble ligands in the presence of surfactants.[41] The highest enantioselectivites (92%) were observed for cetyl-trimethylammonium hydrogen sulfate as the surfactant and chiral atropoiso-meric diphosphines such as BINAP derivatives **13**.

7.2.3 Allylation

Particular attention has been focused on the allylation of carbonyls and imines to corresponding homoallylic alcohols or amines. While a number of metals, such as Zn,[42] Sn,[43] Sb,[44] Mn,[45] Mg,[46] or even Hg,[47] have been reported to be useful for this purpose, indium has showed the most promise due to its unique properties that make it particularly suitable for use in water. This metal is: (i) stable even in boiling water; (ii) resistant to oxidation by air; and (iii) has an unusually low first ionization potential. The literature on indium-mediated reactions in aqueous media up to 1998 has been reviewed[48] and will not be discussed here. The first enantioselective indium-mediated allylations of alde-hydes have been demonstrated by Loh and Zhou in 1999.[49] Later, the authors reported a catalytic enantioselective version of the same transformation using allyltributyltin in place of allylindium and a modified version of Yamamoto-Yanagisawa's catalyst.[50] Excellent asymmetric induction has been also achieved by grafting a chiral auxiliary onto the substrate: Cho and co-workers studied the indium-mediated allylation of α-ketoimide derived from Oppolzer's sultam with various allyl bromides. High diastereoselectivity was observed in a homogeneous water/THF (3:1) solution (Scheme 7.4).[51]

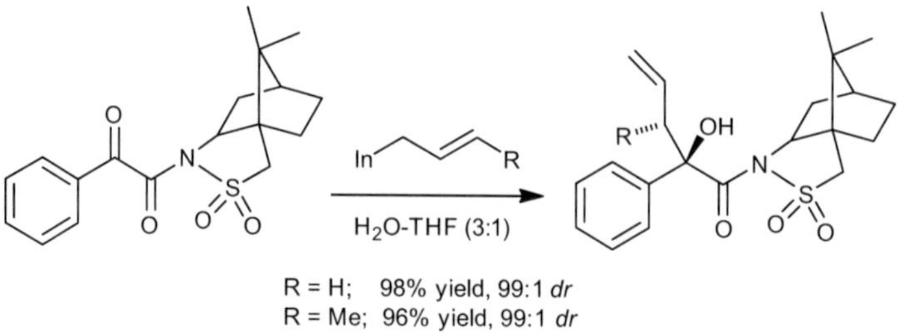

R = H;   98% yield, 99:1 *dr*
R = Me; 96% yield, 99:1 *dr*

Scheme 7.4   Asymmetric indium-mediated allylation.

In 2001 Lubineau re-examined this methodology, in an effort to find efficient entry to the synthetically useful *C*-branched sugars. In pure water containing powdered indium metal the reaction afforded the addition product as a single stereoisomer in 95% yield.[52]

Aryl ketones can be allylated by a mixture of tin derivatives and optically pure monothiobinaphthol (MTB) ligand with high ee (Scheme 7.5).[53] Aliphatic ketones gave complex mixtures of products while a similar reaction using cyclohexyl methyl ketone gave α-methyl-α-(2-propenyl)cyclohexanemethanol with only 59%.

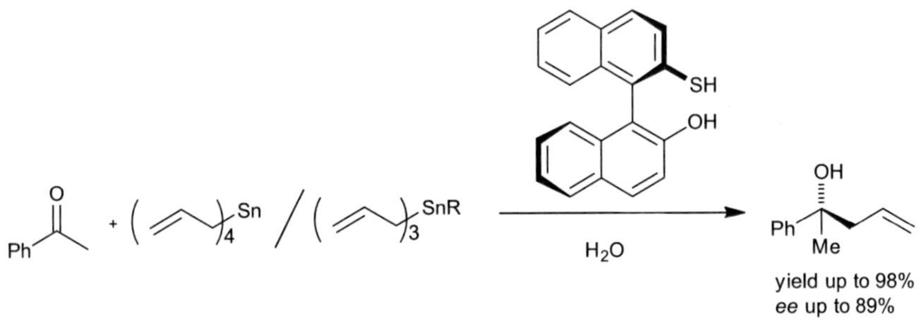

Scheme 7.5 Asymmetric tin-mediated allylation.

Catalytic asymmetric allylation of aldehydes with allyltributyltin in aqueous media has also been realized using combinations of cadmium bromide and chiral diamine ligands.[54]

7.2.4 Cyclopropanation

The ruthenium-catalyzed asymmetric cyclopropanation of styrene with diazoacetates in aqueous media was recently disclosed by Nishiyama and co-workers.[55] The water-soluble, chiral *pybox*-type ligand **19** was used in these transformations (Scheme 7.6). While the reaction proceeded with low enantioselectivity (8%) in pure THF or toluene, the addition of water dramatically increased the selectivity, producing cyclopropane in 78% ee in THF/water (2:1) and 94% ee in biphasic toluene/water (1:1) mixture.

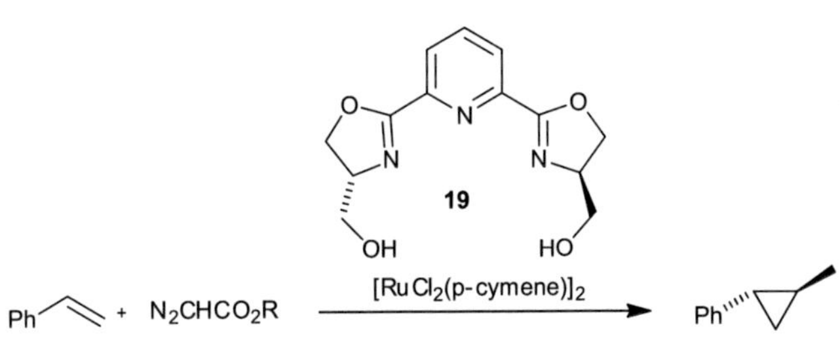

Scheme 7.6 The ruthenium-catalyzed asymmetric cyclopropanation.

7.2.5 Asymmetric Formation of Chlorohydrins

Chlorohydrins have been obtained asymmetrically in the palladium(II)-catalyzed oxidation of various alkenes in aqueous solution using pyridine as a ligand and $CuCl_2$. However, an enantioselectivity of up to 76% ee was obtained in the presence of chelating sulfonated diphosphines such as **3c** and **13b** (Scheme 7.7). The selectivity jumped up to 94% ee while using bimetallic catalysts containing a β-triketone and bridging chiral diphosphine and diamines.[56]

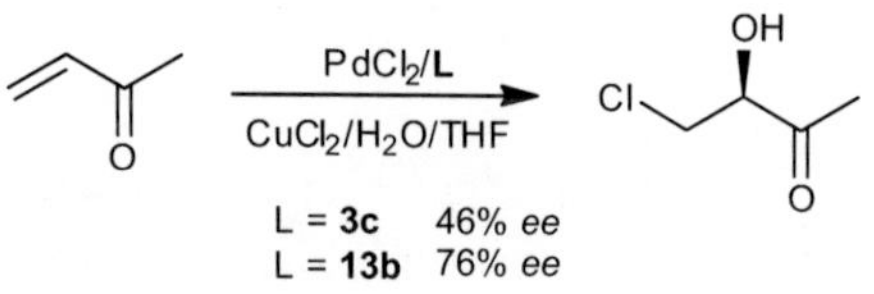
PdCl2/L
CuCl2/H2O/THF
L = 3c 46% ee
L = 13b 76% ee
OH
Cl
O
O

Scheme 7.7 Asymmetric synthesis of chlorohydrins.

7.2.6 Epoxidation

Optically pure epoxides are very useful synthetic intermediates and building blocks for total synthesis. A catalytic asymmetric version of the highly stereospecific, dioxirane-mediated epoxidation of alkenes in aqueous media was developed by Yang and co-workers.[57] The authors observed good to high enantioselectivity (71–95% ee) in the epoxidation of disubstituted and trisubstituted olefins using BINAP-derived ketones (*R*)-**20a–c** as catalysts and Oxone® as the stoichiometric oxidant. Almost simultaneously, Shi and co-workers found that the fructose-derived chiral ketone **21** was an efficient catalyst in the epoxidation of conjugated dienes leading to synthetically useful vinyl epoxides in high enantiomeric excess (89–97%) (Figure 7.6).[58,59]

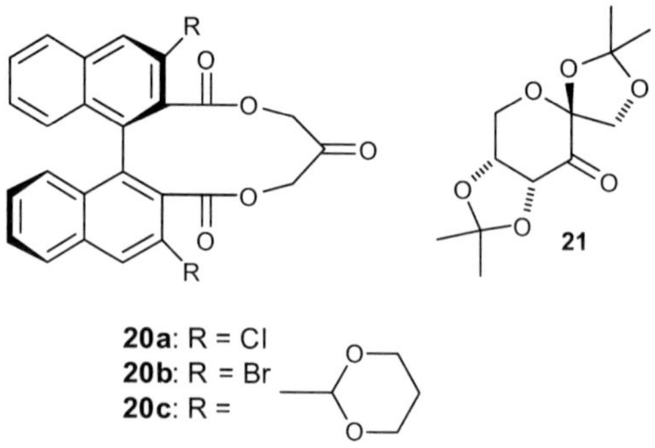
R
O
O
O
O
O
R
O
O
O
O
O
O
O
21
20a: R = Cl
20b: R = Br
20c: R =

Figure 7.6 Asymmetric epoxidation catalyst components.

Malkov and Bourhani developed asymmetric epoxidation of allylic alcohols with the *in situ* generated vanadium catalyst (Scheme 7.8).[60] These oxygen atom

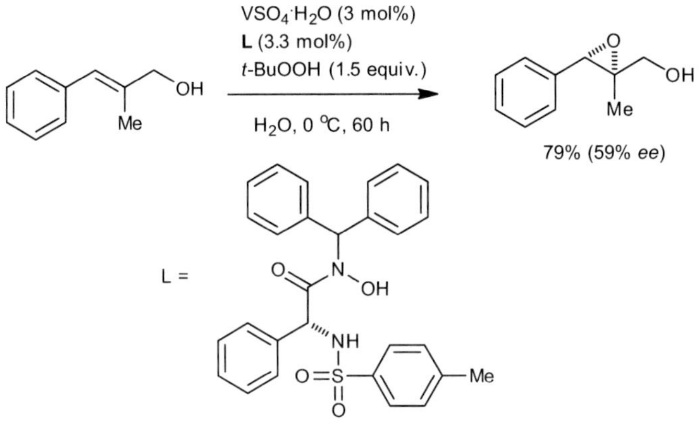

Scheme 7.8 Epoxidation of allylic alcohols.

transfers were successfully performed on water at 0 °C with moderate to good yields of epoxides.

Ring opening of *meso*-epoxides have also been well studied in aqueous solutions. The asymmetric hydrogenolysis of epoxides delivers enantiomerically enriched monoalcohols. Chan *et al.*[61] obtained disodium hydroxysuccinate in rhodium-catalyzed hydrogenations of the corresponding water-soluble epoxides with ee values up to 30%. The reaction was carried out in a presence of chiral phosphine ligands in aqueous methanol. Sinou[62] tested the same reaction with the water-soluble complex of sulfonated rhodium-SKEWPHOS (analogues of **3**) in water or in aqueous methanol. They found a maximum ee of 39%.

Kobayashi and co-workers reported asymmetric ring opening of *meso*-epoxides using 1 mol% of Sc(SDS)₃ and 1.2 mol% of a chiral bipyridine ligand **22** in water. The reaction provided β-amino alcohols in high yields and with excellent enantioselectivities (Scheme 7.9).[63,64]

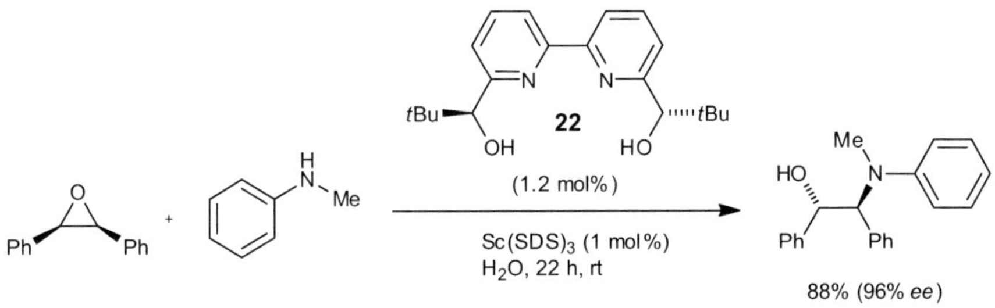

Scheme 7.9 Enantioselective synthesis of β-amino alcohols from *meso*-epoxides.

7.2.7 Diels–Alder Reaction

The benefits of water as an additive acting as an accelerating factor were observed for the very first time for the Diels–Alder reaction. In fact, the discoveries made by pioneers in the early 1980s opened the door for future development in the field of water-assisted chemistry.[65] Further, Engberts *et al.*[66] have shown the beneficial effect of water on both the activity and enantioselectivity in the condensation of cyclopentadiene with an unsaturated ketone in the presence of copper complexes of various chiral (*S*)-amino acids. The Diels–Alder adduct was obtained in yields generally exceeding 90%, and enantioselectivity of up to 74% ee. The enantioselectivity level was higher than those observed for organic solvents. It is to be noticed that the catalyst/water solution can be reused without any decrease in enantioselectivity.

7.2.8 Asymmetric Aldol Reaction

The catalytic aldol reaction is a powerful carbon–carbon bond formation process leading to useful chiral β-hydroxy ketones and esters. Pioneering experiments by Lubineau and co-workers demonstrated that the uncatalyzed aldol reaction of silyl enol ethers (Mukaiyama donor) in water could proceed, though yields were poor.[67] Good chemical and optical yields (up to 73% ee) have been obtained in the addition of various silyl enol ethers to aldehydes in wet DMF in the presence of a palladium(II)-BINAP catalyst.[68]

Copper(II)-catalyzed aldol reactions have also been reported, performed in the presence of bisoxazolines **23** as chiral ligands.[69,70] However, the most successful work on aldol reactions in aqueous media have been initially focused on lanthanides as catalysts. Various lanthanide triflates were complexed with the chiral crown ether **25** and screened for their ability to induce diastereo- and enantioselectivity (up to 85% ee) in the reaction between benzaldehyde and silyl enol ethers in EtOH/water (9:1).[71] It was shown that for larger lanthanide cations, such as Ce^{II}, La^{II}, Pr^{II} and Nd^{II}, which may have a better size fit for the cyclic ligand than smaller ions, both diastereomeric and enantiomeric selectivity were good. Kobayashi also found that in EtOH/water (9:1) lead(II) triflate efficiently catalyzed the aldol reaction of benzaldehyde upon complexation to the chiral crown ligand **24**.[72] Again, good *syn/anti* selectivities and enantioselectivities (up to 87% ee) were obtained.

One of the major drawbacks of using water as a solvent is the low solubility of most organic substrates in water. For the reason of solubility, organic co-solvents have been used in previously mentioned examples. Another solution is the use of surfactants, which solubilize organic substances in water, or the use of a new type of catalyst, 'Lewis acid-surfactant combined catalyst' (LASC).[73] Instructive examples have been elaborated by the same Japanese group. Boron compounds have been developed as catalysts for the reaction of ketone equivalents in water. In the presence of a catalytic amount of a boron source, SDS and benzoic acid, the reaction of benzaldehyde with silyl enol ether gave highly diastereoselective aldol product in water at ambient temperature.[74]

Some other metals that have also been used with some success as catalysts in aqueous Mukaiyama aldol reactions are bismuth[75] and indium.[76] Interestingly, the predominance of *syn* aldol products in the water-based reactions discussed above is in contrast with the analogous reactions run under anhydrous conditions where the *anti* isomer is usually the major product (Scheme 7.10).

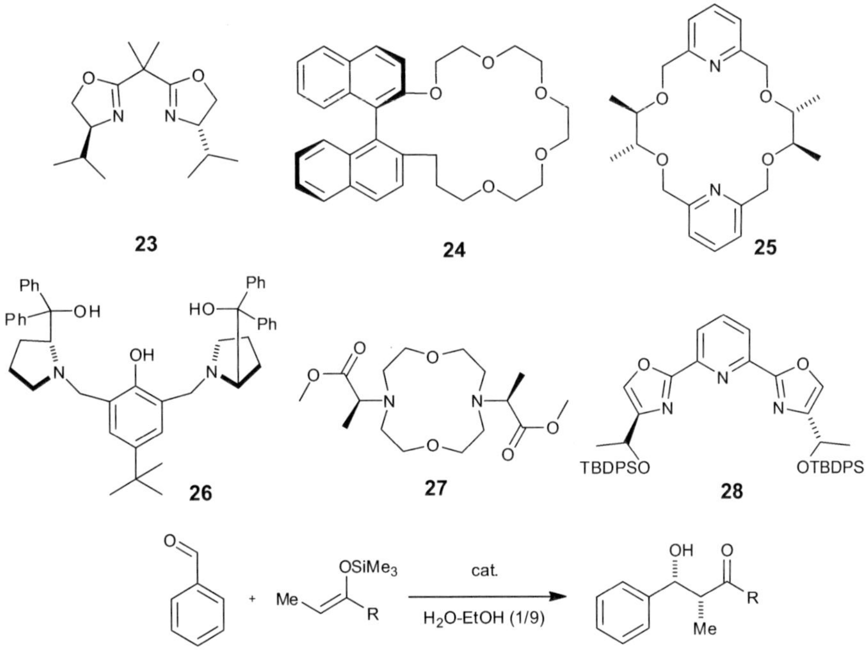

Catalyst	R	yield [%]	ee [%]
Cu(OTf)$_2$ + **23**	Et	81	81
	Ph	74	67
	i-Pr	95	77
Pb(OTf)$_2$ + **24**	Ph	62	55
Ln(OTf)$_2$ + **25**	Ph	85	78
Ga(OTf)$_3$ + **26**	Ph	89	88
Eu(OTf)$_3$ + **27**	Ph	92	93

Scheme 7.10 Enantioselectivities in aldol reaction.

More recently, Li developed a highly efficient asymmetric Mukaiyama reaction by using chiral gallium catalysts with Trost's chiral semi-crown ligands **26**.[77] The combination of Ga(OTf)$_3$ and the chiral ligand made the aqueous reaction proceed smoothly with good yield (89%), diastereoselectivity (*syn/anti* 89:11), and enantioselectivity of *syn* product (ee 87%). It has been noted that gallium salt is known to decompose in the presence of water. However, in the presence of chiral ligand it is stable enough to perform the Mukaiyama aldol reaction.

Recently, a highly efficient asymmetric Mukaiyama reaction by using $Eu(OTf)_3$ and macrocyclic ligand **27** have been presented.[78]

Most of the successful catalysts applied for the enantioselective Mukaiyama aldol reaction are composed of heavy or rare earth metals, which creates drawbacks in their applications because of the toxicity or high price. In 2006 Mlynarski's group focused on application of zinc[79] and iron.[80] Iron is the most widespread metal on earth. It is also cheap and environmentally benign. In this respect enantioselective reactions promoted by iron complexes constitute interesting and valuable *green* alternatives. An iron(II) complex with designed *pybox* ligands **28** showed very good catalytic activity and enantioselectivity in aqueous media. This water-stable iron-based chiral Lewis acid promotes condensation of aromatic silylenol ethers with a range of aldehydes with good yields, excellent *syn* diastereoselectivity and ee up to 92%.[81]

A few instructive examples in neat water have also been presented. A chiral zirconium catalyst generated from $Zr-(O-t-Bu)_4$ and $[(R)-3,3'-I_2-BINOL]$ catalyzed the aldol reaction in high yields under mild conditions.[82] Sulfonate derivatives of chiral 1,1'-binaphthol were used as chiral anionic surfactants in asymmetric aldol-type reactions to give aldol adducts with moderate to good diastereo- and enantioselectivities; $Ga(OTf)_3$ and $Cu(OTf)_2$ provided better results than $Sc(OTf)_3$ as the Lewis acid catalyst in this system.[83]

The aldol reaction of trimethylsilylenol ethers with aqueous formaldehyde proceeded moderately well using tetrabutylammonium fluoride as an activator. Asymmetric hydroxymethylation of trimethoxysilylenol ethers using (R)-BINAP-AgOTf as Lewis acid and KF as Lewis base has been achieved in aqueous media (up to 57% ee).[84] Chiral bis(oxazoline) ligands disubstituted at the carbon atom linking the two oxazolines by Frechet-type polyether dendrimers coordinated with copper(II) triflate were found to provide good yields and moderate enantioselectivities for Mukaiyama aldol reactions in water, which are comparable with those resulting from the corresponding smaller catalysts.[85] $AgPF_6$-BINAP is very active in this reaction and the addition of a small amount of water enhanced the reactivity.[86]

Kobayashi and co-workers have developed a new methodology for asymmetric hydroxymethylation of silicon enolates with aqueous solution of formaldehyde in water that do not require organic solvents. Chiral ligand **29** (Scheme 7.11) in association with $Sc(SDS)_3$/NaSDS provides desired products in good to high yields. A wide range of silicon enolates react smoothly and give high stereoselectivities ($>90\%$ ee).[87]

7.2.9 Mannich Reaction

The development of Mannich-type reactions in aqueous media has been made with the incentive of finding a milder and more convenient approach toward the construction of β-amino ketones or esters. Classical protocols for Mannich reactions are sometimes of limited synthetic potential as they often involve harsh reaction conditions and are plagued with severe side reactions as well as

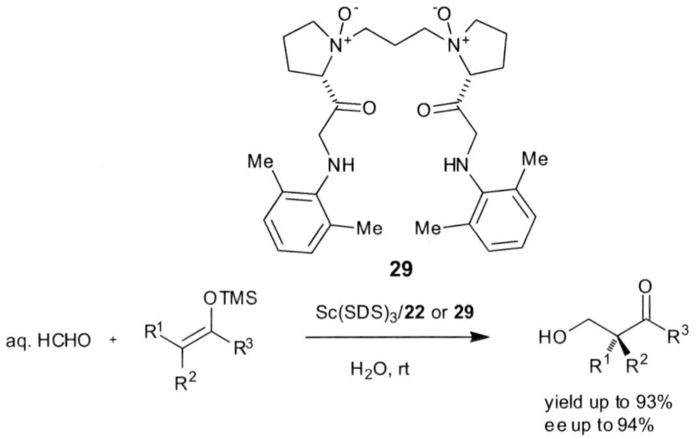

Scheme 7.11 Catalytic hydroxymethylation in water.

poor regio- and stereocontrol.[88] Nonetheless, several research groups have recently reported on the one-pot Mannich-type reaction in water to give β-amino carbonyl compounds using either Lewis acid[89] or Brønsted base catalysis,[90] with or without the addition of surfactants. Both types of reactions generally proceed smoothly in good yields, albeit with diastereoselectivities that are usually moderate at best.

In the one of the smartest examples, Kobayashi showed that a diastereo- and enantioselective Mannich-type reaction of a hydrazono ester with silyl enol ethers in aqueous media can be successfully achieved with ZnF$_2$, a chiral diamine ligand and trifluoromethanesulfonic acid (Scheme 7.12).[91,92] Diastereoselective Mannich-type reactions of chiral aldimines with 2-silyloxybutadienes in the presence of zinc salt and water led to products with 74–90% ee.[93]

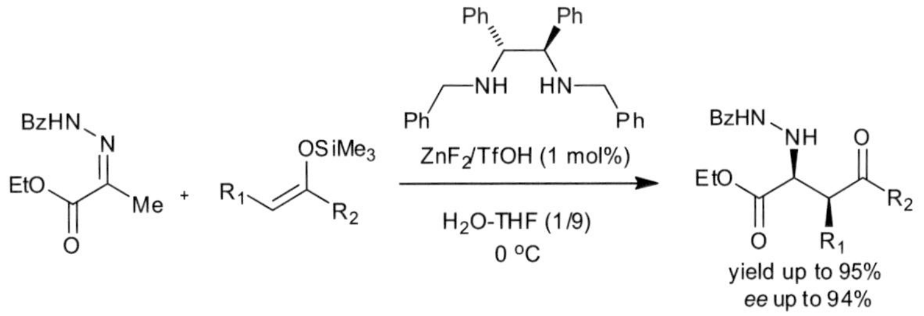

Scheme 7.12 Diastereoselective Mannich-type reactions of chiral aldimines.

7.2.10 Substitution Reactions

Little attention has been devoted to the development of substitution reactions in water. A likely explanation for this is that in many types of nucleophilic reactions in water, hydrolysis of the electrophile may affect the desired reaction pathway. Nevertheless, a few reports of palladium-catalyzed allylic substitutions in water have surfaced in recent years.[39,94] Moreover, Uemura and co-workers used a carbohydrate-based phosphinite-oxazoline ligand 30 in the palladium-catalyzed substitution of 1,3-diphenyl-3-acetoxyprop-1-ene with both carbon and nitrogen nucleophiles.[95] The substitution products were obtained in moderate to high yields (66–95%) and in good enantiomeric excess when performed in water alone or in water/acetonitrile mixes (Scheme 7.13).

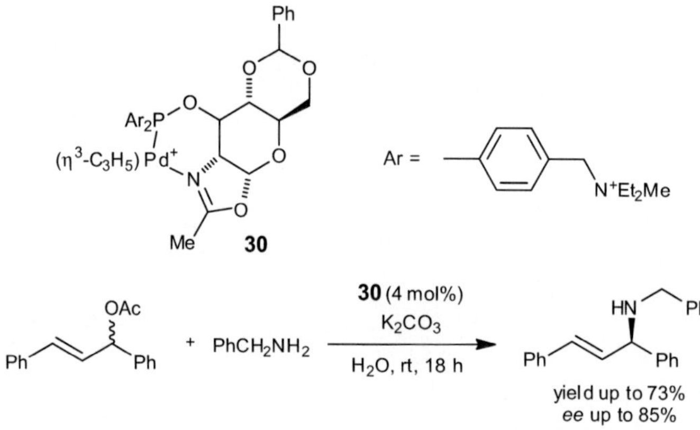

Scheme 7.13   Carbohydrate-based phosphinite-oxazoline ligand.

A year later, Uozumi and Shibatomi reported an immobilized palladium complex of a *P,N*-chelate chiral ligand, which catalyzed the asymmetric alkylation of allylic esters in 0.9 M aqueous Li_2CO_3 with up to 99% enantioselectivity.[40] The catalyst was found to be effective for both cyclic and acyclic substrates. Conveniently, the catalyst could be recovered by simple filtration and reused without any loss of activity or stereoselectivity.

7.2.11   Alkyne with Imine Coupling

Recently, Li and co-workers reported a highly efficient asymmetric aldehyde-alkyne-amine coupling in water.[96] Use of the tridentate bis(oxazolinyl) pyridines 31 with Cu(OTf) afforded the product with both high yield and enantioselectivity up to 99.6% ee in organic solvent and 84% ee in water. In most cases, imines were formed *in situ* and the addition was very simple: mixing an aldehyde, an aniline and an alkyne with the catalyst in one pot (Scheme 7.14).

Tu and co-workers recently reported that a three-component coupling of aldehyde, alkyne and amine *via* C–H activation catalyzed by CuI in water can be greatly accelerated using microwave irradiation. Using (*S*)-proline methyl ester as the source of chirality, a direct and highly diastereoselective method for the construction of chiral propargylamines was achieved.[97]

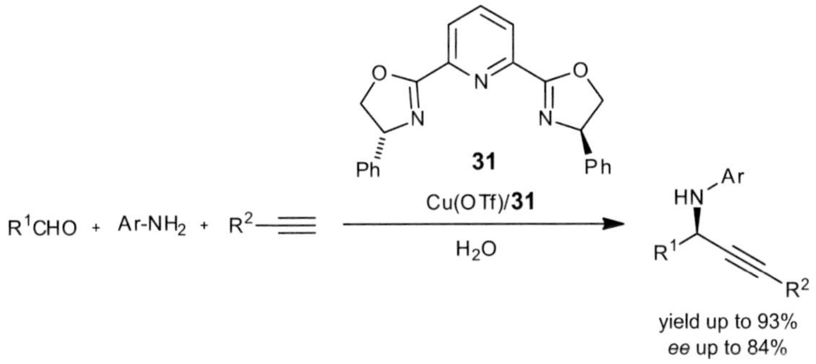

Scheme 7.14 Highly efficient Cu-catalyzed three-compound coupling.

Che and co-workers reported a gold-catalyzed three-component coupling of aldehyde, alkyne and amine in water using gold(III)-salen complexes as precursors.[98] This coupling reaction has been applied to the synthesis of propargylamine-modified artemisinin derivatives with the delicate *endo* peroxide moieties remaining intact.

7.2.12 Asymmetric Suzuki Coupling

Although aqueous asymmetric cross couplings have not been intensively studied, asymmetric Suzuki coupling using a chiral palladium catalyst supported on an amphiphilic resin has been described. Detected high enantioselectivities (99% ee) are promising for further development in the field (Scheme 7.15).[99]

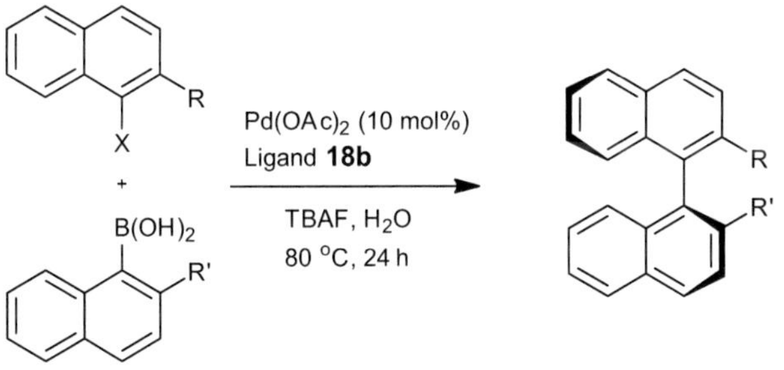

Scheme 7.15 Asymmetric Suzuki coupling.

7.3 Organocatalytic Reactions

Near the end of the 20th century, small metal-free organic molecules attracted attention as organocatalysts.[100] Organic compounds, as compared to metals, are more stable towards air and water, less expensive, possibly non-toxic, readily available, and as a consequence considered as environmentally friendly.

Initially, organocatalytic reactions have been carried out in conventional organic solvents, but soon water solutions were being used by first practitioners.[3,8] Here we present highly hydrophobic catalysts in the presence of water and small organic molecules that could act as enantioselective catalysts in homogeneous solution. While the former is more similar to enzymes protected by shielded active sites inside hydrophobic pockets, latter is extremely interesting as real homogeneous catalysis. Whether or not these reactions are carried out with a catalytic amount of water in truly aqueous medium, the most important aspect was the positive water influence on the stereoselectivity of the catalysts.

7.3.1 Diels–Alder Reactions

In 1999 MacMillan's group developed the concept of iminium activation, which is based on the capacity of chiral amines to function as enantioselective LUMO-lowering catalysts for a broad range of synthetic transformations. For this purpose, chiral secondary amines based on the imidazolidinone ring were developed which incorporated chiral amino acid motifs. To date, this activation strategy based on LUMO lowering has led to the development of over 30 different enantioselective transformations for asymmetric synthesis.[101] In 2002, MacMillan and Northrup presented the first catalytic Diels–Alder reaction in water based on a metal-free imidazolidinone catalyst. They reported the activation of acyclic and cyclic enones for enantioselective catalytic [4 + 2] cycloaddition using catalyst **32**.[102] The chirality of the catalyst originated from an amino acid precursor; **32** was prepared from (S)-phenylalanine methyl amide. The reaction of 4-hexen-3-one and cyclopentadiene provided the Diels–Alder product in 89% yield and with good stereoselectivity (90% ee for the *endo* isomer; Scheme 7.16).

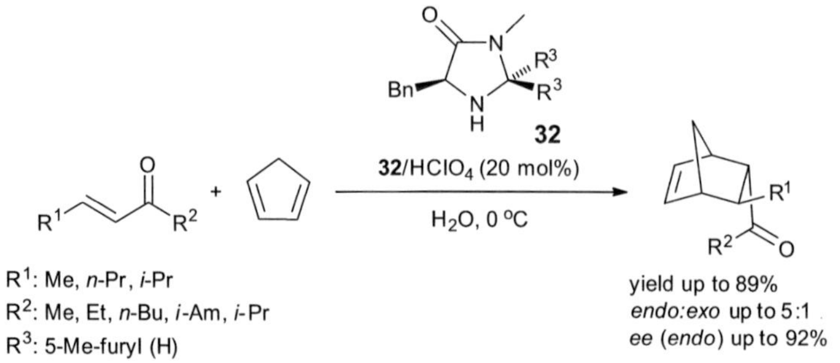

R^1: Me, n-Pr, i-Pr
R^2: Me, Et, n-Bu, i-Am, i-Pr
R^3: 5-Me-furyl (H)

yield up to 89%
endo:exo up to 5:1
ee (endo) up to 92%

Scheme 7.16 Organocatalyzed Diels–Alder cycloadditions between cyclopentadiene and acyclic enones.

7.3.2 Aldol Reactions

Stereoselective formation of the carbon–carbon bond in nature is assisted by enzymes named lyases, which catalyze the usually reversible addition of carbon nucleophiles to carbonyl groups. Aldolases belong to the group of lyases and have evolved to catalyze the anabolism and catabolism of highly oxygenated metabolites. They are essential for many biosynthetic pathways of carbohydrates, keto acids and some amino acids.[103] It is highly desirable to develop chemical systems that can mimic the action of enzymes and perform organic reactions, particularly aldol additions, in water with perfect efficiency and stereoselectivity. Some early studies reported water-tolerant aldolase-type organocatalysts, but it was not possible to achieve high yield and stereoselectivity in the aqueous direct aldol reactions without addition of any organic solvents.[104] The first catalysts which were used in the aldol reaction in aqueous organic solvents were (S)-proline **33**,[105] various proline derivatives **34–36**[106] and some other amino acids (Figure 7.7).[107]

Though asymmetric aldol reactions catalyzed by the proline-derived amides **37**[108] and **38**,[109] diamide **39**[110] and tryptophan[111] have been developed in water without organic co-solvents, only moderate enantioselectivities were observed. In 2006, Barbas III[112] and Hayashi[113] independently presented two examples of far more efficient, modified proline-based catalysts for aldol reaction in the presence of water, diamine **40** and siloxyproline **41** (Scheme 7.17). Although siloxyproline **41** showed better substrate scope, application of this catalyst for water-soluble ketones (acetone, hydroxyacetone) is limited. Both catalysts are of hydrophopic nature, thus reactions must be regarded as more like the 'on water'-type with catalyst operating mostly in organic phase.

Other examples of highly enantioselective aldol reactions in the presence of water have been presented, although limited to cyclic ketones and activated aromatic aldehydes.[114] The efficient and nearly quantitative reaction of cyclohexanone in the presence of a large amount of water was described to be promoted by protonated proline amides **42**,[115] prolinethioamide **43**[114a] and *t*-butylphenoxyproline **44**.[114h]

Studies of enantioselective organocatalytic reactions promoted by primary amino acids and their derivatives[107] provided new and interesting results.[116] The most promising applications of siloxythreonine **45**[114c] and serine **46**[114b] must be seen, however, as reactions 'in the presence of water' rather than 'in water'.

In the 2007, Singh and co-workers[114d] described proline-derived **47** which can be used as an organocatalyst for the direct asymmetric aldol reaction of ketones with aldehyde acceptors in high enantioselectivities and with low catalyst loading (0.5 mol%) in aqueous medium. When the reaction occurred in brine it gave better yields and ee values. The proline amide catalyst was an excellent catalyst not only for the reaction between aromatic aldehydes but also for some α-substituted aliphatic aldehydes. In all cases good yields along with excellent diastereo- and enantioselectivities (>99% ee) were reported.

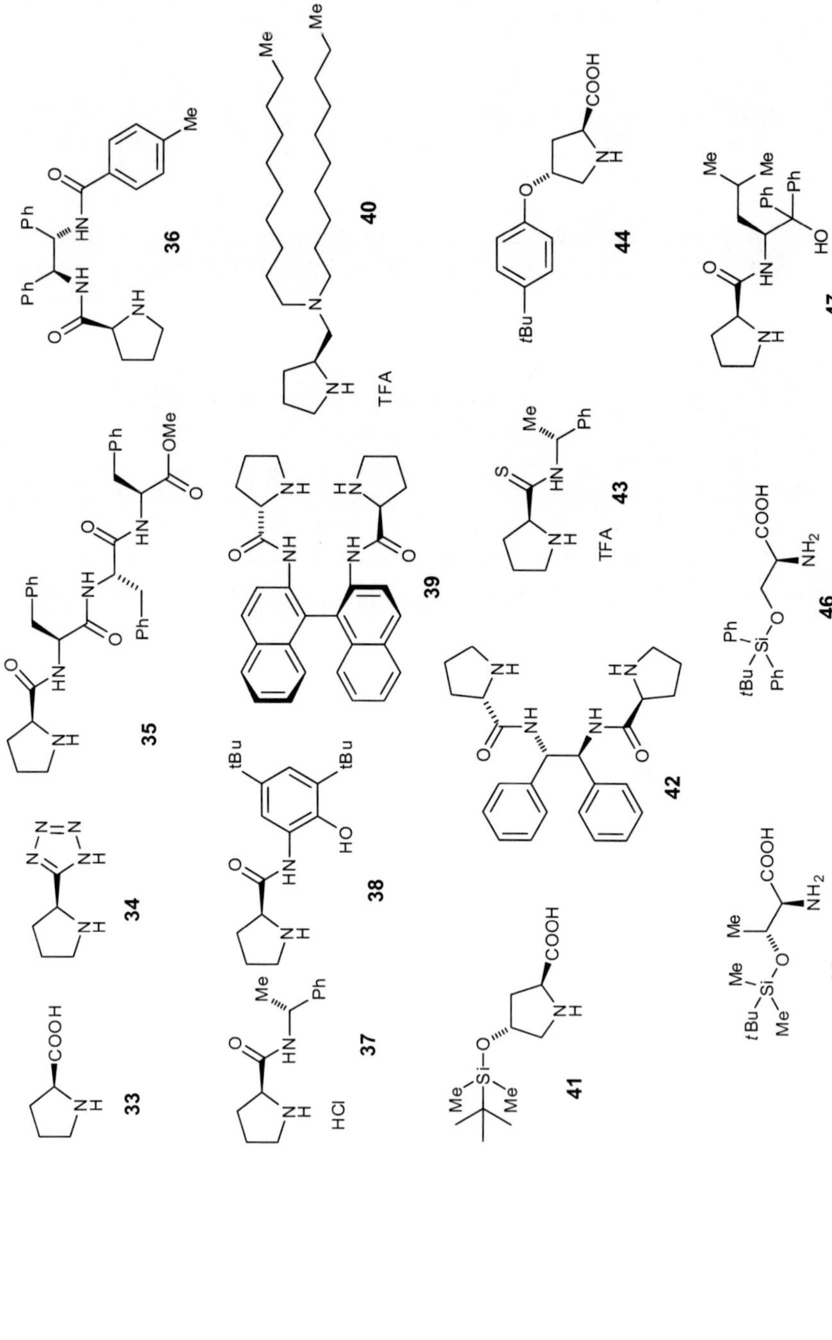

Figure 7.7 Catalysts derived from amino acids for asymmetric aldol reactions in aqueous organic solvents.

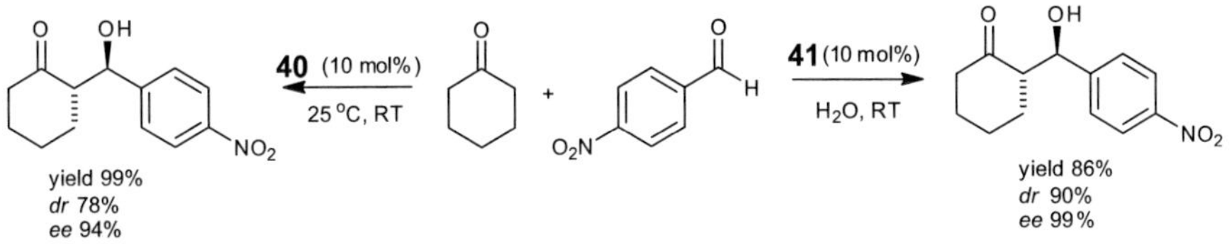

Scheme 7.17 Aldol reaction catalyzed by organocatalyst **40** and **41**.

In 2007 Hayashi's group observed that proline amide **48** acts efficiently in homogeneous solution with water for the enantioselective self-aldol reaction of propanal (Scheme 7.18).[117] They observed high dr and ee. This result was explained by the activation of the carbonyl group of the substrate by an amide proton of the catalyst in the same way that the carboxylic acid proton of proline would.

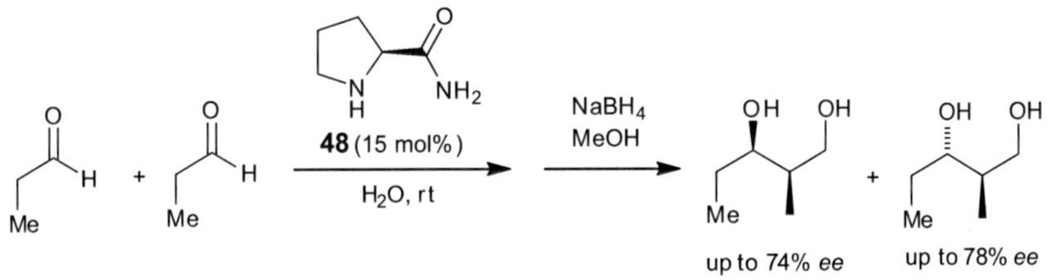

Scheme 7.18 Cross aldol reaction catalyzed by prolinamide.

Simple peptides and their analogues having a primary amino group as the catalytic residue mediate the direct asymmetric intermolecular aldol reaction in aqueous organic solvents. Gong and co-workers[118] used (*S*)-proline-based small peptides **35** as efficient catalysts for the asymmetric direct aldol reactions of hydroxyacetone with aldehydes (Scheme 7.19).

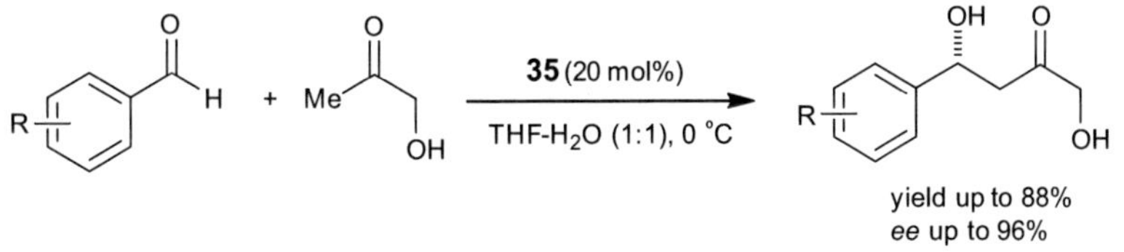

Scheme 7.19 Direct aldol reactions of hydroxyacetone with aromatic aldehydes catalyzed by small peptides.

The abilities of peptides **35** to catalyze the direct aldol reactions of hydroxyacetone with a variety of aldehydes were examined under optimal conditions. Aldol products were obtained with high yields (up to 88%) and enantioselectivity (up to 96% ee). The best results were observed for aromatic aldehydes bearing electron-withdrawing groups.

The research groups led by Lu[114c] and Barbas[119] independently reported that hydrophobic threonine derivatives could catalyze direct aldol reactions between *O*-protected hydroxyacetone and various aromatic aldehydes in the presence of small amounts of water.

In the 2008 Barbas III and co-workers[120] described a novel threonine-based organocatalyst **49** as an effective catalyst for the reaction of protected dihydroxyacetone with a variety of aldehydes in brine. They obtained *syn* aldols with high yield and enantioselectivity (Scheme 7.20).

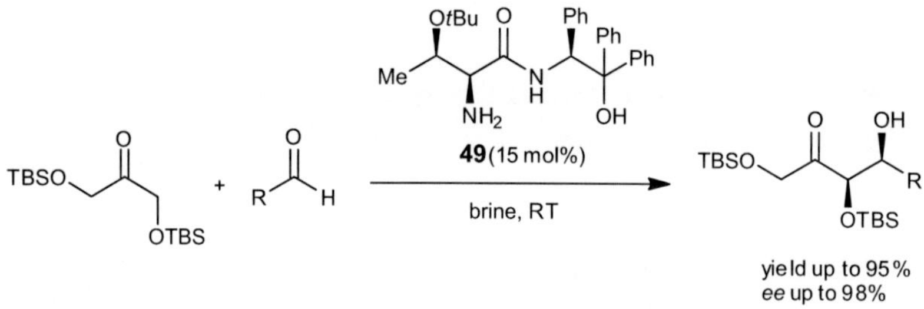

yield up to 95%
ee up to 98%

Scheme 7.20   Aldol reaction of hydroxyacetone in brine.

7.3.3   Michael Reactions

The Michael addition reaction is one of the most important carbon–carbon bond-forming reactions in organic synthesis. In recent years, an intense research effort has been made to find chiral organic molecules as catalysts for this enantioselective reaction.[121] Recent progress in the field of asymmetric organocatalytic 1,4-conjugate addition reactions, regarded as belonging among the more synthetically important carbon–carbon bond-forming reactions, is described. The focus is on some recent advances in the following selected reactions: additions of various nucleophiles to α,β-unsaturated cyclic and acyclic enols, enones, vinyl sulfones and nitro olefins, addition of malonates and/or ketones to acyclic enones, or of aldehydes and ketones to vinyl ketones and nitro olefins, together with some multicomponent domino reactions.[122]

In 2006 Barbas and co-workers reported new organocatalysts **40** (Scheme 7.21) for the asymmetric Michael reaction of aldehydes and ketones to nitro olefins.[112] They developed a catalytic direct asymmetric Michael reaction that can be performed in brine without addition of organic solvents. The diamine **40** bifunctional catalyst system demonstrated excellent reactivity, diastereoselectivity and enantioselectivity in brine. The best result was observed for the reaction in brine because in these electrolyte-rich solutions the anion intermediate undergoes complexation with metal cations which decreases the polymer propagation responsible for the side products.

Liang's group reported triazole **50** as catalyst asymmetric Michael reaction of ketones to nitrostyrene.[123] They developed a new substituted triazole organocatalysts **50** and demonstrated their potential use for Michael reactions

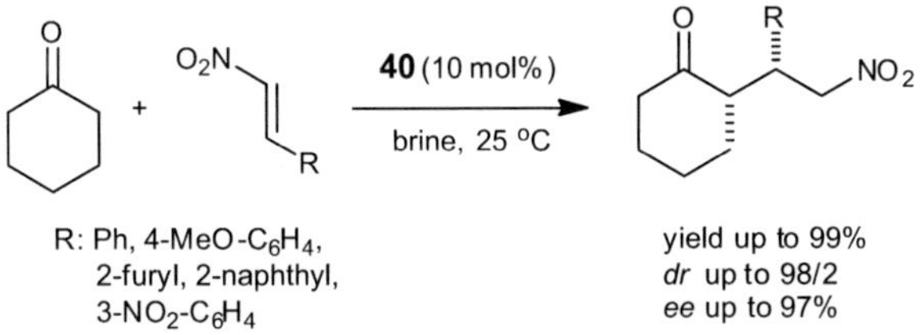
O
O2N
+
R
40 (10 mol%)
brine, 25 °C
O
R
NO2
R: Ph, 4-MeO-C6H4,
2-furyl, 2-naphthyl,
3-NO2-C6H4
yield up to 99%
dr up to 98/2
ee up to 97%

Scheme 7.21 Asymmetric Michael reaction in water.

(Scheme 7.22). The pyrrolidine–triazole catalysts show several interesting features: (i) they can efficiently catalyze the Michael additions with high yields, excellent enantioselectivity and a very good diastereoselectivity; (ii) the Cu^{I}-catalyzed 1,3-dipolar 'click' azide–alkyne cycloaddition provides the modular and tuneable features for the present catalyst; (iii) the triazole moiety cannot only act as a phase tag to complete the reaction in a broad range of solvents (including water), but can also serve as an efficient chiral-induction group to ensure a high selectivity.

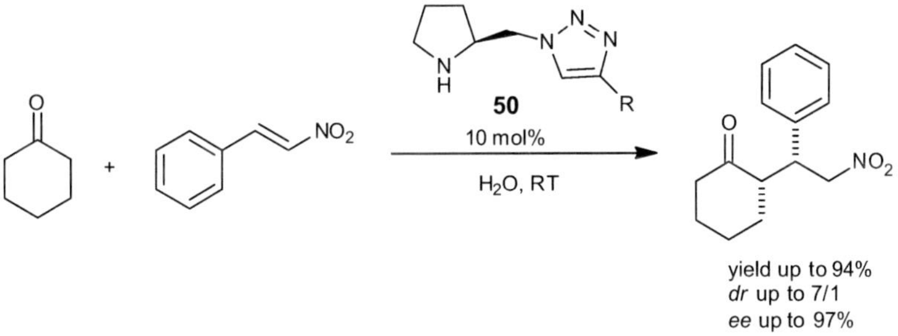
N
H
N
N
N
R
50
10 mol%
H2O, RT
O
+
NO2
O
NO2
yield up to 94%
dr up to 7/1
ee up to 97%

Scheme 7.22 Catalytic asymmetric Michael addition of cyclohexanone to nitrostyrene.

7.3.4 Mannich Reactions

The catalytic, asymmetric Mannich reaction is one of the most powerful methods for the construction of chiral nitrogen-containing molecules. The direct asymmetric Mannich reaction catalyzed by organic molecules plays a huge role in the synthesis of the bioactive molecules. Organocatalysis-mediated, direct, enantioselective Mannich reactions have been developed over the past few years that proceed in polar organic solvents such as DMSO and NMP.

Hayashi and co-workers[124] used a siloxytetrazole **51** as organocatalyst in direct Mannich reaction in the presence of large excess of water (Scheme 7.23). They obtained the product with high yields and enantioselectivities. The best results occurred for the one-pot reaction of dimethoxy aldehyde with *p*-anisidine and cyclohexanone.

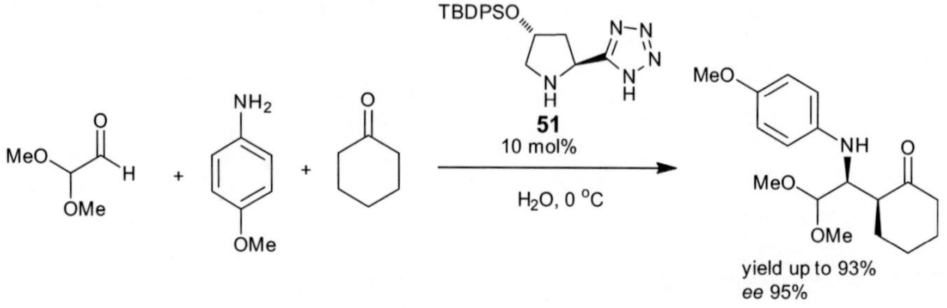

Scheme 7.23   Direct asymmetric Mannich reaction in the presence of water.

The first example of a three-component direct Mannich reaction that can be promoted by a primary amino acid in water was presented by Lu's group.[125] They presented asymmetric reaction of *O*-benzyl hydroxyacetone with *p*-anisidine and aromatic or aliphatic aldehydes in the presence of an (*S*)-threonine-derived catalyst **52** which afforded *anti*-1,2-amino alcohols in good to excellent yields and with enantioselectivities of up to 97% (Scheme 7.24).

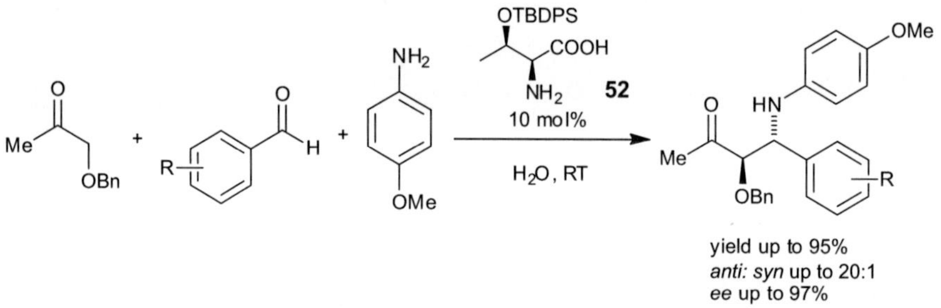

Scheme 7.24   Direct three-component Mannich reactions with aromatic aldehydes in water.

7.4   Conclusions

Organic reactions in water are currently of great interest, and asymmetric catalysis in aqueous media is an established new tool in organic chemistry. A tremendous amount of effort has been applied to mimic enzymes, as they act with high efficiency in the aqueous environment of living cells. Although the field is still in its infancy, the development of organic reactions in aqueous media should benefit academia and industry, including green chemistry. Nevertheless, this does not mean that other strategies are less important. Rather, they are complementary and facilitate the synthesis of the desired target molecules as well as learning about and developing a real understanding of life and nature.

With this review we were primarily focusing on the most instructive examples and aspects of the 'in water' and 'on water' effects. There is now so much organic literature with organic reactions using water as a solvent/medium (over 3000 references so far) that a fully comprehensive review of this area is simply not feasible. We tried to be quite selective but readers can consult all other reviews cited in this chapter.

Acknowledgements

Our team is financed within the Foundation for Polish Science TEAM Programme co-financed by the EU European Regional Development Fund.

References

1. Albert Szent-Gyorgyi (1893–1986), Hungarian biochemist, 1937 Nobel Prize for Medicine.
2. Ch.-J. Li and L. Chen, *Chem. Soc. Rev.*, 2006, **35**, 68.
3. (a) J. Mlynarski and J. Paradowska, *Chem. Soc. Rev.*, 2008, **37**, 1502; (b) N. Mase and C. F. Barbas, III, *Org. Biomol. Chem.*, 2010, **8**, 4043.
4. U. M. Lindström, *Chem. Rev.*, 2002, **102**, 2751.
5. *Organic Reactions in Water: Principles, Strategies and Applications*, ed. U. M. Lindström, Blackwell, Oxford, 2007.
6. (a) K. Manabe and S. Kobayashi, *Chem. Eur. J.*, 2002, **8**, 4094; (b) S. Kobayashi and C. Ogawa, *Chem. Eur. J.*, 2006, **12**, 5954.
7. D. Sinou, *Adv. Synth. Catal.*, 2002, **344**, 221.
8. (a) M. Gruttadauria, F. Giacalone and R. Noto, *Adv. Synth. Catal.*, 2009, **351**, 33; (b) M. Raj and V. K. Singh, *Chem. Commun.*, 2009, 6687.
9. R. N. Butler and A. G. Coyne, *Chem. Rev.*, 2010, **110**, 6302.
10. (a) A. P. Brogan, T. J. Dickerson and K. D. Janda, *Angew. Chem., Int. Ed. Engl.*, 2006, **45**, 8100 and references cited therein; (b) Y. Hayashi, *Angew. Chem., Int. Ed. Engl.*, 2006, **45**, 8103; (c) D. G. Blackmond, A. Armstrong, V. Coombe and A. Wells, *Angew. Chem., Int. Ed. Engl.*, 2007, **46**, 3798.
11. J. Paradowska, M. Stodulski and J. Mlynarski, *Angew. Chem., Int. Ed. Engl.*, 2009, **48**, 4288.
12. R. W. Eckl, T. Priermeier and W. A. Herrmann, *J. Organomet. Chem.*, 1997, **532**, 243.
13. I. Toth, B. E. Hanson and M. E. Davis, *Cat. Lett.*, 1990, **5**, 183.
14. I. Toth, B. E. Hanson and M. E. Davis, *Tetrahedron: Asymmetry*, 1990, **1**, 913.
15. Y.-Y. Yan and T. V. RajanBabu, *J. Org. Chem.*, 2001, **66**, 3277.
16. G. Oehme, E. Paetzold and R. Selke, *J. Mol. Catal.*, 1992, **71**, L1.
17. K. Yonehara, T. Hashizume, K. Mori, K. Ohe and S. Uemura, *J. Org. Chem.*, 1999, **64**, 5593.
18. S. Shin and T. V. RajanBabu, *Org. Lett.*, 1999, **1**, 1229.

19. J. Holz, R. Stürmer and A. Börner, *Tetrahedron Lett.*, 1999, **40**, 7059.
20. I. Grassert, E. Paetzold and G. Oehme, *Tetrahedron*, 1993, **49**, 6605.
21. I. Grassert, V. Vill and G. Oehme, *J. Mol. Catal. A Chem.*, 1997, **116**, 231.
22. G. Oehme, I. Grassert, S. Ziegler, R. Meisel and H. Fuhrmann, *Catal. Today*, 1998, **42**, 459.
23. T. Dwars, U. Schmidt, C. Fischer, I. Grassert, R. Kempe, R. Fröhlich, K. Drauz and G. Oehme, *Angew. Chem., Int. Ed. Engl.*, 1998, **37**, 2851.
24. R. Selke, J. Holz, A. Riepe and A. Börner, *Chem. Eur. J.*, 1998, **4**, 769.
25. K. Yonehara, K. Ohe and S. Uemura, *J. Org. Chem.*, 1999, **64**, 9381.
26. K.-T. Wan and M. E. Davies, *Tetrahedron: Asymmetry*, 1993, **4**, 2461.
27. W. Li, Z. Zhang, D. Xiao and X. Zhang, *J. Org. Chem.*, 2000, **65**, 3489.
28. A. E. S. Gelpke, H. Kooijman, A. L. Spek and H. Hiemstar, *Chem. Eur. J.*, 1999, **5**, 2472.
29. T. Lamouille, C. Saluzzo, R. ter Halle, F. Le Guyader and M. Lemaire, *Tetrahedron Lett.*, 2001, **42**, 663.
30. P. Guerreiro, V. Ratovelomanana-Vidal, J.-P. Genet and P. Dellis, *Tetrahedron Lett.*, 2001, **42**, 3423.
31. T. Thorpe, J. Blacker, S. M. Brown, C. Bubert, J. Crosby, S. Fitzjohn, J. P. Muxworthy and J. M. J. Williams, *Tetrahedron Lett.*, 2001, **42**, 4037.
32. T. Thorpe, J. Blacker, S. M. Brown, C. Bubert, J. Crosby, S. Fitzjohn, J. P. Muxworthy and J. M. J. Williams, *Tetrahedron Lett.*, 2001, **42**, 4041.
33. C. Pinel, N. Gendrean-Diaz, A. Breheret and M. Lemaire, *J. Mol. Catal. A Chem.*, 1996, **112**, L157.
34. (a) C. de Bellefon, N. Tauchoux and S. Caravieilhes, *J. Organomet. Chem.*, 1998, **567**, 143; (b) N. Tanchoux and C. de Bellefon, *Eur. J. Inorg. Chem.*, 2000, 1495; (c) D. Sinou, M. Safi, C. Claver and A. Masdeu, *J. Mol. Catal.*, 1991, **68**, L9.
35. A. M. d'A. Rocha Gonsalves, J. C. Bayon, M. M. Pereira, M. E. S. Serra and J. P. R. Pereira, *J. Organomet. Chem.*, 1998, **553**, 199.
36. D. Carmona, F. J. Lakos, R. Atencio, L. A. Oro, M. P. Lamata, F. Viguri, E. San Jose, C. Vega, J. Reyes, F. Joo and A. Katho, *Chem. Eur. J.*, 1999, **5**, 1544.
37. X. G. Li, X. F. Wu, W. P. Chen, F. E. Hancock, F. King and J. L. Xiao, *Org. Lett.*, 2004, **6**, 3321.
38. C. Wang, C. Q. Li, X. F. Wu, A. Pettman and J. L. Xiao, *Angew. Chem., Int. Ed. Engl.*, 2009, **48**, 6524.
39. Y. Uozumi, H. Danjo and T. Hayashi, *Tetrahedron Lett.*, 1998, **39**, 8303.
40. Y. Uozumi and K. Shibatomi, *J. Am. Chem. Soc.*, 2001, **123**, 2919.
41. C. Rabeyrin, C. Nguefack and D. Sinou, *Tetrahedron Lett.*, 2000, **41**, 7461.
42. (a) B. Alcaide, P. Almendros, C. Aragoncillo and R. Rodrıguez-Acebes, *J. Org. Chem.*, 2001, **66**, 5208; (b) W. Lu and T. H. Chan, *J. Org. Chem.*, 2000, **65**, 8589; (c) Y. S. Cho, J. E. Lee, A. N. Pae, K. I. Choi and H. Y. Koh, *Tetrahedron Lett.*, 1999, **40**, 1725.

43. (a) H.-M. Chang and C.-H. Cheng, *Org. Lett.*, 2000, **2**, 3439; (b) T. H. Chan, Y. Yang and C. J. Li, *J. Org. Chem.*, 1999, **64**, 4452; (c) T. P. Loh and J.-R. Zhou, *Tetrahedron Lett.*, 2000, **41**, 5261.
44. L. H. Li and T. H. Chan, *Tetrahedron Lett.*, 2000, **41**, 5009.
45. C.-J. Li, Y. Meng, X.-H. Yi, J. Ma and T.-H. Chan, *J. Org. Chem.*, 1998, **63**, 7498.
46. W.-C. Zhang and C.-J. Li, *J. Org. Chem.*, 1999, **64**, 3230.
47. T. H. Chan and Y. Yang, *Tetrahedron Lett.*, 1999, **40**, 3863.
48. C.-J. Li and T.-H. Chan, *Tetrahedron*, 1999, **55**, 11149.
49. T.-P. Loh and J.-R. Zhou, *Tetrahedron Lett.*, 1999, **40**, 9115.
50. T. P. Loh and J.-R. Zhou, *Tetrahedron Lett.*, 2000, **41**, 5261.
51. J. A. Shin, J. H. Cha, A. N. Pae, K. I. Choi, H. Y. Koh, H. Y. Kang and Y. S. Cho, *Tetrahedron Lett.*, 2001, **42**, 5489.
52. Y. Canac, E. Levoirier and A. Lubineau, *J. Org. Chem.*, 2001, **66**, 3206.
53. A. Cunningham and S. Woodward, *Synlett*, 2002, 43.
54. S. Kobayashi, N. Aoyama and K. Manabe, *Chirality*, 2003, **15**, 124.
55. S. Iwasa, F. Takezawa, Y. Tuchiya and H. Nishiyama, *Chem. Commun.*, 2001, 59.
56. (a) A. El-Qisairi, O. Hamed and P. M. Henry, *J. Org. Chem.*, 1998, **63**, 2790; (b) O. Hamed and P. M. Henry, *Organometallics*, 1998, **17**, 5184; (c) A. El-Qisairi and P. M. Henry, *J. Organomet. Chem.*, 2000, **603**, 50.
57. D. Yang, M.-K. Wong, Y.-C. Yip, X.-C. Wang, M.-W. Tang, J.-H. Zheng and K.-K. Cheung, *J. Am. Chem. Soc.*, 1998, **120**, 5943.
58. M. Frohn, M. Dalkiewicz, Y. Tu, Z.-X. Wang and Y. Shi, *J. Org. Chem.*, 1998, **63**, 2948.
59. Z.-X. Wang and Y. Shi, *J. Org. Chem.*, 1998, **63**, 3099.
60. Z. Bourhani and A. V. Malkov, *Chem. Commun.*, 2005, 4592.
61. A. S. C. Chan and J. P. Coleman, *J. Chem. Soc., Chem. Commun.*, 1991, 535.
62. J. Bakos, A. Orosz, S. Cserepi, I. Toth and D. Sinou, *J. Mol. Catal. A Chem.*, 1997, **116**, 85.
63. S. P. Azoulay, K. Manabe and S. Kobayashi, *Org. Lett.*, 2005, **7**, 4593.
64. M. Kokubo, T. Naito and S. Kobayashi, *Tetrahedron*, 2010, **66**, 1111.
65. (a) D. C. Rideout and R. Breslow, *J. Am. Chem. Soc.*, 1980, **102**, 7816; (b) R. Breslow, U. Maitra and D. C. Rideout, *Tetrahedron Lett.*, 1983, **24**, 1901; (c) R. Breslow and U. Maitra, *Tetrahedron Lett.*, 1984, **25**, 1239; (d) P. A. Grieco, P. Garner and Z.-m. He, *Tetrahedron Lett.*, 1983, **24**, 1897; (e) P. A. Grieco, K. Yoshida and P. Garner, *J. Org. Chem.*, 1983, **48**, 3137.
66. (a) S. Otto, G. Boccaletti and J. B. F. N. Engberts, *J. Am. Chem. Soc.*, 1998, **120**, 4238; (b) S. Otto and J. B. F. N. Engberts. *J. Am. Chem. Soc.*, 1999, **121**, 6798.
67. A. Lubineau, *J. Org. Chem.*, 1986, **51**, 2142.
68. M. Sodeoka, K. Ohrai and M. Shibasaki, *J. Org. Chem.*, 1995, **60**, 2648.
69. S. Kobayashi, S. Nagayama and T. Busujima, T. *Chem. Lett.*, 1999, 71.

70. S. Kobayashi, Y. Mori, S. Nagayama and K. Manabe, *Green Chem.*, 1999, **1**, 175.
71. S. Kobayashi, T. Hamada, S. Nagayama and K. Manabe, *Org. Lett.*, 2001, **3**, 165.
72. S. Nagayama and S. Kobayashi, *J. Am. Chem. Soc.*, 2000, **122**, 11531.
73. S. Kobayashi and K. Manabe, *Acc. Chem. Res.*, 2002, **35**, 209.
74. Y. Mori, K. Manabe and S. Kobayashi, *Angew. Chem., Int. Ed. Engl.*, 2001, **40**, 2816.
75. C. Le Roux, L. Ciliberti, H. Laurent-Robert, A. Laporterie and J. Dubac, *Synlett*, 1998, 1249.
76. T.-P. Loh, G.-L. Chua, J. J. Vittal and M.-W. Wong, *Chem. Commun.*, 1998, 861.
77. H. J. Li, H. Y. Tian, Y. J. Chen, D. Wang and C. J. Li, *Chem. Commun.*, 2002, 2994.
78. Y. Mei, P. Dissanayake and M. J. Allen, *J. Am. Chem. Soc.*, 2010, **132**, 12871.
79. J. Jankowska and J. Mlynarski, *J. Org. Chem.*, 2006, **71**, 1317.
80. J. Jankowska, J. Paradowska and J. Mlynarski, *Tetrahedron Lett.*, 2006, **47**, 5281.
81. J. Jankowska, J. Paradowska, B. Rakiel and J. Mlynarski, *J. Org. Chem.*, 2007, **72**, 2228.
82. Y. Yamashita, H. Ishitani, H. Shimizu and S. Kobayashi, *J. Am. Chem. Soc.*, 2002, **124**, 3292.
83. H.-J. Li, H.-Y. Tian, Y.-J. Chen, D. Wang and C.-J. Li, *J. Chem. Res., Synop.*, 2003, 153.
84. N. Ozasa, M. Wadamoto, K. Ishihara and H. Yamamoto, *Synlett*, 2003, 2219.
85. B.-Y. Yang, X.-M. Chen, G.-J. Deng, Y.-L. Zhang and Q.-H. Fan, *Tetrahedron Lett.*, 2003, **44**, 3535.
86. M. Ohkouchi, M. Yamaguchi and T. Yamagishi, *Enantiomer*, 2000, **5**, 71.
87. S. Kobayashi, C. Ogawa and M. Kokubo, *Angew. Chem., Int. Ed. Engl.*, 2008, **47**, 6909.
88. N. Risch, M. Arend and B. Westermann, *Angew. Chem., Int. Ed. Engl.*, 1998, **37**, 1044.
89. (a) K. Manabe, Y. Mori, T. Wakabayashi, S. Nagayama and S. Kobayashi, *J. Am. Chem. Soc.*, 2000, **122**, 7202; (b) T.-P. Loh, S. B. K. W. Liung, K.-L. Tan and L.-L Wei, *Tetrahedron*, 2000, **56**, 322; (c) S. Kobayashi, T. Busujima and S. Nagayama, *Synlett*, 1999, 545; (d) T.-P. Loh and L.-L Wei, *Tetrahedron Lett.*, 1998, **39**, 323.
90. (a) T. Akiyama, J. Takaya and H. Kagoshima, *Tetrahedron Lett.*, 2001, **42**, 4025; (b) K. Manabe, Y. Mori and S. Kobayashi, *Tetrahedron*, 2001, **57**, 2537; (c) K. Manabe and S. Kobayashi, *Org. Lett.*, 1999, **1**, 1965.
91. S. Kobayashi, T. Hamada and K. Manabe, *J. Am. Chem. Soc.*, 2002, **124**, 5640.
92. T. Hamada, K. Manabe and S. Kobayashi, *J. Am. Chem. Soc.*, 2004, **126**, 7768.

93. K. Ishimaru and T. Kojima, *J. Org. Chem.*, 2003, **68**, 4959.
94. (a) S. Dos Santos, Y. Tong, F. Quignard, A. Choplin, D. Sinou and J. P. Dutasta, *Organometallics*, 1998, **17**, 78; (b) M. Feuerstein, D. Laurenti, H. Doucet and M. Santelli, *Tetrahedron Lett.*, 2001, **42**, 2313.
95. T. Hashizume, K. Yonehara, K. Ohe and S. Uemura, *J. Org. Chem.*, 2000, **65**, 5197.
96. (a) C. Wei and C.-J. Li, *J. Am. Chem. Soc.*, 2002, **124**, 5638; (b) C. Wei, J. T. Mague and C.-J. Li, *Proc. Natl. Acad. Sci. USA*, 2004, **101**, 5748.
97. L. Shi, Y.-Q. Tu, M. Wang, F.-M. Zhang and C.-A. Fan, *Org. Lett.*, 2004, **6**, 1001.
98. V. K.-Y. Lo, Y. Liu, M.-K. Wong and C.-M. Che, *Org. Lett.*, 2006, **8**, 1529.
99. Y. Uozumi, Y. Matsuura, T. Arakawa and Y. M. A. Yamada, *Angew. Chem., Int. Ed. Engl.*, 2009, **48**, 2708.
100. (a) A. Berkessel and H. Gröger, *Asymmetric Organocatalysis: From Biomimetic Concepts to Applications in Asymmetric Synthesis*, Wiley-VCH, 2005; (b) A. Dondoni and A. Massi, *Angew. Chem., Int. Ed. Engl.*, 2008, **47**, 4638.
101. G. Lelais and D. W. C. MacMillan, *Aldrichim. Acta*, 2006, **39**, 79.
102. A. B. Northrup and D. W. C. MacMillan, *J. Am. Chem. Soc.*, 2002, **124**, 2458.
103. T. D. Machajewski and C. H. Wong, *Angew. Chem., Int. Ed. Engl.*, 2000, **39**, 1352.
104. (a) A. Bogevig, N. Kumaragurubaran and K. A. Jørgensen, *Chem. Commun.*, 2002, 620; (b) A. Cordova, W. Notz and C. F. Barbas, III, *Chem. Commun.*, 2002, 3024; (c) Y.-Y. Peng, Q.-P. Ding, Z. Li, P. G. Wang and J.-P. Cheng, *Tetrahedron Lett.*, 2003, **44**, 3871; (d) A. Hartikka and P. I. Arvidsson, *Tetrahedron: Asymmetry*, 2004, **15**, 1831; (e) S. S. Chimni, D. Mahajan and V. V. S. Babu, *Tetrahedron Lett.*, 2005, **46**, 5617; (f) Y.-S. Wu, Y. Chen, D.-S. Deng and J. Cai, *Synlett*, 2005, 1627; (g) K. Akagawa, S. Sakamoto and K. Kudo, *Tetrahedron Lett.*, 2005, **46**, 8185.
105. (a) J. Casas, H. Sunden and A. Cordova, *Tetrahedron Lett.*, 2004, **45**, 6117; (b) D. E. Ward and V. Jheengut, *Tetrahedron Lett.*, 2004, **45**, 8347; (c) Y. Hayashi, S. Aratake, T. Itoh, T. Okano, T. Sumiya and M. Shoji, *Chem. Commun.*, 2007, 957.
106. (a) H. Torii, M. Nakadai, K. Ishihara, S. Saito and H. Yamamoto, *Angew. Chem.*, 2004, **116**, 2017, *Angew. Chem Int. Ed.*, 2004, **43**, 1983; (b) Z. Tang, Z. H. Yang, L. F. Cun, L. Z. Gong, A. Q. Mi and Y. Z. Jiang, *Org. Lett.*, 2004, **6**, 2285; (c) J. R. Chen, X. Y. Li, X. N. Xing and W. J. Xiao, *J. Org. Chem.*, 2006, **71**, 8198; (d) X. H. Chen, S. W Luo, Z. Tang, L. F. Cun, A. Q. Mi, Y. Z. Jiang and L. Z. Gong, *Chem. Eur. J.*, 2007, **13**, 689.
107. (a) I. Ibrahem and A. Cordova, *Tetrahedron Lett.*, 2005, **46**, 3363; (b) M. Amedjkouh, *Tetrahedron Asymmetry*, 2005, **16**, 1411; (c) A. Cordova, W. Zou, I. Ibrahem, E. Reyes, M. Engqvist and W. W. Liao,

Chem. Commun., 2005, 3586; (d) P. Dziedzic, W. Zou, J. Hafren and A. Cordova, *Org. Biomol. Chem.*, 2006, **4**, 38.

108. (a) S. S. Chimni, D. Mahajan and V. V. S. Babu, *Tetrahedron Lett.*, 2005, **46**, 5617; (b) S. S. Chimni and D. Mahajan, *Tetrahedron: Asymmetry*, 2006, **17**, 2108; (c) Y.-S. Wu, Y. Chen, D.-S. Deng and J. Cai, *Synlett*, 2005, 1627.

109. Y.-Q. Fu, Z.-C. Li, L.-N. Ding, J.-C. Tao, S.-H. Zhang and M.-S. Tang, *Tetrahedron: Asymmetry*, 2006, **17**, 3351.

110. G. Guillena, M. del Carmen Hita and C. Nájera, *Tetrahedron: Asymmetry*, 2006, **17**, 1493.

111. (a) Z. Jiang, Z. Ling, X. Wu and X. Lu, *Chem. Commun.*, 2006, 2801; (b) M. Amedjkouh, *Tetrahedron: Asymmetry*, 2007, **18**, 390.

112. N. Mase, Y. Nakai, N. Ohara, H. Yoda, K. Takabe, F. Tanaka and C. F. Barbas III, *J. Am. Chem. Soc.*, 2006, **128**, 734.

113. Y. Hayashi, T. Sumiya, J. Takahashi, H. Gotoh, T. Urushima and M. Shoji, *Angew. Chem.*, 2006, **118**, 972.

114. (a) D. Gryko and W. J. Saletra, *Org. Biomol. Chem.*, 2007, **5**, 2148; (b) Y.-C. Teo, *Tetrahedron: Asymmetry*, 2007, **18**, 1155; (c) X. Wu, Z. Jiang, H.-M. Shen and Y. Lu, *Adv. Synth. Catal.*, 2007, **349**, 812; (d) V. Maya, M. Raj and V. K. Singh, *Org. Lett.*, 2007, **9**, 2593; (e) C. Wang, Y. Jiang, X.-X. Zhang, Y. Huang, B.-G. Li and G.-L. Zhang, *Tetrahedron Lett.*, 2007, **48**, 4281; (f) M. Lei, L. Shi, G. Li, S. Chen,W. Fang, Z. Ge, T. Cheng and R. Li, *Tetrahedron*, 2007, **63**, 7892; (g) S. Guizzetti, M. Benaglia, L. Raimondi and G. Celentano, *Org. Lett.*, 2007, **9**, 1247; (h) J. Huang, X. Zhang and D. W. Armstrong, *Angew. Chem.*, 2007, **119**, 9231; *Angew. Chem., Int. Ed. Engl.*, 2007, **46**, 9073.

115. J. Paradowska, M. Stodulski and J. Mlynarski, *Adv. Synth. Catal.*, 2007, **349**, 1041.

116. L.-W. Xu and Y. Lu, *Org. Biomol. Chem.*, 2008, **6**, 2047.

117. S. Aratake, T. Itoh, T. Okano, T. Usui, M. Shoji and Y. Hayashi, *Chem. Commun.*, 2007, 2524.

118. Z. Tang, Z. H. Yang, L. F Cun, L. Z. Gong, A. Q. Mi and Y. Z. Jiang, *Org. Lett.*, 2004, **6**, 2285.

119. N. Utsumi, M. Imai, F. Tanaka, S. S. V. Ramasastry and C. F. Barbas III, *Org. Lett.*, 2007, **9**, 3445.

120. S. S. V. Ramasastry, K. Albertshofer, N. Utsumi and C. F. Barbas III, *Org. Lett.*, 2008, **10**, 1621.

121. O. M. Berner, L. Tedeschi and D. Enders, *Eur. J. Org. Chem.*, 2002, 1877.

122. S. B. Tsogoeva, *Eur. J. Org. Chem.*, 2007, 1701.

123. Z. Y. Yan, Y. N. Niu, H. L. Wei, L. Y. Wu, Y. B. Zhao and Y. M. Liang, *Tetrahedron: Asymmetry*, 2006, **17**, 3288.

124. Y. Hayashi, T. Urushima, S. Aratake, T. Okano and K. Obi, *Org. Lett.*, 2008, **10**, 21.

125. L. Cheng, X. Wu and Y. Lu, *Org. Biomol. Chem.*, 2007, **5**, 1018.

CHAPTER 8
Non-covalent Immobilization

J. M. FRAILE,[a] J. I. GARCÍA,[a] C. I. HERRERIAS,[b]
J. A. MAYORAL[b] AND E. PIRES[b]

[a] Instituto de Ciencia de Materiales de Aragón (ICMA), Universidad de Zaragoza, Spain; [b] Instituto de Catálisis Homogénea (IUCH), Universidad de Zaragoza, Spain

8.1 Introduction

A method to allow simple separation of a chiral catalyst from the reaction medium is to make it insoluble, that is to convert it into a heterogeneous chiral catalyst. Immobilization of chiral complexes on solid supports is the most frequently used approach to prepare chiral heterogeneous catalysts.[1-4] Immobilization requires the existence of any type of support–catalyst interaction that prevents the leaching of catalyst to the reaction medium. The formation of a covalent bond between the chiral ligand and the support is by far the most frequently used method to retain the chiral complex on the insoluble support, because it guarantees the absence of leaching of the valuable chiral ligand. The main drawback of this method is the need for an additional modification of the ligand and/or the support which requires a synthetic effort, with a concomitant increase in cost, and whose influence on the stereochemical course of the reaction is difficult to predict. Due to these facts the interest in methods that do not require the covalent link between the chiral ligand and the support has increased in the last years.[5] In principle any type of interaction, from the weak van der Waals forces to the strong coordinative bonds or electrostatic interactions, can be used to link the chiral complex and the insoluble supports. In some cases the interaction occurs between the support and a liquid phase in

RSC Green Chemistry No. 15
Enantioselective Homogeneous Supported Catalysis
Edited by Radovan Šebesta
Published by the Royal Society of Chemistry, www.rsc.org

which the chiral complex is dissolved. In this chapter we present representative examples of all the different existing strategies for non-covalent immobilization of chiral catalysts.

8.2 Supported Liquid Phases

The use of supported liquid phases (SLP) is one simple and original solution for the immobilization of homogeneous catalysts that presents some advantages: it is applicable to any catalyst soluble in the immobilized liquid phase without needing any interaction between this catalyst and the support, the catalyst keeps the mobility in the solvent phase, and the SLP has a larger surface area improving in this way the diffusion of reagents. The limitations of this system are the same as those of the biphasic liquid–liquid systems, mainly the solubility of the catalyst in the second liquid phase, as well as the need for a specific solvent–support interaction able to retain the liquid phase.

The technique of supported aqueous phase (SAP) uses the hydrophilic nature of a surface to adhere water or another polar solvent containing a soluble catalyst. This strategy was used in the asymmetric synthesis of naproxen through asymmetric hydrogenation promoted by Ru-BINAP-4SO$_3$Na (Figure 8.1), made soluble in water by the incorporation of anionic sulfonic groups.[6] An aqueous solution of the catalyst was adsorbed on a controlled pore glass. The SAP catalyst was 50 times more active than the biphasic system AcOEt/H$_2$O, leading to 70% ee without leaching of ruthenium. The low enantioselectivity, in comparison with the 90% ee obtained with Ru-BINAP in MeOH, was due to the reaction of the catalyst with water. In order to overcome this limitation a solution of the catalyst in ethylene glycol instead of water was supported.[7] Using a mixture of cyclohexane/chloroform as the organic phase the enantioselectivity increased up to 96% ee and the catalyst was fully recoverable without leaching of ruthenium. It is important to note that when the

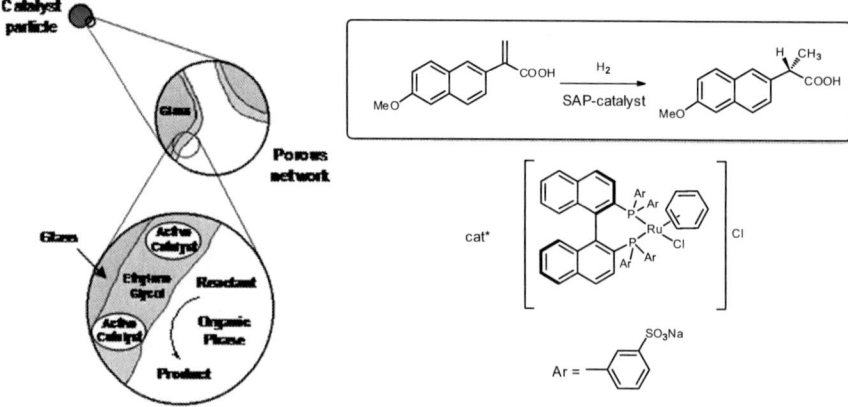

Figure 8.1   Enantioselective synthesis of naproxen using SAP.

reaction was carried out in a similar biphasic system the conversion was very low (<2%) due to the small interfacial area.

Solubility in water of Rh-diphosphine complexes can be increased by incorporation of cationic ammonium groups to the ligand (Figure 8.2).[8] In general the SAP on a controlled pore glass was more active than the biphasic water/(AcOEt:benzene) system but slightly less enantioselective in the hydrogenation of (Z)-2-benzamidocinnamic acid. The recovered catalyst was less active and some Rh leaching and phosphine oxidation were observed.

In the recent past, supported ionic liquid phases (SILP)[9] have appeared as a new class of supported liquid phases with a high potential for enantioselective catalysis.

In an attempt to enhance the affinity between purely siliceous materials and the ionic liquid phase, 1-methyl-3-(3-trimethoxysilylpropyl)imidazolium salts have been grafted to different supports. One example is the immobilization of a solution of Mn^{III}-salen in [bmim]PF$_6$ on modified MCM-48 and its use in the epoxidation of several alkenes with m-CPBA (Figure 8.3).[10] The supported catalyst led, in most cases, to improved enantioselectivities, which were

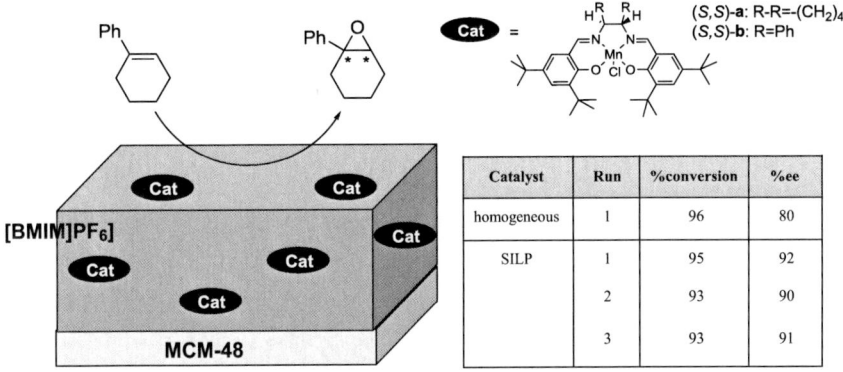

| n=0 | *am*BDPP | SAP 16% ee | biphasic 50% ee |
| n=1 | *am*Chiraphos | SAP 55% ee | biphasic 65% ee |

Figure 8.2 Enantioselective hydrogenation of (Z)-2-benzamidocinnamic acid using SAP.

Catalyst	Run	%conversion	%ee
homogeneous	1	96	80
SILP	1	95	92
	2	93	90
	3	93	91

Figure 8.3 m-CPBA epoxidation of alkenes with a SILP containing salen-Mn^{III}.

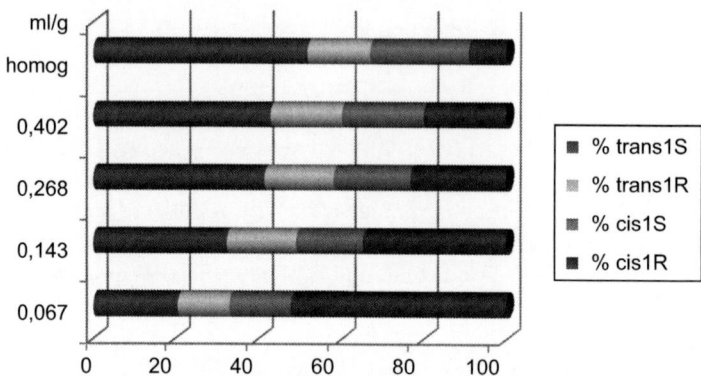

Figure 8.4   Effect of the thickness of the ionic liquid phase on the stereoselectivity of
the cyclopropanation of styrene with ethyl diazoacetate.

suggested to be due to spatial restrictions. The catalyst was recovered by evaporation of dichloromethane and hexane washing. No leaching of either catalyst or ionic liquid was observed and the recovered catalyst led to similar
results in both activity and selectivity.

However, this strategy implies the modification of the support to improve the
affinity towards the ionic liquid phase. Silica pre-treated with K_3PO_4 was shown
to be suitable for the immobilization of phosphonium ionic liquid phases.[11] In
that case the immobilization had a positive effect in enantioselectivity in ketone
hydrogenation with Rh and Ru complexes. This positive effect was ascribed to
the formation of a 'ionic liquid cage' around the organometallic complex.

Naturally charged supports should be suitable for immobilization of ionic liquid
phases without the need for further modification. This is the case of clays, on which
[bmim]PF_6 containing Box-Cu catalysts was supported. This solid was used in the
cyclopropanation of styrene with ethyl diazoacetate.[12] After the reaction the solid
was washed with hexane to extract the products, and reused. Using a decreasing
amount of ionic liquid the thickness of the liquid phase is reduced, leading to the
appearance of an effect on the *trans/cis* ratio and the enantioselectivity (Figure 8.4)
that can be ascribed to a surface effect, as will be explained in Section 8.7.

In spite of their great potential, the use of supported liquid phases needs
further studies to understand the nature of the interactions between support,
liquid and catalyst, and their consequences for the reaction. The main limitation of this strategy is the importance of the reaction solvent choice to prevent
leaching of catalyst, whereas the most attractive methods are those that do not
require any modification of the ligand and the support.

8.3   Immobilization by Physisorption

Physisorption is a general term to describe the retention of a molecule on a solid
surface due to unspecific interactions, usually weak van der Waals forces.

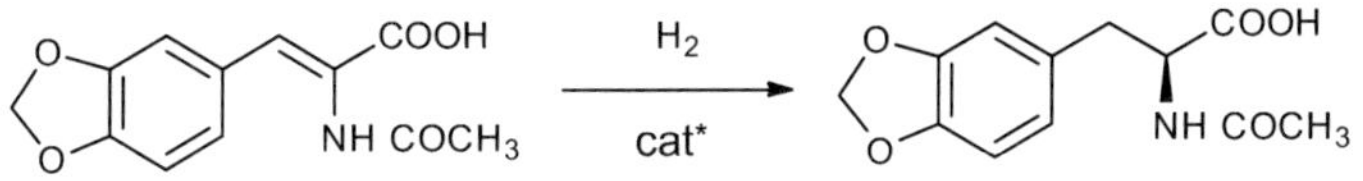

Figure 8.5 Hydrogenation of a substituted acetamidocinnamic acid, precursor of DOPA.

In many cases immobilization is carried out on pre-treated supports with polar or even ionic groups on their surfaces, so that the existence of coordinative or electrostatic interactions between the catalyst and the support cannot be discarded.

In one of the first examples of this approach, a neutral [(BPPM)Rh(cod)Cl] complex was immobilized on methylated silica by adsorption from aqueous solutions, and used in the hydrogenation of a precursor of DOPA (Figure 8.5).[13] Freshly prepared catalyst led to good results (88% ee), but enantioselectivity decreased upon recovery (52 and 30% ee in consecutive runs). The adsorption method and the unsuitability of non-methylated silica seem to indicate the hydrophobic character of the immobilization, although the improved recoverability of the cationic [(BPPM)Rh(cod)]$^+$ClO$_4^-$ points to a possible participation of other mechanisms of adsorption.

Adsorption through hydrophobic interactions can be improved by using surfactant-modified supports. Surfactants, both ionic and non-ionic, covalently bonded to silica, were compared with sodium dodecyl sulfate adsorbed on alumina as support for [(BPPM)Rh(cod)]BF$_4$. The catalysts were used in the hydrogenation of methyl (Z)-α-acetamidocinnamate in water. The results were compared with those reached in solution, both in the presence and in the absence of surfactants (Figure 8.6).[14] The positive effect of surfactants was similar both in homogeneous and heterogeneous phases. The better behaviour of alumina was ascribed to a favourable bilayer structure. SiO$_2$ did not show any leaching of rhodium after 10 cycles and the recovered catalyst led to the same enantioselectivity and slight decrease in catalytic activity, probably due to the participation of electrostatic interactions.

[(BINAP)Ru(p-cymene)Cl]Cl and [(BPPM)Rh(cod)Cl] were absorbed from benzene into mesoporous silica HMS, whose pores (average size of 26 Å) fit well with the dimensions of those complexes.[15] Both catalysts were used in the hydrogenation of the sodium salt of the (Z)-2-acetamidocinnamic acid in water. The enantioselectivities (near 50% ee) were lower than those reached in the homogenous hydrogenations of the sodium salt in water or the corresponding acid in organic solvents. Although the Rh catalyst was more active, an important decrease in activity was observed with the recovered catalyst, together with 4% rhodium leaching. The Ru-catalyst was reused with the same results and less than 2% leaching. When these catalysts were used in the hydrogenation of itaconic acid in aqueous media, enantioselectivities lower than 20% ee were obtained with both catalysts. [(BINAP)Ru(p-cymene)Cl]Cl was also incorporated on mesoporous silicas with a narrow range of pore size (26 and 37 Å) and amorphous silica (average pore size 68 Å), whose external

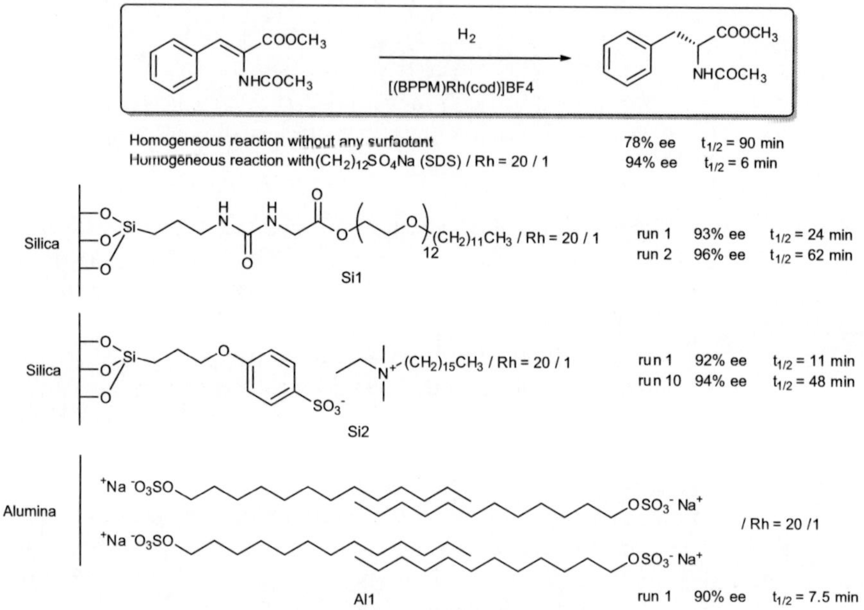

Figure 8.6 Results obtained from the hydrogenation of methyl (Z)-α-acet-amidocinnamate with surfactants.

surfaces were passivated with Ph$_2$SiCl$_2$ and the internal surfaces modified with amino groups for potential tethering of the catalyst.[16] Results with large pore modified and unmodified silicas ranged between 80–100% of conversion and 40–50% ee in the aqueous hydrogenation of sodium (Z)-α-acet-amidocinnamate. However, the catalyst immobilized onto 26 Å pore silica showed a low catalytic activity due to the restricted access of the reagent to the pores. Catalysts were reused with only small decreases in activity and enan-tioselectivity. These results showed that the van der Waals interactions between the catalyst and the pores are strong enough to retain the catalyst in aqueous solvents and results were not improved by the use of an internal tether. The use of these systems is limited to reactions carried out in water, which is probably due to the low solubility of the catalytic complexes in this solvent.

Cr^{III}-salen-complexes were impregnated on silica from diethyl ether solutions and used as catalysts for the asymmetric ring-opening reaction of epoxides in hexane (Figure 8.7).[17] Low selectivities were obtained in the opening of styrene, cyclopentene and cyclohexene oxides with trimethylsilyl azide. However, kinetic resolution of 1,2-epoxyhexane and 1,2-epoxyheptane resulted in enan-tioselectivities higher than 97% ee at conversions of about 50%. Catalyst reuse was studied during 10 consecutive cycles for the reaction of the 1,2-epox-yhexane, showing that leaching was somewhat important. A less soluble dimeric catalyst was impregnated from THF and used in the same reaction carried out in hexane in an attempt to reduce leaching (Figure 8.7).[18] In the

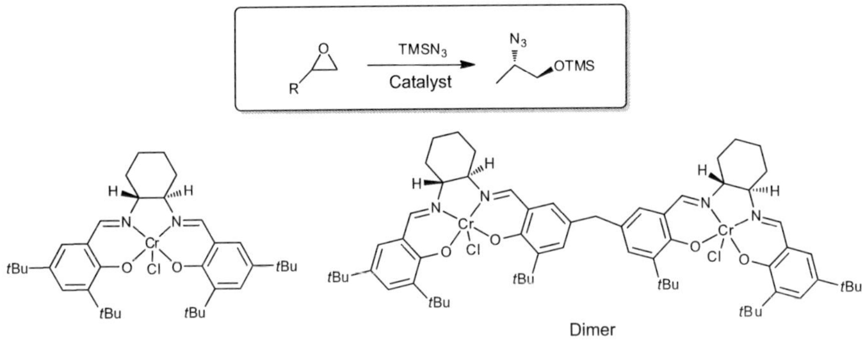

Figure 8.7 Kinetic resolution of epoxides with TMSN$_3$ using monomeric and dimeric salen-Cr^{III} catalysts.

kinetic resolution of 1,2-epoxyhexane, enantioselectivity (up to 98% ee) was maintained for 12 batch reactions with an overall leaching of only 1% after 10 runs, and 3.7% in a continuous system four times less than leaching observed with the monomeric catalysts, confirming the influence of catalyst solubility.

Co^{III}-salen complexes were impregnated at the interface between the macroporous matrix and the ZSM-5 layer of a membrane reactor and used to promote the hydrolytic kinetic resolution of epoxides.[19] High enantioselectivity was reached with salen-Co(OAc) (up to 98% ee) but the catalyst was not recoverable. Complexes with less coordinating anions (*e.g.* salen-CoPF$_6$) were more active, with up to 99.8% ee, and better recoverability. The same catalyst was used in a continuous membrane reactor. The organic phase, containing the epoxide, was circulated along the ZSM-5 side and the aqueous phase was circulated along the alumina side. The complex is insoluble in water and remains at the interface between layers. The hydrophilic diol diffused to the aqueous phase through the macropores whereas the unreacted epoxide remained in the organic phase. Enantioselectivities were high in both diol (up to 98% ee) and epoxide (up to 99% ee) with epoxide yields in the range of 38–45%.

It can be concluded that physisorption involves very week catalyst–support interactions and a low solubility of the complex in the reaction medium is crucial to prevent leaching.

8.4 Immobilization by Hydrogen Bonding

Most inorganic supports present different types of accessible hydroxyl groups that can be used to immobilize chiral catalysts by hydrogen bonding between the catalyst and the support. This method may be also considered as an example of the physisorption method.

Complexes of (*S,S*)-1,2-diphenylethylenediamine with [Rh(C$_6$H$_{10}$)Cl]$_2$ (2:1 ratio) were adsorbed onto different supports and it was proposed that immobilization was due to the formation of hydrogen bonds between the amino

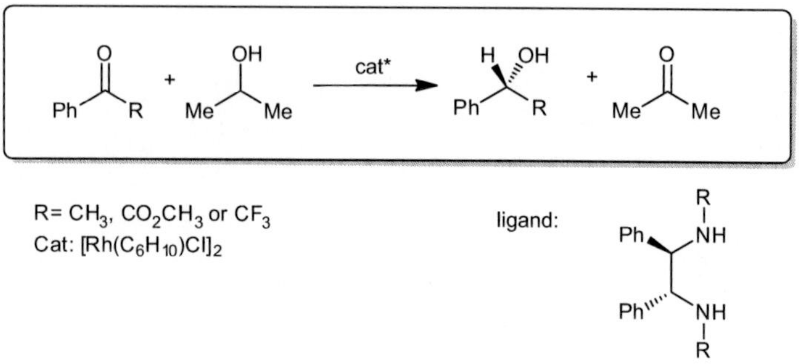

R= CH$_3$, CO$_2$CH$_3$ or CF$_3$
Cat: [Rh(C$_6$H$_{10}$)Cl]$_2$

Figure 8.8 Transfer hydrogenation of prochiral ketones, catalyzed by a Rh complex.

groups of the chiral ligand and the support. The catalyst was better retained on silica than on alumina when used in the transfer hydrogenation of prochiral ketones using a continuous flow reactor (Figure 8.8).[20] Activity was noticeably higher with supports of larger particle size, TON = 300 with silica of 36–70 μm particle size, compared with TON = 20 when particle size is 15–40 μm.

A complementary strategy consists in the immobilization through hydrogen bonding between the support and the counter ion of a cationic catalyst. Using this approach, several diamine-Rh complexes were immobilized on a series of commercial silicas with different average pore diameter (3.8, 6.0 and 25 nm).[21] The catalysts were tested in the hydrogenation of methyl 2-oxo-2-phenylacetate (Figure 8.9). The spatial restrictions imposed by lowering the pore diameter of the support showed to be markedly positive for the enantioselectivity, reaching values much higher than in solution. The stability of the catalysts was proven in recovery experiments.

The same strategy was used to bond some rhodium catalysts onto the surface of MCM-41 silica.[22] The positive confinement effect with this mesoporous crystalline material was shown in the hydrogenation of (*E*)-2-phenylcinnamic acid, with a noticeable increase in the enantioselectivity using the immobilized catalyst (Figure 8.10).

Hydrogen bonds can be formed with triflate counter anions of cationic complexes or with sulfonic groups attached to the ligand. Both strategies have been used for rhodium catalysts (Figure 8.11).[23] Catalysts are leached by hydrogen bond acceptor solvents so that homogenous reactions take place in methanol and heterogeneous in *n*-heptane. With BDPP and BINAP catalysts the results with homogeneous and supported catalysts were similar. With DIOP enantioselectivity decreased with the supported catalyst, which may be due to an interaction of the oxygen atoms of the ligand with the surface.

The same strategy has been applied to immobilization of cationic metal-bis(oxazoline) complexes with triflate counter ions, able to act as catalysts in Lewis acid-promoted reactions such as Diels–Alder reactions (Figure 8.12).[24,25] Catalysts based on Box-*t*-Bu was quite efficient and a slight decrease in conversion was observed upon recovery, but without modification in the enantioselectivity. The use of supported catalysts based on BoxPh led to a reversal

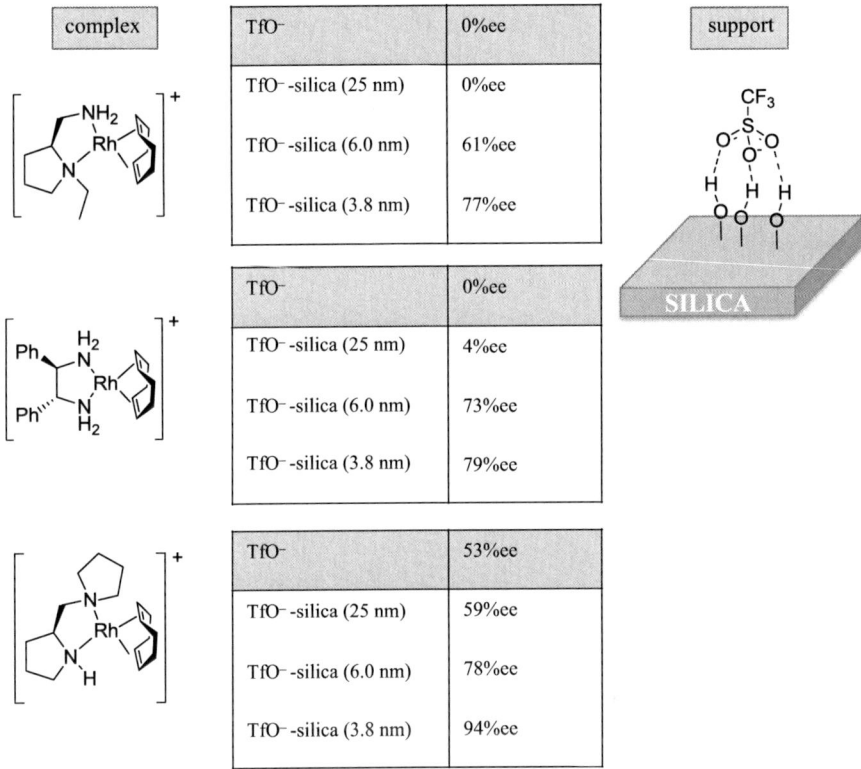

Figure 8.9 Asymmetric hydrogenation of a β-keto ester promoted by chiral catalysts immobilized by hydrogen bonding of the triflate counter ion with silica.

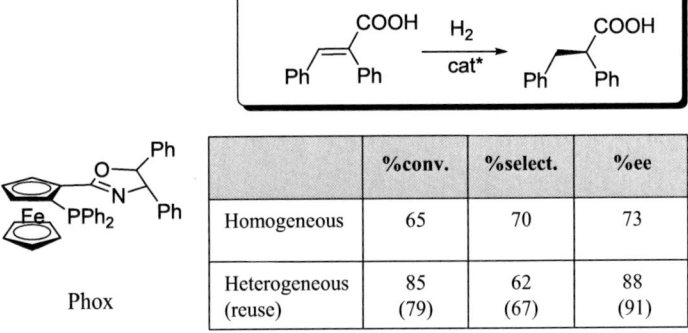

	%conv.	%select.	%ee
Homogeneous	65	70	73
Heterogeneous (reuse)	85 (79)	62 (67)	88 (91)

Phox

Figure 8.10 Results obtained from the hydrogenation of (*E*)-2-phenylcinnamic acid with Phox-Rh.

Si-(SO_3)BDPP-Rh Si-TfO-Rh-DIOP Si-TfO-Rh-BINAP

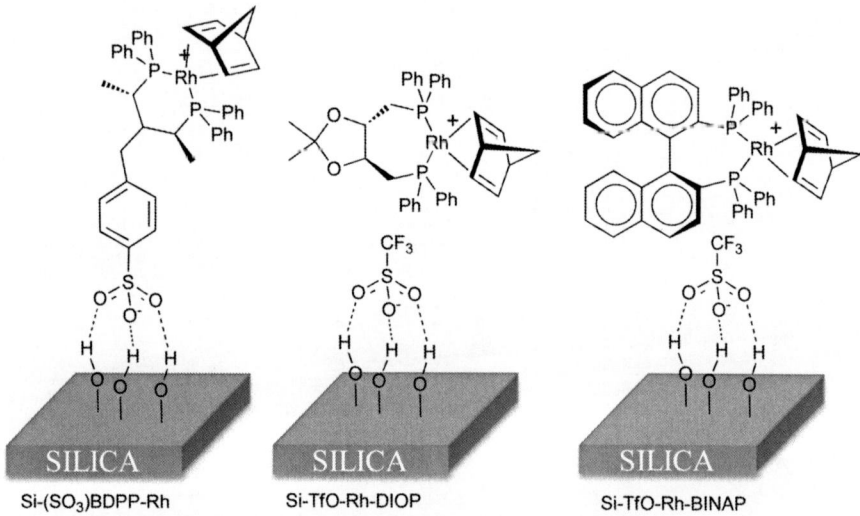

Figure 8.11 Rh complexes immobilized on silica by hydrogen bond formation.

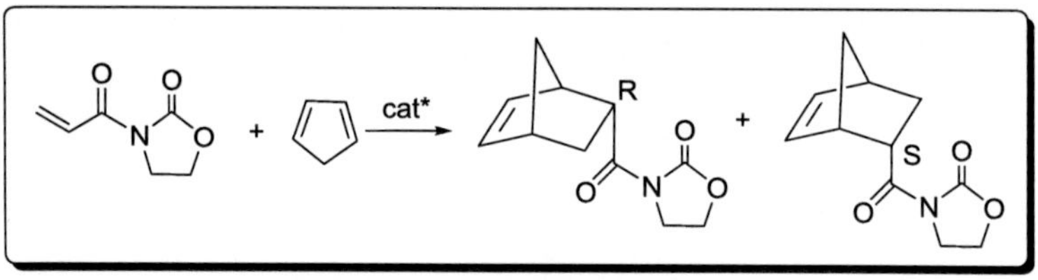

R	Metal /phase	%ee
^{t}Bu	Cu	91% (*S*)*
Ph	Cu / homogeneous	20% (*S*)
	Cu / heterogeneous	46% (*R*)
	Zn / homogeneous	22% (*S*)
	Zn / heterogeneous	24% (*R*)
	Mg / homogeneous	60% (*S*)
	Mg / heterogeneous	30% (*R*)

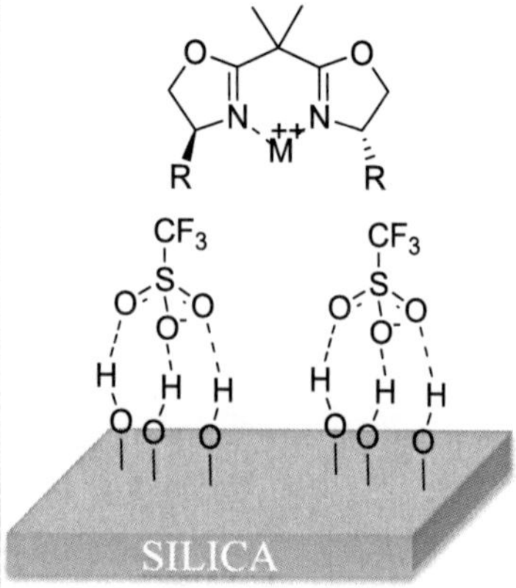

Figure 8.12 Diels–Alder reactions promoted by [Box-metal cation]TfO complexes immobilized by hydrogen bonding onto silica.

of the absolute configuration of the major *endo* enantiomer in comparison with that preferably obtained in solution. In the case of Mg and Zn catalysts a similar effect was observed in solution when going from triflate to less coordinating anions, such as ClO_4^- or SbF_6^-, due to changes in the geometry of the complex. It was suggested that the formation of the hydrogen bond with the support reduces the coordinating ability of the triflate, causing the similar effect. On the contrary the counter ion has no influence on the stereochemical course of the reaction catalyzed by copper complexes in solution, so that the steric effect of the support may be responsible for this behaviour in this case.

Heteropolyacids (HPA) are strong acids able to generate weakly coordinating anions. When these anions are used as counter ions of cationic complexes the catalysts obtained are normally soluble in the reaction medium. These HPAs can be immobilized by weak, normally hydrogen bond, interactions with inorganic supports.

Pioneering work[26] used phosphotungstic acid (PTA) over different supports to immobilize diphosphine-Rh complexes. In general, supported catalysts were more active and enantioselective than homogeneous ones in the hydrogenation of methyl 2-acetamidoacrylate. Although both supported catalysts were efficient, results were better with alumina (Figure 8.13).

The influence of the nature of the HPA was studied[27] using several chiral and non-chiral catalysts and alumina as the support. The heteropolyacids tested were phosphotungstic acid (PTA), phosphomolybdic acid (PMA),

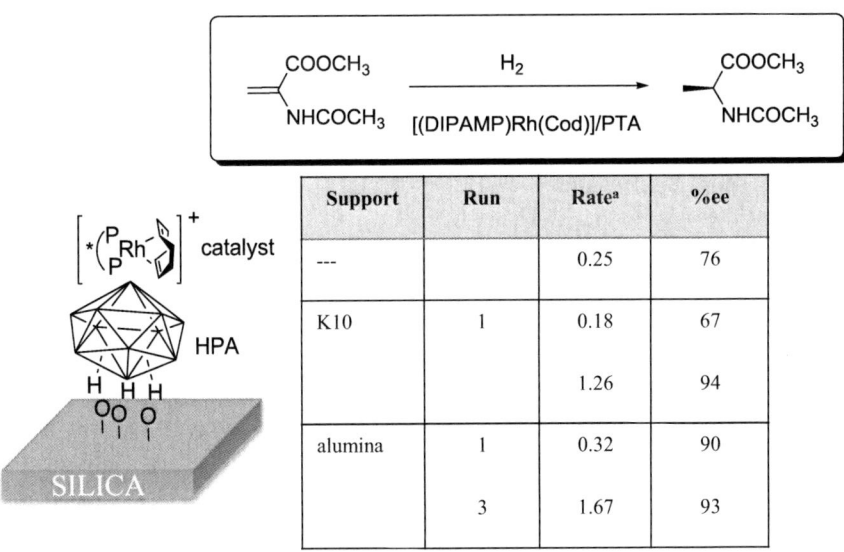

Support	Run	Rate[a]	%ee
---		0.25	76
K10	1	0.18	67
		1.26	94
alumina	1	0.32	90
	3	1.67	93

a) mol of H_2 (mol of Rh)$^{-1}$min^{-1}

Figure 8.13 Enantioselective hydrogenation promoted by [(DIPAMP)Rh(Cod)/PTA] immobilized on different supports.

silicotungstic acid (STA) and silicomolybdic acid (SMA). In the hydrogenation of methyl 2-acetamidoacrylate promoted by immobilized [(DIPAMP)Rh(Cod)] HPA the use of different heteropolyacids resulted in a different influence of pressure and temperature. In general higher TOF and enantioselectivities were reached with PMA, although the best TOF was obtained with STA at 40 °C and 30 psi (1 psi = 6895 N m^{-2}). In the hydrogenation of dimethyl itaconate promoted by [(BINAP)Rh]HPA/alumina, enantioselectivity in the first run was not noticeably dependent of the HPA, but SMA allowed the most efficient recovery both in activity and enantioselectivity. These results show that the nature of the HPA has an influence more significant than that expected for a mere change of counter ion.

Looking for an even greener methodology, the continuous hydrogenation of dimethyl itaconate was tested using [(BDPP)Rh(nbd)]BF$_4$/PTA/alumina as catalyst.[28] Conversion was not dependent on the pressure, but higher enantioselectivities were reached at 10 MPa. Temperature dependence was more important and the best results were obtained at 60 °C; under these conditions conversion was stable over a period of 8 h with a total absence of metal leaching (< 1 ppm), although leaching of Rh (7–16 ppm) and W (1 ppm) were observed when the reaction was carried out at 100 °C.

This methodology was compared with cationic exchange in hectorite (see Section 8.7) for the hydrogenation of methyl 2-acetamidoacrylate,[29] using AcOEt as the solvent. A TOF of 2300 h^{-1}, higher than that reached with the exchanged catalyst, was obtained, with 97% ee. Furthermore, the catalyst was recoverable.

More recently this type of methodology has been successfully extended to the cyclopropanation reaction of styrene with ethyl diazoacetate catalyzed by Box and AzaBox-Cu complexes.[30]

8.5   Immobilization by Entrapment

In many cases, van der Waals forces and hydrogen bonding are too weak as interactions to avoid leaching and it is necessary to enclose the catalyst inside the structure of the support, either a flexible organic polymer or a rigid inorganic matrix. For this purpose the size of the catalyst must be larger than the pore window during all of the catalytic process. In a broad sense, it can be considered that entrapment is based on the use of repulsive van der Waals interactions.

It is possible to entrap a catalyst into a flexible polymer by mixing the catalyst with a prepolymer and then forming the polymers. [(BINAP)Ru(*p*-cymene)Cl]Cl[31] and [(MeDuPhos)Rh(cod)]OTf[32] were immersed in a mixture of dimethylsilicone prepolymer and a tetrasilane cross-linker in a Petri dish, and a silicone membrane (PDMS) containing the catalyst was formed after curing (Figure 8.14). Both catalysts were used in the hydrogenation of methyl acetylacetate in alcohols. Using the Ru-BINAP catalyst enantioselectivities were in the range of 61–70% ee in ethylene glycol, PEG and glycerol. Leaching

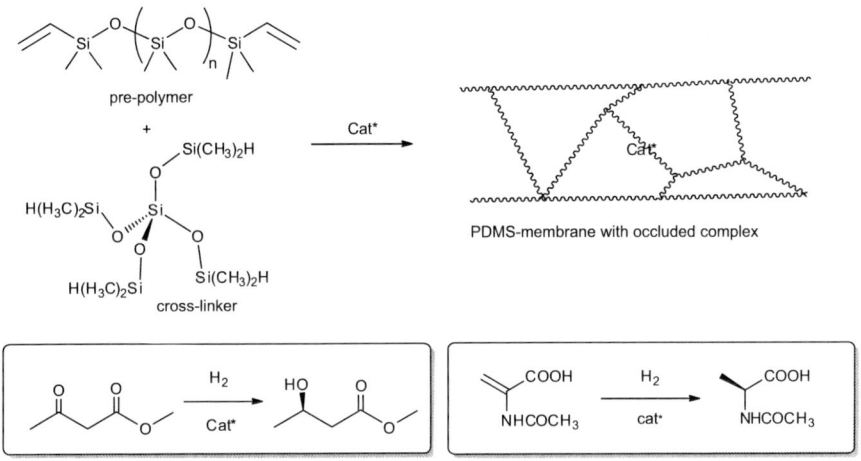

Figure 8.14 Preparation of catalysts occluded into PDMS membranes and enantio-selective hydrogenations in which they have been applied.

was not detected in any of these solvents. With the Rh catalyst, activity was lower than in the homogenous phase in methanol and ethylene glycol, and enantioselectivity was reduced from 99% ee to 90–93% ee. However leaching was not detected and the catalyst was recovered once. The Rh catalyst was also used in the hydrogenation of methyl 2-acetamidoacrylate[33] by adding the membrane cut into pieces to the reaction medium. Under these conditions leaching was controlled by the solubility of the complex and the swelling of the membrane, being extensive in methanol (31%) and dichloromethane (82%) and low (<1%) in water or *n*-heptane. The catalyst was again less active than in solution leading to ca. 97% ee. The use of poly(vinylalcohol) (PVA) instead of PDMS did not improve the results of the reaction and leaching was higher.[34]

Mn-salen catalysts occluded in PDMS membranes[35] were used in the alkene epoxidation using NaOCl, with similar results to those obtained in solution (18% ee with *trans*-β-methylstyrene and 52% ee with styrene) using the Jacobsen's catalyst (Figure 8.15). The main problem was leaching, ranging from 12% in *n*-hexane to 100% in chlorobenzene from the Jacobsen's catalyst, 56% in chlorobenzene for the dimeric complex, and 2.7% in *n*-heptane and 65.9% in chloroform for the Katsuki's complex.

Another strategy to entrap a catalyst into a flexible polymeric network is microencapsulation.[36] Pioneering work in this field[37] described the micro-encapsulation of OsO_4 in a linear acrylonitrile–butadiene–styrene copolymer by cooling a THF solution of the osmium derivative and the support, and final hardening of the capsules in methanol. These capsules were used in the dihydroxylation of alkenes with the addition of 1,4-bis(9-*O*-dihydrox-yquinidyl)phthalazine (DHQD)₂PHAL as the chiral ligand. Catalyst and chiral ligand were separately recovered, the first one by filtration of the capsules and the ligand by extraction. Using this methodology, *trans*-β-methylstyrene was

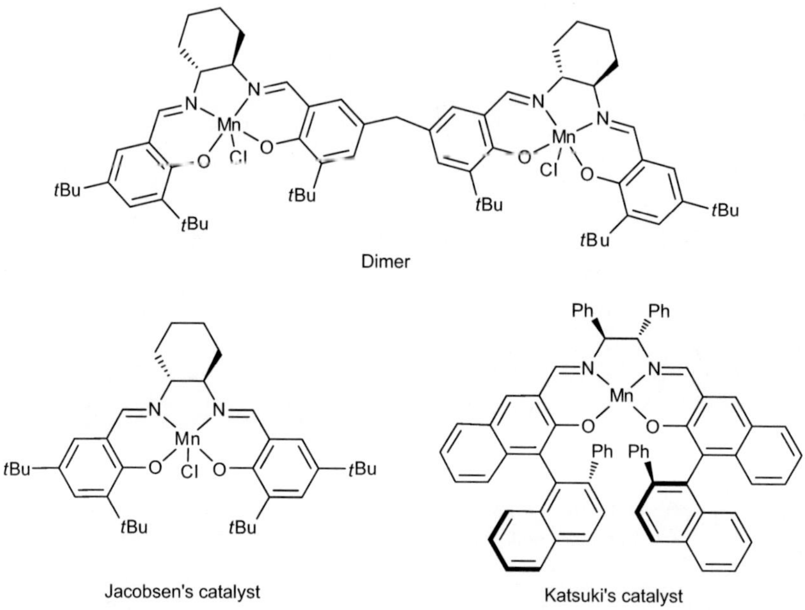

Figure 8.15   Salen-Mn complexes immobilized in PDMS.

dihydroxylated with *N*-methylmorpholine oxide, with yields of 88, 75, 97, 81 and 88% and enantioselectivities of 84, 95, 94, 96 and 95% ee in five consecutive reactions. Different oxidants and polymers,[38] including a flexible cross-linked polystyrene (1% divinylbenzene) have been tested,[39] but in none of these examples recycling of the whole chiral complex was described.

Reversibility of the microencapsulation process was shown in the immobilization of *i*-PrPyBox-RuCl$_2$ into polystyrene capsules (Figure 8.16).[40] The cyclopropanation of styrene with ethyl diazoacetate was carried out in dichloromethane, a solvent where the catalyst and the support were soluble, so the reaction took place in solution. After the reaction, the capsules were formed again by addition of a non-polar solvent, such as hexane or cyclohexane. These capsules were filtered, washed and reused up to three times without loss of catalytic activity and enantioselectivity.

Cu(acac)$_2$ encapsulated in polystyrene was modified by the addition of Box ligands and used in enantioselective sulfimidation reactions (Figure 8.17).[41] It was necessary to use an excess of chiral ligand (Box:Cu = 2:1) to obtain a significant, though still low, enantioselectivity (23% ee).

It is clear that leaching is the main problem associated to the use of flexible polymeric supports. Polymer swelling and complex solubility in the reaction medium are key parameters. These problems might be solved by the use of rigid supports to entrap the chiral catalyst.

It is possible to build the support around the catalytic complex if this complex is stable under the synthetic conditions. Using this approach

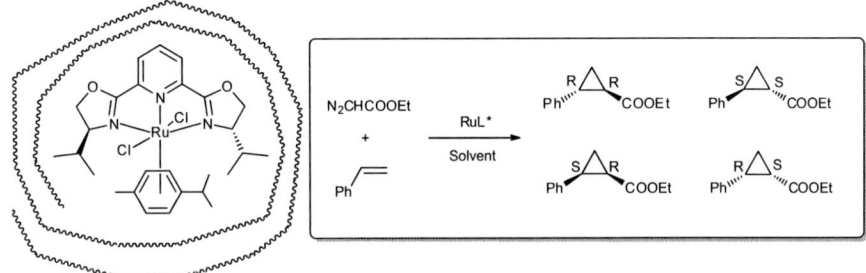

Solvent	Run	Yield	Trans/cis	%ee trans	%ee cis
Cyclohexane	1	31	94/6	85	68
	2	12	91/9	66	35
CH$_2$Cl$_2$	1	44	88/12	84	50
	2	41	89/11	86	61
	3	25	88/12	84	54

Figure 8.16 Cyclopropanation of styrene promoted by microencapsulated PyBox-Ru complexes.

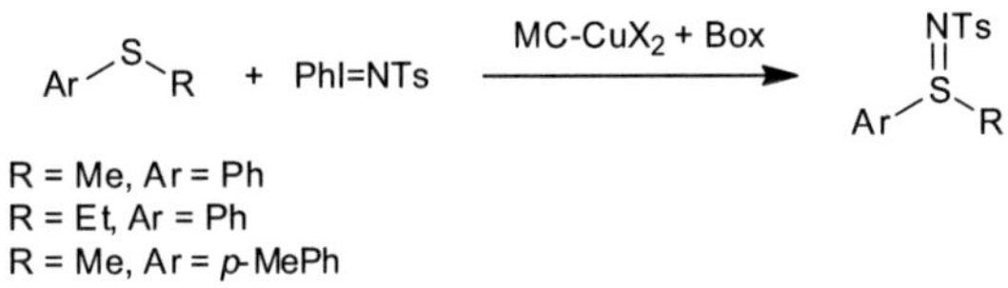

Ar$-$S$-$R + PhI=NTs $\xrightarrow{\text{MC-CuX}_2 + \text{Box}}$ Ar$-$S$(=$NTs$)-$R

R = Me, Ar = Ph
R = Et, Ar = Ph
R = Me, Ar = p-MePh

Figure 8.17 Enantioselective sulfimidation reactions.

[(BINAP)Ru(p-cymene)Cl]Cl, [(DIOP)Rh(cod)Cl] and [(BPPM)Rh(cod)Cl] were entrapped into porous silica prepared by sol-gel synthesis from Si(OEt)$_4$.[42] In this way it is not possible to control the position of the catalyst in the solid: some molecules may be placed in non-accessible internal sites, so that they are not active, some others may be placed on the external surface and they are only retained by physisorption, and finally some molecules may be placed in internal but accessible sites. Hydrogenation of itaconic acid in THF led to an important leaching, which indicates that a great part of the catalyst was placed on the external surface.

Jacobsen's catalyst was entrapped in alumina prepared by reaction of AlCl$_3$ with diisopropyl ether.[43] Although the catalyst was more active and selective than the homogeneous one with several oxidants and it was recycled, enantioselectivity was not described.

These examples show that the structure of the amorphous solids does not allow controlling the position of the catalytic sites and the access of the reagents to these sites. In order to get a better control, crystalline solids with regular pore systems should be used and, in this regard, zeolites with supercages in their

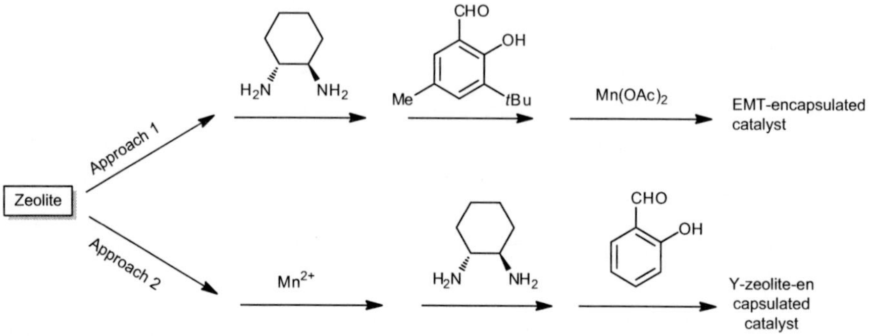

Figure 8.18   Zeolite-immobilized salen-Mn catalysts.

structure were considered the best alternative. The catalyst is introduced in portions, small enough to enter through the pore window and the complex is formed inside the supercages. As this complex is bigger than the pore opening, it is retained inside the zeolite. This approach is known as 'ship-in-a-bottle' synthesis. In most cases the support is an anionic zeolite and electrostatic interactions may participate in the immobilization. It is important to avoid the presence of non-complexed catalytic sites, able to promote competitive non-enantioselective reactions and at the same time the preparation method must prevent the possible formation of species different from the desired one.

EMT[44] and Y-zeolites[45] have been used to immobilize salen-Mn^{III} catalysts using two different strategies (Figure 8.18). In the first one, the chiral diamine and the aldehyde were sequentially introduced in the zeolite and made to react under reflux, the salen ligand was then complexed with Mn(AcO)$_2$ and the solid washed by Soxhlet extraction. In the second case, Mn exchange was previous to ligand synthesis and complexation. The catalyst immobilized in EMT (Figure 8.19) was tested in the epoxidation of alkenes with aqueous NaOCl. The reduced activity and selectivity, which were also observed when the support was Y-zeolite, as well as the lack of activity with bulkier alkenes or oxidants, were due to the severe diffusion restrictions imposed by these supports, which make an important part of the catalytic centres non-accessible to the reagents.

The small size of the pore system of zeolites is a limitation for the methodology that has been tried to be solved by using dealuminated faujasites,[46] which have mesopores surrounded by micropores (Figure 8.20). Several salen-metal complexes were occluded and used in the epoxidation of (R)-limonene and α-pinene with O$_2$/pivalaldehyde (Figure 8.21).[47] With Mn^{III}-salen it was necessary to use high O$_2$ pressure to obtain moderate epoxide selectivities (32–70%) and diastereoselectivities (32–55% de). The best result, 100% conversion, 96% epoxide selectivity and 91% de, slightly better than in solution, was obtained with the Co-salen catalyst.

The cationic character of salen-Mn^{III} makes the electrostatic interaction with the zeolite possible, so some of the complex may remain on the external surface in spite of Soxhlet extraction, which would not happen using neutral

Figure 8.19 View of Mn(salen) encapsulated in EMT. (Reprinted from S. B. Ogun-wumi and T. Bein, *Chem. Commun.*, 1997, 901, with permission from RSC.)

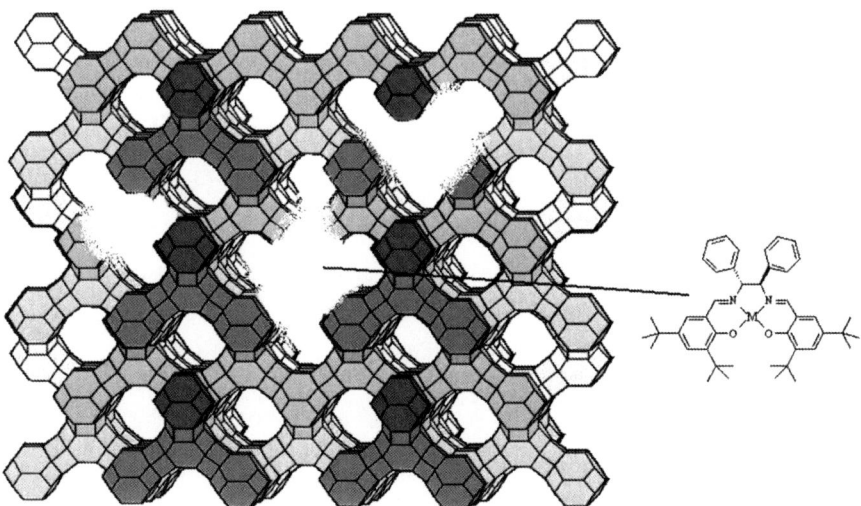

Figure 8.20 Salen-metal complex occluded into dealuminated faujasite. (Reprinted from C. Schuster and W. F. Hölderich, *Catal. Today*, 2000, **60**, 193, with permission from Elsevier.)

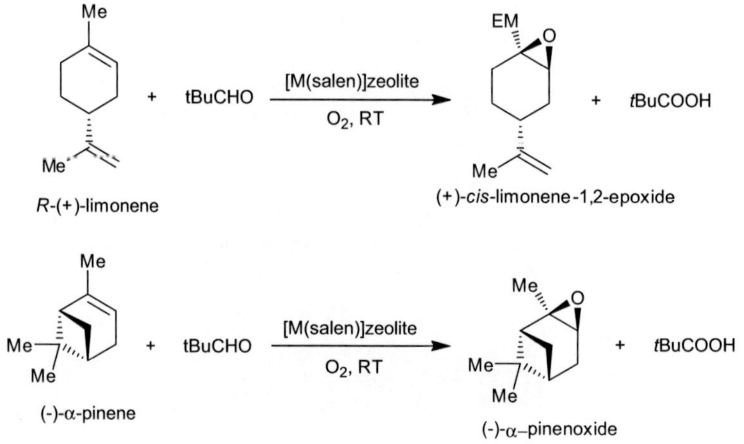

Figure 8.21 Epoxidation of terpenes with entrapped Mn(salen) catalysts.

complexes. Thus, Co^{II}-salen was encapsulated into ultrastable Y-zeolite (USY) by the 'flexible ligand' method, by sublimation of the ligand on Co-exchanged USY.[48] This method considers that the ligand is flexible enough to enter the pore aperture, but once the complex is formed the conformation is blocked and it cannot exit. The catalyst showed very little enantioselectivity in the reduction of acetophenone with NaBH$_4$.

The situation of the complexes in the solid and their participation in the catalytic reactions remained unclear and this point was deeply studied by several techniques[49] for occluded Mn-salen complexes. From these studies it was demonstrated that only 15–20% of Mn^{2+} present in the starting Y-zeolite was oxidized to Mn^{3+} and hence complexed with the ligand. Given that the oxidized manganese was more abundant in the external part of the zeolite most of the catalysis may be due to the minor fraction of complex residing in the outermost layers.

Trying to overcome this problem, Mn complexes of tetradentate ligands (Figure 8.22) were entrapped into Y-zeolite by the flexible ligand method, external and non-complexed cationic species were eliminated by exchange with Na$^+$.[50] The solids were used in the oxidation of sulfides with NaOCl in the presence of *N*-methylmorpholine oxide (Figure 8.22). Compared with the homogenous catalyst, the solid was less active and enantioselective (up to 19% ee) in the oxidation of methyl phenyl sulfide but more active and enantioselective (up to 21% ee in comparison with <5% ee) in the oxidation of 2-ethylbutyl phenyl sulfide, and the catalyst could be reused four times.

It can be concluded that the 'ship-in-a-bottle' methodology has serious limitations regarding complex size, due to the microporous character of the zeolites. Moreover, the limited accessibility to some catalytic sites and the participation of external and non-complexed species are also important limitations of this strategy.

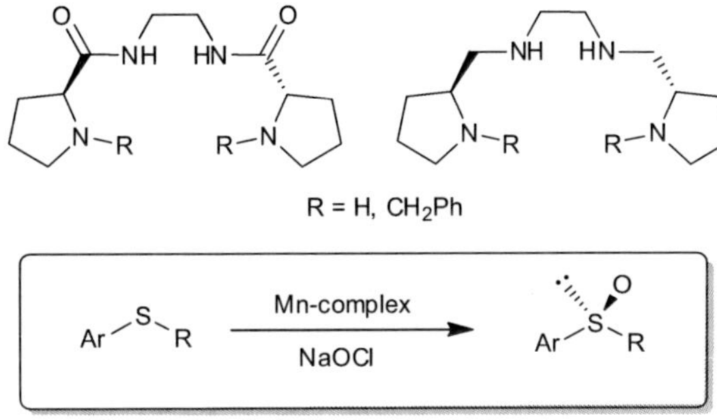

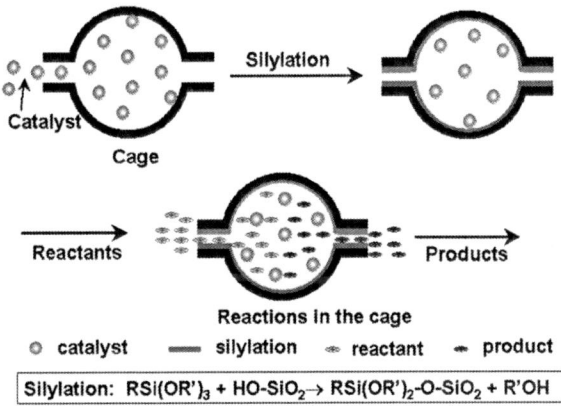

R = H, CH₂Ph

Figure 8.22 Tetradentate ligands used in the enantioselective sulfide oxidation.

Silylation

Catalyst

Cage

Reactants → Products

Reactions in the cage

○ catalyst ▬ silylation ⚬ reactant ⚬ product

Silylation: $RSi(OR')_3 + HO\text{-}SiO_2 \rightarrow RSi(OR')_2\text{-}O\text{-}SiO_2 + R'OH$

Figure 8.23 Entrapping of homogenous catalysts into mesoporous silicas. (Reprinted from H. Yang, J. Li, J. Yang, Z, Liu, Q. Yang and C. Li, *Chem. Commun.*, 2007, 1086, with permission from RSC.)

A new approach using a mesoporous crystalline silica,[51] allows overcoming these limitations. The pores of silica SBA-16 are large enough to allow the entrance of the complex in the adsorption process, a subsequent treatment with alkylsilane reduces the pore opening, entrapping the complex but allowing the diffusion of reagents and products (Figure 8.23). In this regard, the size of the silane was essential as was shown for the occlusion of salen-Co complexes. SBA was modified with methyl, propyl, benzyl, octyl and dodecylsilanes. Only C_8 and C_{12} were able to prevent diffusion of the complex, but C_{12} blocked completely the porosity of the solid. The solids were tested in the asymmetric ring opening of epoxides (Figure 8.24) with excellent results and activity retained for 12 runs in the reaction of epichlorohydrin, and 13 runs in the reaction of propylene oxide (51% yield and 95% ee after 20 hours in the first run and 46% yield and 97% ee after 28 hours in the thirteenth run).

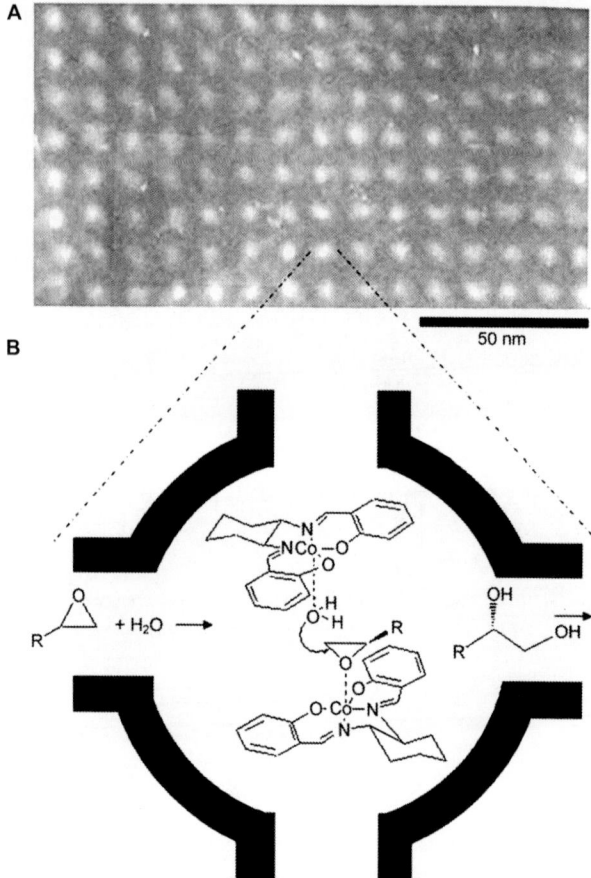

Figure 8.24 (a) TEM image of the (100) projection of cubic SBA-16 (*Im3m*); (b) (salen)Co complex promoting the hydrolytic kinetic resolution of epoxides in the cage of mesoporous crystalline silica. (Reprinted from H. Yang, L. Zhang, L. Zhong, Q. Yang and C. Li, *Angew. Chem., Int. Ed. Engl.*, 2007, **46**, 6861, with permission from Wiley.)

This strategy is particularly suitable in those reactions involving two molecules of catalyst in the transition state for which the isolation of the catalytic sites makes most of the immobilized catalysts useless. This effect was shown in the hydrolytic kinetic resolution of epoxides.[52] Results in solution and in heterogeneous phase are only similar when the average number of catalyst molecules per hypercage is higher than 2. With lower Co content, catalyst molecules are isolated due to the closing of the pores connecting the hypercages. In solution, dilution has a clear detrimental effect because the cooperative effect is more difficult. Given the confinement in the hypercages of the support, the entrapped catalyst is not sensitive to this dilution effect and it is much more active.

8.6 Immobilization by Coordinative Bonds

In this strategy a group of the support enters into the coordination sphere of the metal responsible for the catalytic activity, compensating at the same time a positive charge of the complex or not. We will consider only those examples in which this group belongs to the structure of the support. In other examples the coordinating group is added to the support, which requires an additional preparation step.

In the first example of this strategy chiral aluminium catalysts were grafted to silica and alumina by means of the formation of O–Al bonds.[53] Immobilization was carried out by reacting AlEt$_2$Cl, (1R,2S,5R)-menthol and silica or alumina in refluxing toluene. The best results in the reaction between cyclopentadiene and methacrolein, comparable to those obtained in homogeneous phase under the same conditions, were obtained using silica as the support (Figure 8.25).

Ti-tartrate catalysts were grafted onto amorphous[54] and MCM-41[55] silica by reaction of the metal with the surface silanol groups, and used in the oxidation of sulfides to sulfoxides with *t*-butylhydroperoxide (TBHP) or hydrogen peroxide (H$_2$O$_2$) (Figure 8.26). The amorphous silica-immobilized catalyst was highly active and selective to sulfoxide production (almost quantitative conversion with 99:1 sulfoxide/sulfone selectivity in the oxidation of *p*-tolyl methyl sulfide), but low ee (13%) was obtained. The MCM-41 catalyst was obtained by the template-ion-exchange method and was tested in the oxidation of methyl phenyl sulfide. Under the best conditions the sulfoxide was obtained with 30% ee and a poor 60:40 sulfoxide/sulfone ratio. The enantioselectivity was not mainly due to the sulfide-sulfoxide oxidation, but to a kinetic resolution in the sulfoxide-sulfone oxidation step.

Using a similar strategy, tantalum-tartrate complexes were supported onto silica and used in the epoxidation of allyl alcohols (Figure 8.27).[56] The larger

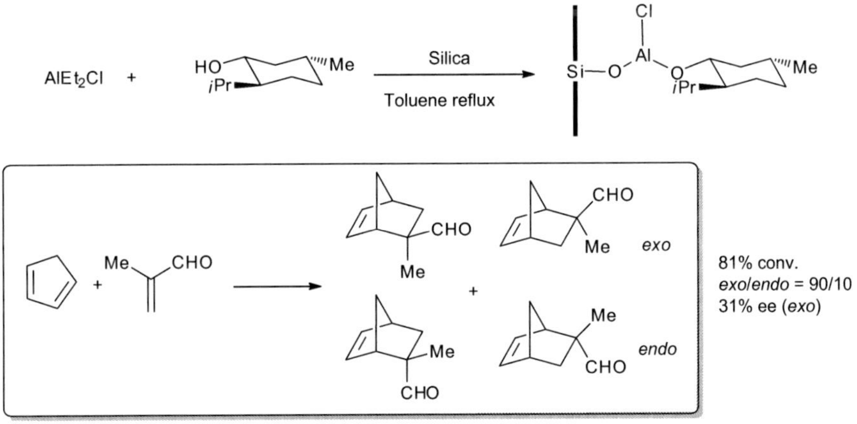

Figure 8.25 Preparation of silica supported Al-menthol catalyst and application to enantioselective Diels–Alder reaction.

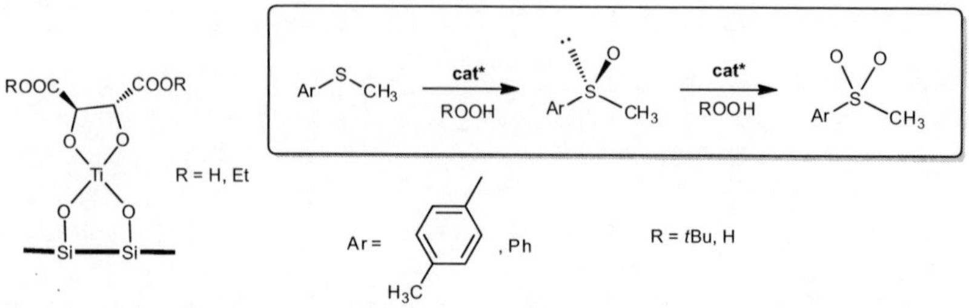

Figure 8.26 Supported Ti-tartrate catalysts and application to enantioselective oxidation of sulfides.

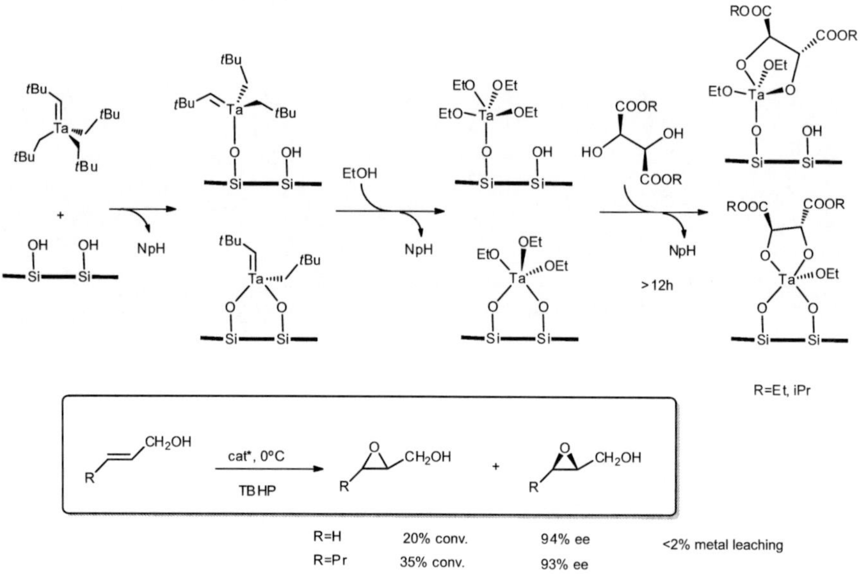

Figure 8.27 Preparation of silica-supported Ta-tartrate catalysts and enantioselective epoxidation.

coordination sphere of tantalum allows the simultaneous coordination of tartrate, support and reagents. Furthermore Ta catalysts display a very low activity in solution, minimizing the concurrence of homogeneous catalysis by possible leached species.

Results depended on several factors related both to the preparation of the catalyst and to the reaction conditions. Regarding the preparation of the catalyst, long times in the treatment of the Ta-silica solid with the chiral modifier are needed to reach the best enantioselectivities. Regarding the reaction conditions, alkyl hydroperoxides must be used because H_2O_2 led to high conversion but no enantioselection. The good results obtained with high concentration of allyl alcohol (98% ee and 45% conversion with 1 M) or higher

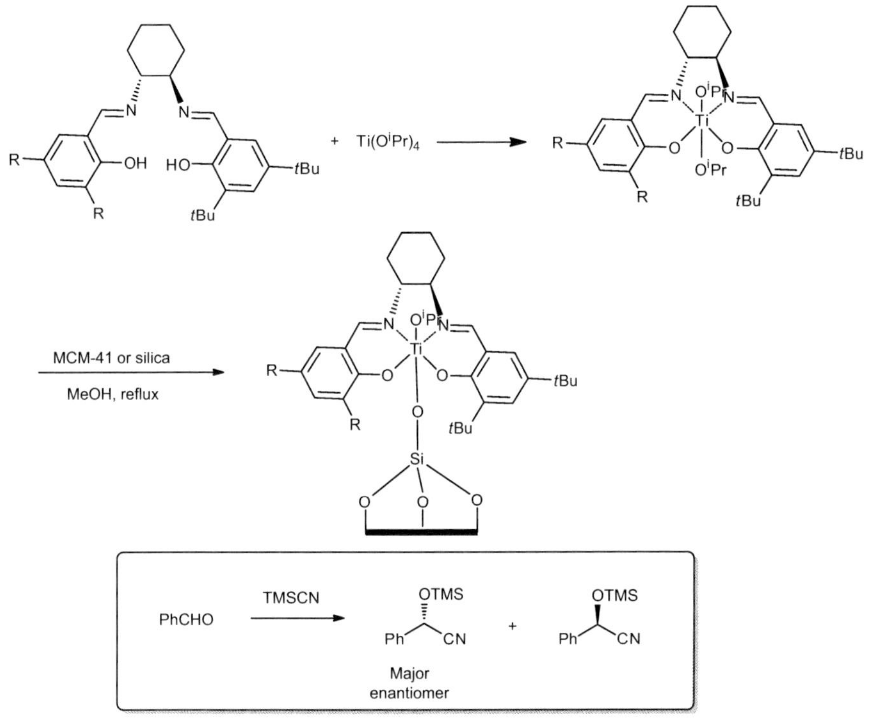

Figure 8.28 Preparation of silica-supported Ti(salen) and enantioselective tri-methylsilylcyanation reaction.

temperatures (91% ee at 10 °C) are accompanied by lower epoxide selectivity, so that a contribution of the kinetic resolution of the epoxide in a subsequent hydrolysis or alcoholysis reaction to the high enantioselectivity obtained cannot be discarded. Unfortunately, recovery experiments were not described.

Ti-salen immobilized through silanol groups to crystalline MCM-41[57] and amorphous silica[58] was used in the trimethylsilylcyanation of benzaldehyde (Figure 8.28). Better results were obtained with the crystalline support. In general, enantioselectivities are better than those obtained in solution with the same ligand and the best results are obtained with one of the non-symmetrical ligands (R = H) (94% ee with heterogeneous and 90% ee with the homogeneous catalyst), whereas with the C_2-symmetric ligand (R = t-Bu) lower enantioselectivities (83% ee in heterogeneous and 77% ee in homogeneous phase) were obtained.

Basic carbons with oxygen atoms on their surface, able to act in the same way as –OH groups of silica or alumina, can be also used in this immobilization strategy. Several Rh-diphosphine complexes were bonded to carbon through Rh–O bonds and used in the asymmetric hydrogenation of dimethyl itaconate and methyl 2-acetamidoacrylate[59] (Figure 8.29), with enantioselectivities

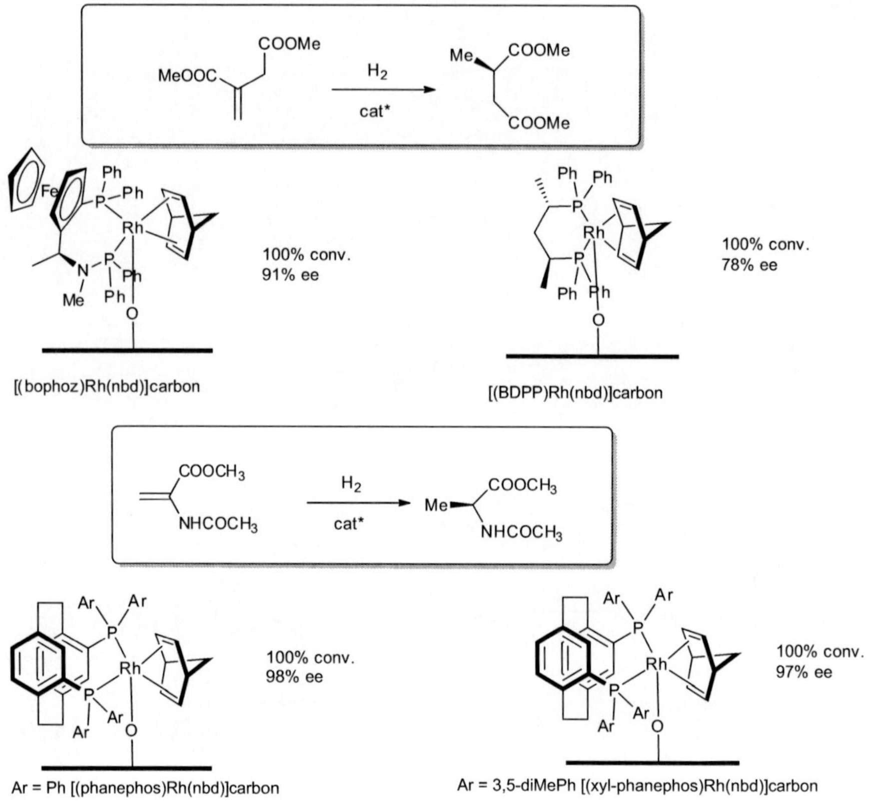

Figure 8.29 Results obtained in enantioselective hydrogenation promoted by [(diphosphine)Rh(nbd)] complexes grafted on basic carbon.

similar or slightly better than those obtained in solution and leaching below the detection limit, even in scaled-up experiments.

Organic polymers bearing coordinating groups can also be used to support metal complexes by this strategy. Polyaniline (PANI) was reacted with OsO_4 and/or $MeReO_3$ to obtain PANI-Os, PANI-Re and PANI-Os-Re complexes that were used in the asymmetric dihydroxylation of alkenes using *N*-methylmorpholine *N*-oxide (NMO) or H_2O_2 as the oxidant and [(DHQD)$_2$PHAL] as the chiral modifier (Figure 8.30).[60] In general, high isolated yields (83–94%) and enantioselectivities (62–99% ee), comparable to those obtained in solution, were reached. PANI-Os and PANI-Os-Re were used in five consecutive reactions with the same results without detection of metal leaching. The chiral modifier can be separately recovered from the aqueous phase in the treatment of the reaction, and fresh chiral modifier should be added in each reuse, which reduces the practical interest of catalyst reuse.

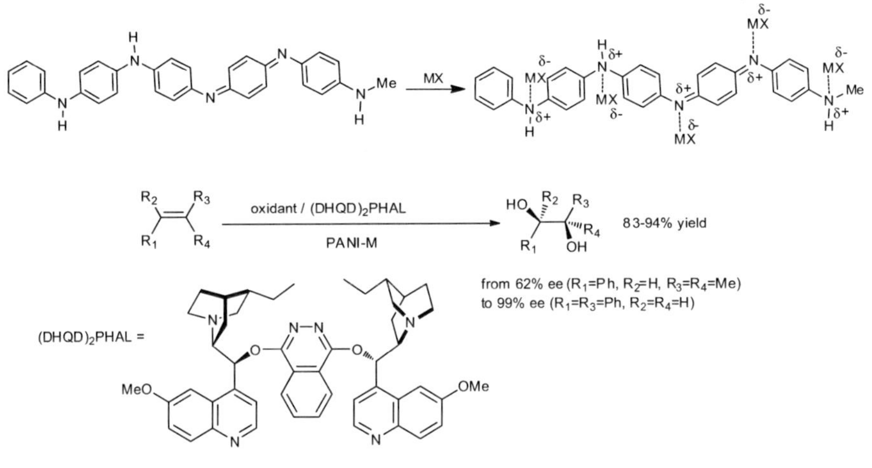

Figure 8.30 Results from enantioselective dihydroxylation of alkenes with PANI-Os and PANI-Os-Re complexes in the presence of (DHQD)$_2$-PHAL.

8.7 Immobilization by Electrostatic Interactions

Solids with negative charges can act as counter ions for cationic complexes, so that the electrostatic interaction links the complex to the support. This produces a chiral solid able to act as a heterogeneous catalyst, provided the complex remains charged all along the catalytic cycle.

Immobilization can be carried out by means of two different approaches (Figure 8.31). The first one consists in the exchange of the compensating cations of the solid by the cationic chiral complex previously formed in solution. In the second one, the exchange is carried out with the catalytic metal cation, which is then complexed by the chiral ligand. Both strategies have advantages and disadvantages and examples of both will be presented later on. The solvent in which the exchange is carried out is important because it should favour the mobility of the ions and may contribute to the stabilization of the different species by solvation.

Any kind of anionic material, inorganic, organic or hybrid, can be used to support cationic complexes. Historically the first examples of non-covalent immobilization used clays, a well-known class of inorganic cation exchangers. These materials are lamellar aluminosilicates or magnesiosilicates, in which isomorphic substitutions (*e.g.* aluminium by silicon, magnesium by aluminium or lithium by magnesium) generate charge defaults in the structure, compensated by the inclusion of hydrated alkaline or alkaline earth cations that can be subsequently exchanged by different cations.

In the first reported example[61] the complex [(PNNP)-Rh(cod)]$^+$ (Figure 8.32) was electrostatically bonded to several clays by cationic exchange of its perchlorate salt in methanol. The solid catalyst was used in the enantioselective hydrogenation of several *N*-acetyl dehydroamino acids with a noticeable

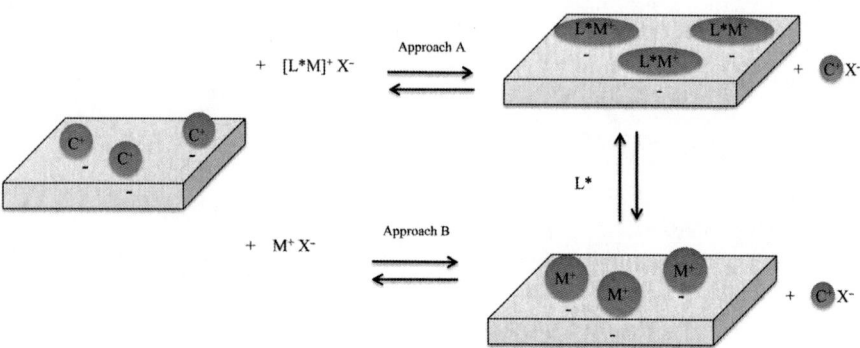

Figure 8.31   Immobilization by cation exchange.

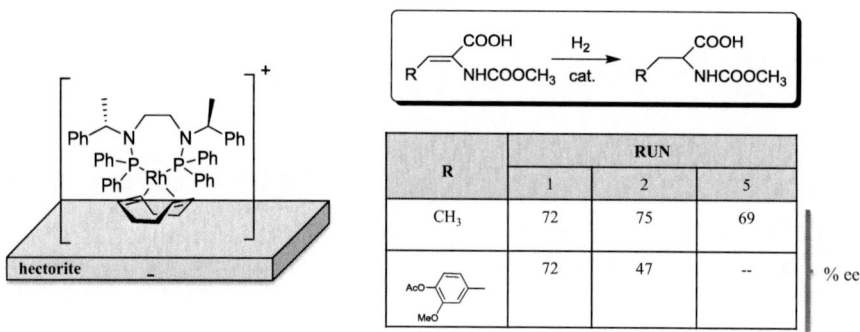

Figure 8.32   Hydrogenation of amino acid precursors with [(PNNP)-Rh(cod)] exchanged onto hectorite.

influence of the nature of the clay support, the best results being obtained with hectorite (Figure 8.32). The hectorite-immobilized catalyst was very active in the hydrogenation of *N*-acetyl dehydroalanine and it was recovered five times with the same enantioselectivity; however, recovery was less efficient for bulkier substrates.

Several chiral diphosphine-rhodium perchlorate complexes (Figure 8.33) were analogously immobilized[62] on a synthetic hectorite by cation exchange in a water/acetonitrile mixture. The catalysts were compared in the hydrogenation of several itaconates showing that the nature of the substrate and the reaction solvent are important in both homogeneous and heterogeneous catalysis, although the effects in both phases are not identical (Figure 8.33). In general the immobilized catalysts were less active than their homogeneous counterparts, with a detrimental effect of less polar solvents, not so evident in homogeneous phase, and justified by the limited access of the reagents due to the spatial restrictions between the clay sheets. Regarding enantioselectivity, the immobilization had in general a positive effect, more important in the case of non-polar solvents and with BPPFA ligand (Figure 8.33).

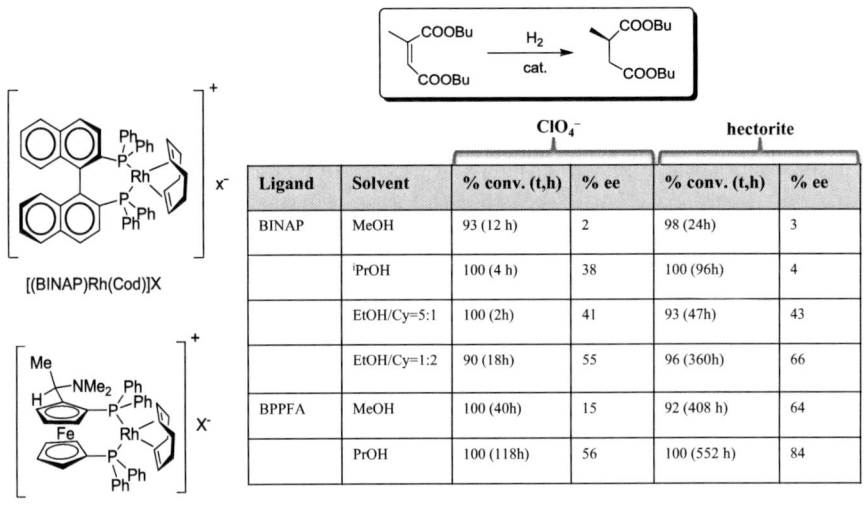

Figure 8.33 Results obtained in the itaconate hydrogenation with homogeneous and supported chiral catalysts.

As commented above (Section 8.5), zeolites are also able to exchange cations, and the role of electrostatic interactions cannot be discarded in the retention mechanism of entrapped catalysts.

Mesoporous materials, mainly Al-MCM41, have also been used for electrostatic immobilization of Mn-salen complexes. The mesoporous character of these materials overcomes the strong diffusional limitation imposed by the use of zeolites. As a consequence it has been possible to compare direct exchange of the (salen)MnPF$_6$ complex (method A) with the complexation of the Mn^{II}-exchanged Al-MCM41, followed by oxidation to Mn^{III} (method B)[63] (Figure 8.34).

The behaviour of both types of immobilized catalysts in the epoxidation of styrene with *m*-CPBA was similar, that prepared by direct exchange being slightly more active and enantioselective. The immobilized catalysts led to better enantioselectivities and one of them was recovered three times without loss of activity.

The influence of the nature of the mesoporous support has been studied using [(MeDuPhos)Rh(cod)]$^+$ and [(MeBPE)Rh(cod)]$^+$ (Figure 8.35), immobilized on different materials as catalyst for the hydrogenation of dimethyl itaconate and methyl 2-acetamidoacrylate.[64]

The amount of complex exchanged was in the range of 25–50% over the theoretical maximum. One reason might be the low accessibility of inner sites once the most external sites are occupied by the exchanged complex. Despite its large pore size, results obtained with SBA-15 were worse than those obtained with MCM materials. The use of MeDuPhos and MCM solids allows carrying out the reactions with very low catalyst/alkene ratio, which together with the efficient recovery results in overall TON > 20 000.

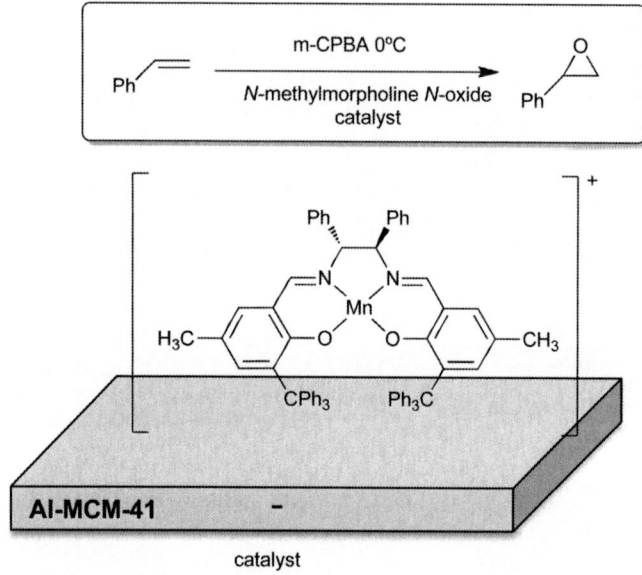

catalyst

Figure 8.34 Epoxidation of styrene promoted by (salen)Mn immobilized on Al-MCM41.

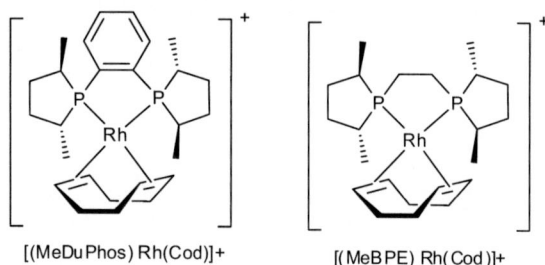

[(MeDuPhos) Rh(Cod)]+   [(MeBPE) Rh(Cod)]+

Figure 8.35 Structure of [(MeDuPhos)Rh(cod)]$^+$ and [(MeBPE)Rh(cod)]$^+$ ligands.

The influence of the reaction solvent was studied by Sheldon and co-workers[65] in the hydrogenation of dimethyl itaconate catalyzed by the complex [(MeDuPhos)Rh(cod)]$^+$ immobilized in the large pore support AITUD-1 (1.5 nm pore diameter with broad distribution) by cationic exchange from alcohol solutions. In general, high enantioselectivities, from 96 to 98% ee, were obtained with higher catalytic activity in 2-propanol and leaching below 1% in less polar solvents such as dichloromethane or ethyl acetate. In the hydrogenation of methyl 2-acetamidoacrylate leaching was again lower in less polar solvents but the influence of the solvent on the enantioselectivity was more relevant going from 75% in 2-propanol to 98% in methanol.

Not only inorganic materials are suitable supports for the electrostatic immobilization of chiral catalysts, in fact the use of organic polymers as cation-exchange resins is well known. Organic polymers offer a large variety of structures and properties, which makes it possible to adapt the support to the reaction conditions. One example is the immobilization of [(DIOP)Rh(Cod)]PF$_6$ and [(Norphos)Rh(Cod)]PF$_6$ on sulfonic and carboxylic polymeric resins (Figure 8.36).[66] Exchange was carried out starting from the acidic and the Na$^+$ forms of the resins, and the supported catalysts were tested in the hydrogenation of (Z)-2-acetamidocinnamic acid and its methyl ester.

Better enantioselectivities (up to 75% ee with DIOP and 80% ee with Norphos) were obtained with the catalysts prepared from the sodium form of the resin. The use of carboxylic resins led to poorer results, in agreement with the influence of the counter ion also observed in solution. Macroporous resins, with permanent porosity, and gel type resins that do not have permanent porosity were compared. The better results obtained with the macroporous Dowex were ascribed to the improved diffusion.

The cross-linking degree of the polymer has also an important influence in the behaviour of catalyst immobilized on gel-type polymeric cation exchanger. Several Rh complexes (Figure 8.36) were immobilized on sulfonated polystyrene-divinylbenzene and compared in the hydrogenation of methyl

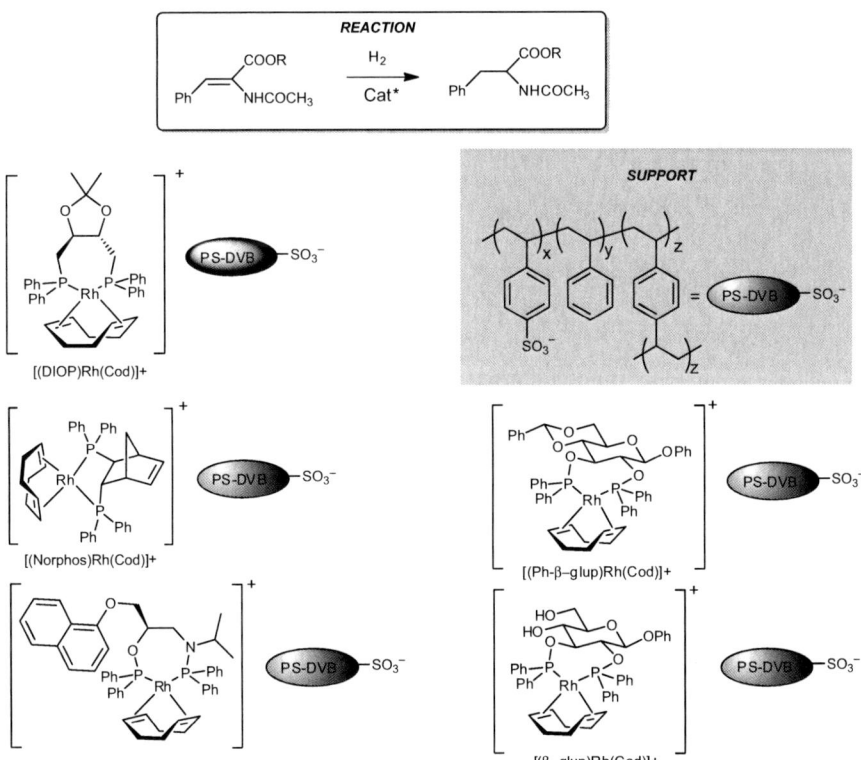

Figure 8.36 Hydrogenation of 2-acetamido cinnamate esters catalyzed by chiral Rh complexes immobilized on sulfonic resins.

(Z)-2-acetamidocinnamate.[67] Lowering the cross-linking degree from 2% to
0.5% increased the catalytic activity, which was ascribed to better diffusion
caused by the improved resin swelling. The nature of the exchangeable
cation was again important with Li^+ leading to more active catalysts. The
best enantioselectivities, independent of the exchange method, were obtained
with β-glup ligand (93-96% ee along 15 cycles) with a low leaching degree
(0.5% per cycle).

As indicated above, the nature of the acidic group of the resin influences the
behaviour of these catalysts as it acts as the counter ion of the cationic complex.
In this regard the anionic form of nafion, a perfluorinated anionic resin, is less
coordinating than the sulfonated anions of normal resins. Nafion NR50 was
used to immobilize aminated derivatives of BDPP and chiraphos[68] (Figure
8.37). The acidic form of nafion was able to protonate the four amino groups of
the ligand leading to species with several positive charges efficiently bonded to
the resin.

Alternatively, the tetra(trimethylammonium)ligand was exchanged from the
Na^+ form of the polymer in water. The catalysts were used in the hydrogenation of several cinnamic acid derivatives with moderated enantioselectivities (56–76% ee), though better than those obtained using the sulfonic resin
Amberlyst-15H. It is not clear that the resin acts as the counter ion of Rh^+,
because of the presence of the ammonium groups; in fact, the absence of Rh
leaching indicates that the interaction between the catalyst and the support
does not depend on the oxidation state of Rh and the possible formation of Rh^0
through reduction.

One of the advantages of polymers, regarding electrostatic immobilization, is
the variability of the anionic centre. However, mechanical fragility makes them
less suitable from a practical point of view. A possible solution is the use of

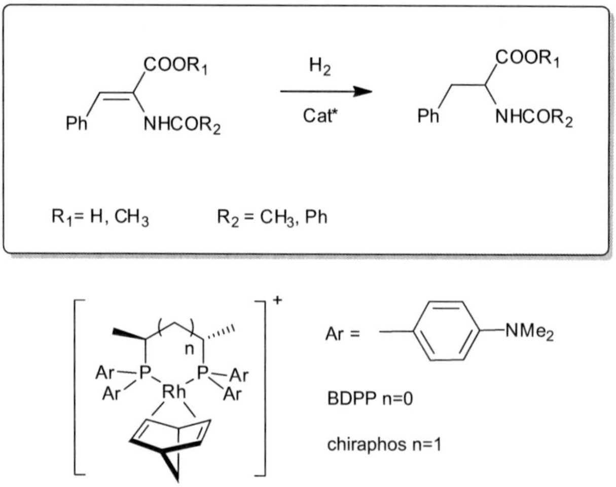

Figure 8.37   Hydrogenation of cinnamic acid and its derivatives with chiral catalysts
immobilized on nafion NR50.

organically modified inorganic supports, which has been tested in hydrogenation reactions using silica grafted benzenesulfonic acid[69] (Figure 8.38).

Trying to get together the advantages of crystalline mesoporous materials, namely mechanical stability and a regular mesoporous system, with those of polymers, several organic moieties bonded to macroporous materials have been used to immobilize Mn-salen complexes (Figure 8.39). A phenoxide group bonded to MCM-41 immobilized the Jacobsen's complex through cation exchange in refluxing ethanol to yield a catalyst able to epoxidize α-methylstyrene with aqueous NaClO with 72% ee, better than the 56% ee reached in solution. The lack of activity in the epoxidation of 1-phenylcyclohexene proved the placement of the catalyst inside the mesopores. Different supports (MCM41, SBA-15 and silica), organic linkers and counter ions were used for the immobilization of the Jacobsen's Mn-salen complex and compared in the epoxidation of 6-cyano-2,2-dimethylchromene with NaClO (Figure 8.39).[70]

Figure 8.38   Silica-grafted benzenesulfonic acid.

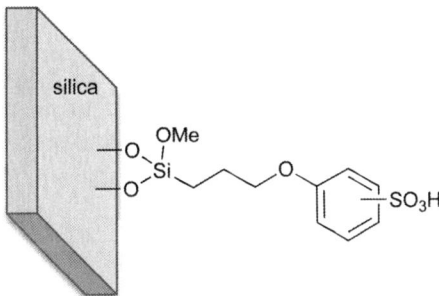

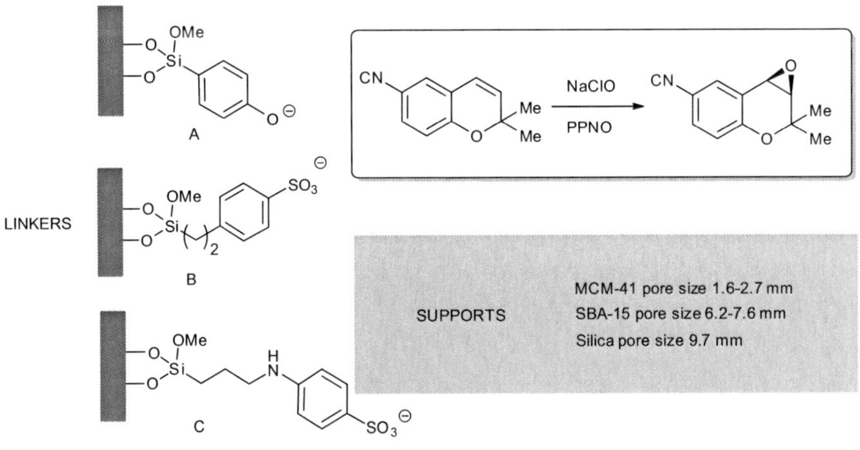

Figure 8.39   Results obtained in the epoxidation of 6-cyano-2,2-dimethylchromene with supported Jacobsen's (salen)Mn complexes.

The pore size is an important parameter and better results were reached with larger pore diameters both in activity (yields over 80%) and selectivity (enantioselectivities over 80% ee). Regarding the anion, sulfonate seems to be better than phenoxide, and the increase in spacer length was always positive, but more dramatically in solids with larger pores. Confinement effects inside the pores were considered to be in the origin of some of the observed effects. Activity and selectivity were improved by using a more hydrophobic methylated surface that favours the adsorption of the reagents.

The influence of the different factors involved in the electrostatic immobilization has been studied in depth in the case of the bis(oxazoline) (Box) and the related azabis(oxazoline) (AzaBox) ligands. The copper complexes of these ligands were used to promote the enantioselective cyclopropanation of styrene with ethyl diazoacetate (Figure 8.40).

BoxPh-CuCl$_2$ was immobilized by cation exchange onto several clays,[71] such as the natural aluminosilicates K10-montmorillonite and bentonite, and the synthetic magnesium silicate laponite. All the catalysts led to very similar results in the cyclopropanation reaction, and in subsequent works the more reproducible synthetic laponite was used as support.

Regarding the catalytic results, the exchanged catalysts were more active than the homogeneous ones, which is due to the detrimental effect of the chloride as a co-ordinating counter ion. In fact when compared with the homogeneous catalysts obtained from Cu(OTf)$_2$ the results are very similar. The immobilized catalyst can be recovered and reused with similar results, i.e. *trans/cis* selectivities about 60/40 and enantioselectivities of 50% ee in *trans*-cyclopropanes and 30% ee in *cis*-cyclopropanes.

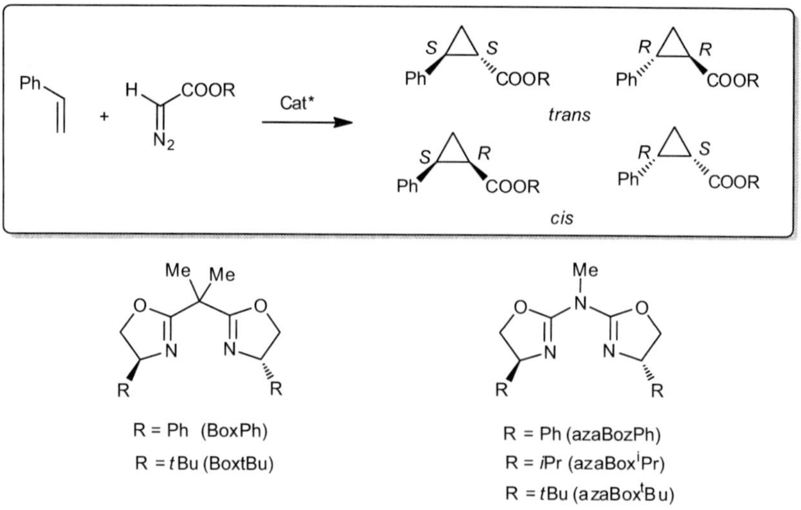

R = Ph (BoxPh)
R = *t*Bu (BoxtBu)

R = Ph (azaBozPh)
R = *i*Pr (azaBox^{i}Pr)
R = *t*Bu (azaBox^{t}Bu)

Figure 8.40 Cyclopropanation of styrene promoted by bis(oxazolines) and azabis(oxazolines) copper complexes.

Given the strong influence of the counter ion, anionic supports less coordinating than clays were also studied. BoxPh-Cu complexes immobilized on nafion and a nafion-silica nanocomposite[72] led to results identical to those obtained in solution with the $Cu(OTf)_2$ complex (ca. 60% ee in *trans*-cyclopropanes) and the catalyst could be recovered and reused with the same results. The better catalytic performance of the catalyst immobilized on the nafion-silica nanocomposite was due to the much higher surface area ($>80\,m^2\,g^{-1}$ as compared to $<0.02\,m^2\,g^{-1}$ of nafion), which allows better accessibility to the catalytic sites.

Other cation-exchange polymeric supports with sulfonic groups, such as Dowex 50W and Delowax1/9, led to low enantioselectivities that were almost completely lost upon reuse, which was ascribed to the effect of the counter ion.[73]

Due to the excellent enantioselectivity results obtained in solution using the complex Cu^{II}-Box-*t*-Bu, the corresponding immobilized catalyst was also tested. After immobilization the catalyst led to lower enantioselectivities that were even lower after recovery. In order to improve these results the immobilization conditions were deeply studied[74] by using different exchange solvents and different supports. Even using the best exchange conditions and support (nafion-silica nanocomposite SAC-13) the results were still far from those reached in solution and recovery of the catalyst was poor, not explained by the influence of the counter ion.

The hypothesis of ligand leaching during the exchange process, due to a lower stability of Cu^{II}-Box-*t*-Bu complex as compared to Cu^{II}-BoxPh, was validated when 91% ee and 88% ee in *trans*- and *cis*-cyclopropanes, respectively, were obtained by adding an excess of Box-*t*-Bu ligand to the heterogeneous reaction.[75] As the exchange equilibrium is accompanied by the equilibrium of the complex formation, if the latest is not completely shifted toward the complexed form, it is possible the formation of free copper sites on the solid (Figure 8.41). Taking into account that these sites are more active, the presence of even a small number of them is enough to justify the important decrease in enantioselectivities. It is worth noting that the even lower enantioselectivity obtained with the recovered catalyst may also be due to the leaching of the chiral ligand.

AzaBox ligands form stronger copper complexes than Box[76] and hence they allowed reproducing the results in homogeneous phase (90% ee in *trans*-cyclopropanes and 88% ee in *cis*-cyclopropanes) in a recoverable manner when Cu-AzaBox complexes were electrostatically immobilized.

This is a clear example of the importance of the strength of the ligand–metal complex in immobilized chiral catalysis, except in ligand-accelerated reactions, to avoid the participation of non-enantioselective catalytic sites. The excess of ligand frequently used in solution to shift the equilibrium of complex formation is not possible with supported catalysts. Regarding catalyst recovery, when the immobilization is carried out by covalent bonding between the ligand and the support, the formation of strong complexes prevents from catalyst deactivation by metal leaching. When the catalyst is immobilized by electrostatic

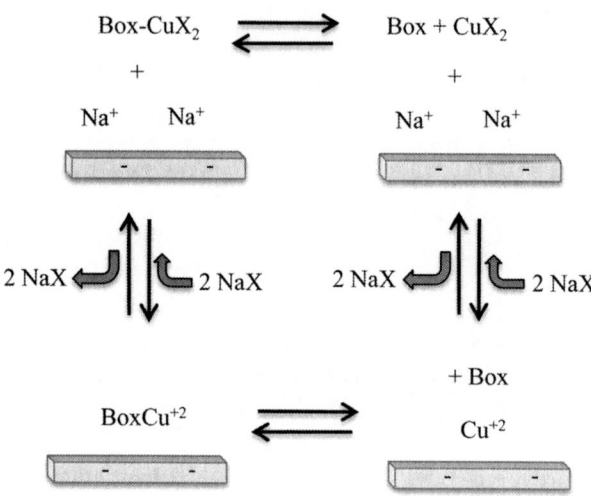

Figure 8.41   Chemical equilibria involved in the Box-Cu^{II} complex exchange process.

interactions the metal cation cannot be leached, unless it loses its charge during the reaction or some other cations take part in the process, but the strong complexes prevent the formation of non-enantioselective catalytic sites by chiral ligand leaching.

Finally some interesting results were found in the study of the influence of the solvent on the reaction results. When cyclopropanation was promoted by Cu-BoxPh complex exchanged on laponite using solvents with low dielectric constant, such as hexane, toluene or styrene, a complete reversal of the diastereoselectivity (*trans/cis* $= 31/69$) was observed, and furthermore the major *cis*-cyclopropane had the absolute configuration opposite to that preferably obtained in solution.[77] The effect on the diastereoselectivity was magnified with C_1-symmetric ligands, reaching *trans/cis* selectivities up to 8/92.[78]

Theoretical studies[79] have shown that the step determining the stereoselectivity is the concerted addition of the alkene to the Cu^I-carbene intermediate formed in the rate-limiting step. Energy differences between the four possible diastereomeric transition states are responsible for the *trans/cis* diastereoselectivity and for the enantioselectivity in *trans*- and *cis*-cyclopropanes. One possible model to explain the solvent effect on the stereoselectivity is based on the formation of a tighter ion pair between the intermediate and the anionic surface, in a solvent of low dielectric constant. The steric interactions with the surface should favour the position of the ester group pointing outside the clay surface. The steric interactions between the surface and the incoming alkene increase the energy of some transition states, so that the transition state free from steric interactions with the clay surface becomes the most stable, leading to the major product obtained in these conditions (Figure 8.42).

Carbene C–H insertion is a reaction mechanistically related to cyclopropanation, also taking place through a metal-carbene complex. The same Box and

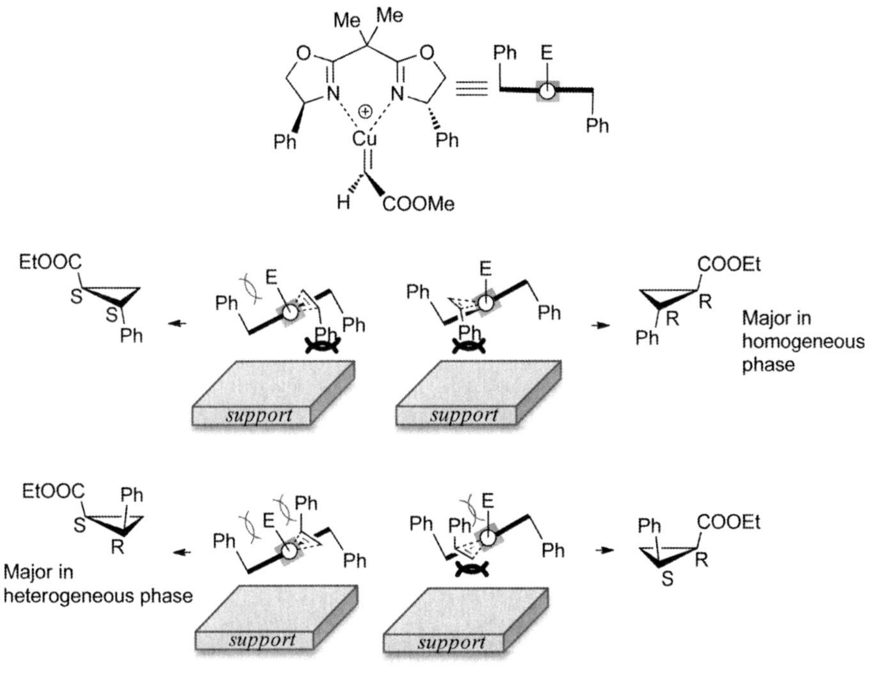

Figure 8.42 Model to explain the surface effect on the stereoselectivities of cyclo-propanation reactions promoted by bis(oxazolines)-copper complexes immobilized onto laponite.

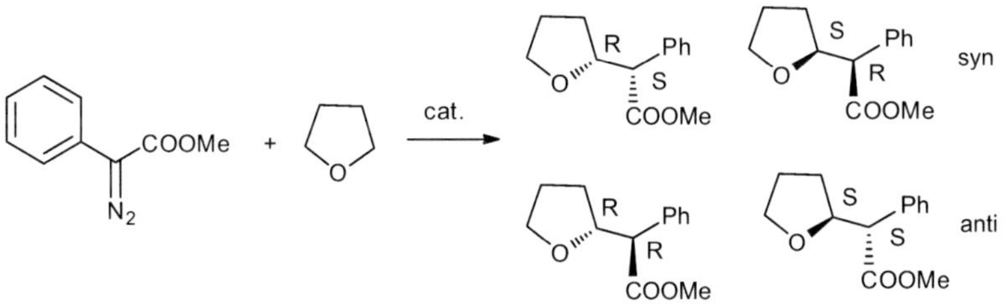

Figure 8.43 C–H insertion reaction.

Cu-AzaBox complexes immobilized on laponite have been used to catalyze the reaction between THF and ethyl 2-phenyldiazoacetate[80] (Figure 8.43). Whereas Cu[II]-BoxPh led only to moderate results in homogeneous phase (around 60% ee), enantioselectivity increased up to 88% ee in the major *syn*-product when immobilized on laponite, in another example of positive support effect. Furthermore, the catalyst was recovered and reused four times with the same results. The observed primary KIE (kinetic isotopic effect) using d_8-THF showed that in this case the limiting step is not the formation of the Cu^I-carbene intermediate but the insertion of this carbene into the C–H bond. The similarity of results between homogeneous and heterogeneous catalysts indicates that this positive effect is not due to a change in the reaction mechanism.

This kind of surface effect is not limited only to carbene-metal mediated reactions, as demonstrated by the results in a Mukaiyama aldol reaction (Figure 8.44),[81] in which the Box-Cu^{II} complex acts as a Lewis acid. The reversion in the diastereoselectivity together with the important increase in the enantioselectivity in the *anti*-product allowed obtaining the major product in enantiopure form for the first time.

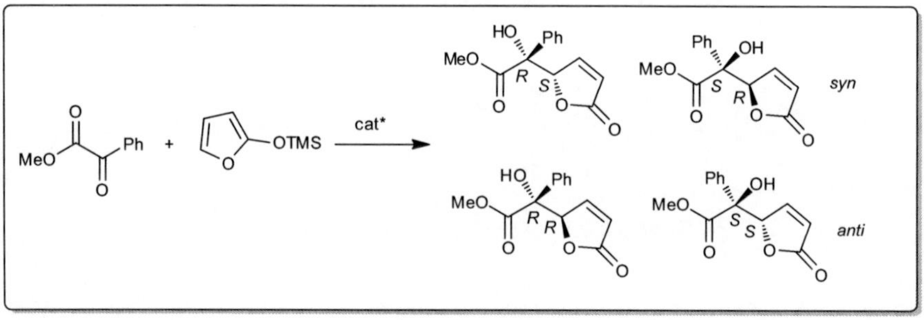

Catalyst	Syn/anti	%ee syn	%ee anti
BoxPhCu(OTf)$_2$	62:38	7	16
BoxPhCu-Laponite	14:86	9	90

Figure 8.44   Results obtained from the Mukaiyama aldol reaction with homogeneous and supported catalysts.

Chiral catalysts bearing anionic groups can be electrostatically bonded to cationic supports. This strategy has been less frequently used because it requires the modification of the ligand to introduce the anion. Layered double hydroxides (LDHs), also known as anionic clays, are among the most popular inorganic cationic supports. In fact they have been used to immobilize a sulfonated Ru-BINAP complex (Figure 8.45).[82] In the hydrogenation of dimethyl itaconate the enantioselectivity depends on the nature of the support and using Mg-Al/Cl LDH the results are similar to those obtained in solution. In the case of geraniol an excellent enantioselectivity was achieved with both supports but the immobilized catalysts are less active than the homogeneous one. Catalysts bonded to anion-exchange resins were not active in this reaction.

Disulfonated Mn-salen complexes have been immobilized onto Zn-Al/LDH by aqueous exchange of the compensating benzoate anion and the supported catalysts have been used in the epoxidation of several alkenes with oxygen using

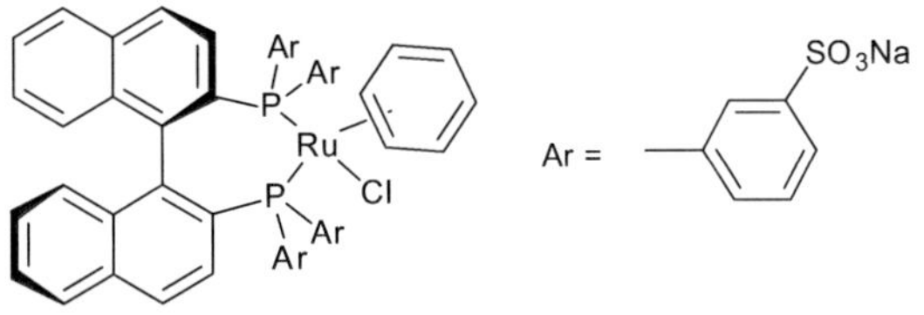

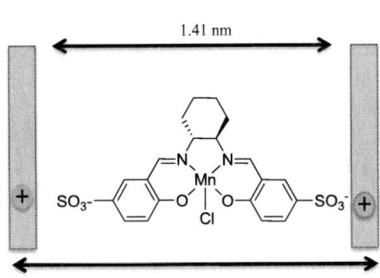

Ar = —⟨benzene⟩—SO₃Na

Substrate	Support	TOF (h⁻¹)	%ee
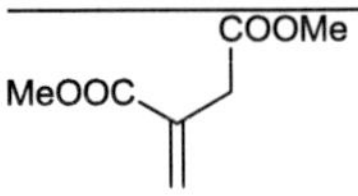 MeOOC / COOMe	---	29	51
	Mg-Al/Cl LDH	27	48
	Zn-Al/NO₃ LDH	25	11
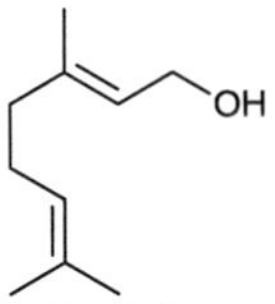 OH (geraniol)	---	4	100
	Mg-Al/Cl LDH	2	100
	Zn-Al/NO₃ LDH	2	100

Figure 8.45 Hydrogenation reaction promoted by chiral complexes immobilized on LDHs.

pivalaldehyde as a sacrificial aldehyde[83] (Figure 8.46). The activity is very high for this kind of reaction but enantioselectivity is moderate and highly dependent on the alkene structure.

A related catalyst bearing only one sulfonate group was immobilized onto Mg-Al-LDH, as well as on a Merrifield resin bearing triethylammonium groups, and used in the epoxidation of various alkenes with *m*-CPBA.[84]

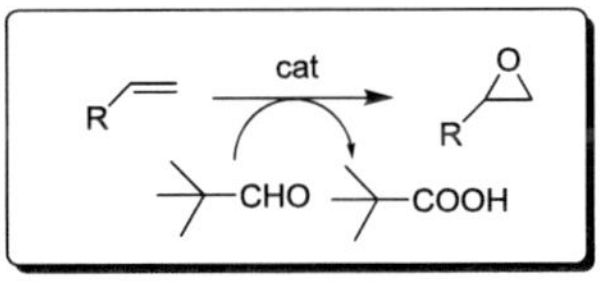

alkene	TOF (h⁻¹)	% epoxide selectivity	% ee
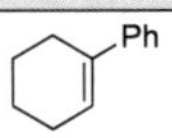 ⟨cyclohexene-Ph⟩	165.0	86	27
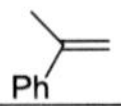 Ph⟨isopropenyl⟩	360.2	70	28
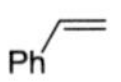 Ph⟨propenyl⟩	121.8	85	67
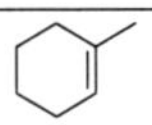 ⟨cyclohexene⟩	234.2	90	68

Figure 8.46 Results obtained of the alkene epoxidation with LDH-supported salen-Mn catalysts.

In general, solids were more active and enantioselective than the homogenous catalyst, with particularly good results for cyclic alkenes. Catalyst recovery was also quite efficient.

8.8 Concluding Remarks

Non-covalent immobilization strategies can lead to efficient catalysts, in some cases even more efficient than their homogeneous counterparts, in a way simpler than covalent bonding of the chiral ligand to the support, provided no modifications of the homogeneous catalyst should be done. As it happens with homogeneous catalysis, the catalysts and the reaction conditions should be optimized to obtain the highest yields and selectivities but, with immobilized catalysts, also to avoid catalyst deactivation and leaching of metal, chiral ligand and/or the whole catalyst.

Comparison between homogeneous and immobilized catalysts shows that the best ligand in one phase may not be the best in the other. In some reactions immobilized catalysts improve or even completely change the stereochemical course of the catalyzed reactions. Although in most cases spatial restrictions imposed by the support and site isolation of the catalytic sites seem to be in the origin of these results, more extensive work is necessary to better understand these results, allowing the design of catalysts better adapted to work in immobilized phases.

References

1. D. E. De Vos, I. F. J. Vankelecom and P. A. Jacobs, *Chiral Catalysts Immobilisation and Recycling*, Wiley-VCH, Weinheim, Germany, 2000.
2. Q. H. Fan, Y. M. Li and A. S. C. Chan, *Chem. Rev.*, 2002, **102**, 3385.
3. D. Mc Morn and G. J. Hutchings, *Chem. Soc. Rev.*, 2004, **33**, 108.
4. M. Heitbaum, F. Glorius and I. Escher, *Angew. Chem., Int. Ed. Engl.*, 2006, **45**, 4732.
5. J. M. Fraile, J. I. García and J. A. Mayoral, *Chem. Rev.*, 2009, **109**, 360.
6. K. T. Wan and M. E. Davis, *J. Catal.*, 1994, **148**, 1.
7. K. T. Wan and M. E. Davis, *J. Catal.*, 1995, **152**, 25.
8. I. Tóth, I. Guo and B. E. Hanson, *J. Mol. Catal. A*, 1997, **116**, 217.
9. C. Van Doorslaer, J. Wahlen, P. Mertens, K. Binnemans and D. De Vos, *Dalton Trans.*, 2010, **39**, 8377.
10. L.-L. Lou, K. Yu, F. Ding, W. Zhou, X. Peng and S. Liu, *Tetrahedron Lett.*, 2006, **47**, 6513.
11. K. L. Fow, S. Jaenicke, T. E. Müller and C. Sievers, *J. Mol. Catal. A*, 2008, **279**, 239.
12. M. R. Castillo, L. Fousse, J. M. Fraile, J. I. García and J. A. Mayoral, *Chem. Eur. J.*, 2007, **13**, 287.
13. N. Ishizuka, M. Togashi, M. Inoue and S. Enomoto, *Chem. Pharm. Bull.*, 1987, **35**, 1686.

14. H. N. Flach, I. Grassert, G. Oehme and M. Capka, *Colloid Polym. Sci.*, 1996, **274**, 261.
15. J. Jamis, J. R. Anderson, R. S. Dickson, E. M. Campi and W. R. Jackson, *J. Organomet. Chem.*, 2000, **603**, 80.
16. J. Jamis, J. R. Anderson, R. S. Dickson, E. M. Campi and W. R. Jackson, *J. Organomet. Chem.*, 2001, **627**, 37.
17. B. M. L. Dioos and P. A. Jacobs, *Tetrahedron Lett.*, 2003, **44**, 8815.
18. B. M. L. Dioos and P. A. Jacobs, *Appl. Catal. A*, 2005, **282**, 181.
19. D. Chois and G. J. Kim, *Catal. Lett.*, 2004, **92**, 35.
20. P. Gamez, F. Fache and M. Lemaire, *Bull. Soc. Chim. Fr.*, 1994, **131**, 600.
21. R. Raja, J. M. Thomas, M. D. Jones, B. F. G. Johnson and D. W. E. Vaugham, *J. Am. Chem. Soc.*, 2003, **125**, 14982.
22. J. Rouzaud, M. D. Jones, R. Raja, B. F. G. Johnson, J. M. Thomas and M. J. Duer, *Helv. Chim. Acta*, 2003, **86**, 1753.
23. C. Bianchini, P. Barbaro, V. Dal Santo, R. Gobetto, A. Meli, W. Oberhauser, R. Psaro and F. Vizza, *Adv. Synth. Catal.*, 2001, **343**, 41.
24. P. O'Leary, N. P. Krosveld, K. P. De Jong, G. van Koten and R. Gebbink, *Tetrahedron Lett.*, 2004, **45**, 3177.
25. H. Wang, X. Liu, H. Xia, P. Liu, J. Gao, P. Ying, J. Xiao and C. Li, *Tetrahedron*, 2006, **62**, 1025.
26. R. Augustine, S. Tanielyan, S. Anderson and H. Yang, *Chem. Commun.*, 1999, 1257.
27. R. Augustine, S. Tanielyan, N. Mahata, Y. Gao, A. Zsigmond and H. Yang, *Appl. Catal., A*, 2003, **256**, 69.
28. P. Stephenson, P. Licence, S. K. Ross and M. Poliakoff, *Green Chem.*, 2004, **6**, 521.
29. C. Simons, U. Hanefeld, I. W. C. E. Arends, T. Maschmeyer and R. A. Sheldon, *J. Catal.*, 2006, **239**, 212.
30. M. R. Torviso, M. N. Blanco, C. V. Cáceres, J. M. Fraile and J. A. Mayoral, *J. Catal.*, 2010, **275**, 70.
31. I. F. J. Vankelecom, D. Tas, R. F. Parton, V. Van de Vyver and P. A. Jacobs, *Angew. Chem., Int. Ed. Engl.*, 1996, **35**, 1346.
32. I. Vankelecom, A. Wolfson, S. Geresh, M. Landau, M. Gottlieb and M. Herskowitz, *Chem. Commun.*, 1999, 2407.
33. A. Wolfson, S. Janssens, I. Vankelecom, S. Geresh, M. Gottlieb and M. Herskowitz, *Chem. Commun.*, 2002, 388.
34. A. Wolfson, S. Geresh, M. Gottlieb and M. Herskowitz, *Tetrahedron: Asymmetry*, 2002, **13**, 465.
35. R. F. Parton, I. F. J. Vankelecom, D. Tas, K. B. M. Janssen, P.-P. Knops-Gerrits and P. A. Jacobs, *J. Mol. Catal. A*, 1996, **113**, 283.
36. R. Akiyama and S. Kobayashi, *Chem. Rev.*, 2009, **109**, 594.
37. S. Kobayashi, M. Endo and S. Nagayama, *J. Am. Chem. Soc.*, 1999, **121**, 11229.
38. S. Kobayashi, T. Ishida and R. Akiyama, *Org. Lett.*, 2001, **3**, 2649.
39. T. Ishida, R. Akiyama and S. Kobayashi, *Adv. Synth. Catal.*, 2005, **347**, 1189.

40. A. Cornejo, J. M. Fraile, J. I. García, M. J. Gil, V. Martínez-Merino and J. A. Mayoral, *Tetrahedron*, 2005, **61**, 12107.
41. M. L. Kantam, B. Kavita, V. Neeraja, Y. Haritha, M. K. Chaudhuri and S. K. Dehury, *Adv. Synth. Catal.*, 2005, **347**, 641.
42. F. Gelman, D. Avnir, H. Schumann and J. Blum, *J. Mol. Catal. A*, 1999, **146**, 123.
43. T. C. O. MacLeod, D. F. C. Guedes, M. R. Lelo, R. A. Rocha, B. L. Caetano, K. J. Ciuffi and M. D. Assis, *J. Mol. Catal. A*, 2006, **259**, 319.
44. S. B. Ogunwumi and T. Bein, *Chem. Commun.*, 1997, 901.
45. M. J. Sabater, A. Corma, A. Domenech, V. Fornés and H. García, *Chem. Commun.*, 1997, 1285.
46. C. Heinrichs and W. F. Hölderich, *Catal. Lett.*, 1999, **58**, 75.
47. C. Schuster and W. F. Hölderich, *Catal. Today*, 2000, **60**, 193.
48. W. Kahlen, H. H. Wagner and W. F. Hölderich, *Catal. Lett.*, 1998, **54**, 85.
49. A. Domenech, P. Formentin, H. García and M. J. Sabater, *J. Phys. Chem. B*, 2002, **106**, 574.
50. M. J. Alcón, A. Corma, M. Iglesias and F. Sánchez, *J. Mol. Catal. A*, 2002, **178**, 253.
51. H. Yang, J. Li, J. Yang, Z, Liu, Q. Yang and C. Li, *Chem. Commun.*, 2007, 1086.
52. H. Yang, L. Zhang, L. Zhong, Q. Yang and C. Li, *Angew. Chem., Int. Ed. Engl.*, 2007, **46**, 6861.
53. J. M. Fraile, J. I. García, J. A. Mayoral and A. J. Royo, *Tetrahedron: Asymmetry*, 1996, **7**, 2263.
54. J. M. Fraile, J. I. García, B. Lázaro and J. A. Mayoral, *Chem. Commun.*, 1998, 1807.
55. M. Iwamoto, Y. Tanaka, J. Hirosumi and N. Kita, *Chem. Lett.*, 2001, 226.
56. D. Meunier, A. de Mallmann and .J. M. Basset, *Top. Catal.*, 2003, **23**, 183.
57. G. J. Kim and J. H. Shin, *Catal. Lett.*, 1999, **63**, 83.
58. J. H. Kim and G. J. Kim, *Catal. Lett.*, 2004, **92**, 123.
59. C. F. Barnard, J. Rouzaud and S. H. Stevenson, *Org. Proc. Res. Dev.*, 2005, **9**, 164.
60. B. M. Choudary, M. Roy, S. Roy, M. L. Kantam, B. Sreedhar and K. V. Kumar, *Adv. Synth. Catal.*, 2006, **348**, 1734.
61. M. Mazzei, W. Marconi and M. Riocci, *J. Mol. Catal.*, 1980, **9**, 381.
62. S. Shimazu, K. Ro, T. Sento, N. Ichikuni and T. Uematsu, *J. Mol. Catal. A*, 1996, **107**, 297.
63. G. J. Kim and J. H. Shin, *Catal. Lett.*, 1999, **63**, 83.
64. (a) A. Crosman and W. F. Hoelderich, *J. Catal.*, 2005, **232**, 43; (b) A. Crosman and W. F. Hoelderich, *Catal. Today*, 2007, **121**, 130.
65. C. Simons, U. Hanefeld, I. W. C. E. Arends, R. A. Sheldon and T. Maschmeyer, *Chem. Eur. J.*, 2004, **10**, 5829.
66. H. Brunner, E. Bielmeier and J. Wiehl, *J. Organomet. Chem.*, 1990, **384**, 223.
67. R. Selke, K. Haüpke and H. W. Krause, *J. Mol. Catal.*, 1989, **56**, 315.

68. I. Tóth, B. E. Hanson and M. E. Davis, *J. Organomet. Chem.*, 1990, **397**, 109.
69. S. Xiang, Y. Zhang, Q. Xin and C. Li, *Chem. Commun.*, 2002, 2696.
70. H. Zhang and C. Li, *Tetrahedron*, 2006, **62**, 6640.
71. J. M. Fraile, J. I. García, J. A. Mayoral and T. Tarnai, *Tetrahedron: Asymmetry*, 1998, **9**, 3997.
72. J. M. Fraile, J. I. García, J. A. Mayoral, T. Tarnai and M. A. Harmer, *J. Catal.*, 1999, **186**, 214.
73. M. J. Fernáindez, J. M. Fraile, J. I. García, J. A. Mayoral, M. I. Burguete, E. García-Verdugo, S. L. Luis and M. A. Harmer, *Top. Catal.*, 2000, **13**, 303.
74. J. M. Fraile, J. I. García, C. I. Herrerías, J. A. Mayoral and M. A. Harmer, *J. Catal.*, 2004, **221**, 532.
75. J. M. Fraile, J. I. García, M. A. Harmer, C. I. Herrerías, J. A. Mayoral, O. Reiser and H. Werner, *J. Mater. Chem.*, 2002, **12**, 3290.
76. J. M. Fraile, J. I. García, C. I. Herrerías, J. A. Mayoral, O. Reiser, A. Socuellamos and H. Werner, *Chem. Eur. J.*, 2004, **10**, 2997.
77. A. I. Fernández, J. M. Fraile, J. I. García, C. I. Herrerías, J. A. Mayoral and L. Salvatella, *Catal. Commun.*, 2001, **2**, 165.
78. J. M. Fraile, J. I. García, J. A. Mayoral, E. Pires and I. Villalba, *Chem. Eur. J.*, 2007, **13**, 8830.
79. J. M. Fraile, J. I. García, V. Martínez-Merino, J. A. Mayoral and L. Salvatella, *J. Am. Chem. Soc.*, 2001, **123**, 7616.
80. J. M. Fraile, J. I. García, J. A. Mayoral and M. Roldán, *Org. Lett.*, 2007, **9**, 731.
81. M. J. Fabra, J. M. Fraile, J. I. García, F. Lahoz, J. A. Mayoral and I. Pérez, *Chem. Commun.*, 2008, 5402.
82. D. Tas, D. Jeanmart, R. F. Parton and P. A. Jacobs, *Stud. Surf. Sci. Catal.*, 1997, **108**, 493.
83. S. Bhattacharjee and J. A. Anderson, *Adv. Synth. Catal.*, 2006, **348**, 151.
84. B. M. Choudary, T. Ramani, H. Maheswaran, L. Prashant, K. V. S. Ranganath and K. V. Kumar, *Adv. Synth. Catal.*, 2006, **348**, 493.

Subject Index